Addressing Environmental Challenges With AI, Robotics, and Augmented Reality

Mourade Azrour
Moulay Ismail University of Meknes, Morocco

Imran Hossain
Varendra University, Bangladesh

A.K.M. Mahmudul Haque
University of Rajshahi, Bangladesh

IGI Global
Scientific Publishing
Publishing Tomorrow's Research Today

Vice President of Editorial Melissa Wagner
Director of Acquisitions Mikaela Felty
Director of Book Development Jocelynn Hessler
Production Manager Mike Brehm
Cover Design Jose Rosado

Published in the United States of America by
IGI Global Scientific Publishing
701 East Chocolate Avenue
Hershey, PA, 17033, USA
Tel: 717-533-8845
Fax: 717-533-7115
Website: https://www.igi-global.com E-mail: cust@igi-global.com

Library of Congress Cataloging-in-Publication Data

Names: Azrour, Mourade, editor. | Hossain, Imran, 1998- editor. | Haque, A. K. M., 1976- editor.
Title: Addressing environmental challenges with AI, robotics, and augmented reality / edited by Mourade Azrour, Imran Hossain, and A.K.M. Mahmudul Haque.
Description: Hershey, PA : IGI Global Scientific Publishing, [2025] | Includes bibliographical references and index. | Summary: "The primary objective of this book is to explore how emerging technologies such as AI, robotics, blockchain, and AR are being leveraged to tackle environmental challenges"-- Provided by publisher.
Identifiers: LCCN 2025009703 (print) | LCCN 2025009704 (ebook) | ISBN 9798337318929 (h/c) | ISBN 9798337318936 (s/c) | ISBN 9798337318943 (ebook)
Subjects: LCSH: Artificial intelligence--Environmental applications. | Robotics. | Blockchains (Databases) | Augmented reality. | Sustainable development--Technological innovations. | Sustainable Development Goals (Project)
Classification: LCC GE45.D37 A34 20254 (print) | LCC GE45.D37 (ebook) | DDC 363.7/060285--dc23/eng/20250421
LC record available at https://lccn.loc.gov/2025009703
LC ebook record available at https://lccn.loc.gov/2025009704

British Cataloguing in Publication Data
A Cataloguing in Publication record for this book is available from the British Library.

Table of Contents

Detailed Table of Contents

This study closes a significant knowledge gap on AI's unique offering to smart cities and provides useful information for those involved in information technology, social and sustainable development, urban planning, climate science, and intelligent urban planning ventures. It functions as a scholarly research output and strategic guideline for improving the layout of smart cities. It also emphasizes how AI may be used to build efficient, ecological, and dynamic metropolitan landscapes. The results showed how AI affects (i) outstanding health care, (ii) availability and effectiveness, (iv) education, (v) security for the citizens, and (iii) public amenities and connectivity. Similar tactics were put forth to encourage the incorporation of AI in prospective upcoming smart cities, including capacity building, digital infrastructure, collaborative ecosystems, and explicit ethics and guidelines.

This study looks into air pollution in developing countries and how Artificial Intelligence (AI) can contribute to air quality monitoring and data-driven interventions in those nations. Using a systematic literature review, it identifies successful AI implementations, like that from South Africa, which demonstrates cost-effective monitoring, and one from India that emphasizes the need for enabling legislation and policy. The major challenges are data integration, infrastructure constraints, and the need for energy sector reform because fossil fuel combustion remains one of the major sources of pollution. The research also suggests the potential of smart

AR is changing how we learn by mixing digital and real-life experiences. This study looks at how AR can help us understand the environment and promote sustainability. The researchers reviewed 86 articles from sources like Scopus, Web of Science, and IEEE Xplore, focusing on studies from 2014 to 2024. They explored how AR is used in varioul topics like biodiversity, climate change, waste management, and disaster preparedness. AR helps students stay engaged, remember information better, and understand complex environmental issues. Using 3D models, real-time simulations, and personalized experiences helps students think more deeply and critically. The research shows that AR can encourage lasting changes in behavior and promote sustainable habits when used in education. Policymakers and teachers are encouraged to add AR to school programs, invest in resources and training for teachers, and create rules to ensure data is used ethically. The study concludes that AR has great potential to help create environmentally aware citizens ready to face sustainability challenges in the 21st century.

The rapid urbanization in Bangladesh has led to significant challenges in solid waste management, with traditional systems struggling to meet the growing demands of urban centers. This study explores the potential of Internet of Things (IoT) technologies in revolutionizing waste management practices, focusing on their role in enhancing efficiency, sustainability, and citizen engagement. By analyzing global case studies from cities such as Barcelona, Singapore, and New York, this research demonstrates how IoT solutions, such as smart bins, real-time data analytics, and waste-to-energy technologies, can optimize waste collection, reduce operational costs, and promote recycling efforts. However, the adoption of these technologies in Bangladesh faces several barriers, including financial constraints, inadequate infrastructure, a lack of technical expertise, and limited public awareness. The study identifies these challenges and provides recommendations for overcoming them, emphasizing the need for pilot projects, cost-effective solutions, and capacity-building initiatives.

The integration of Artificial Intelligence (AI) and robotics is transforming urban infrastructure maintenance, addressing challenges associated with traditional methods. Predictive maintenance, real-time monitoring, and collaborative frameworks enhance efficiency and resource management. Robotics, including cobots and swarm technologies, play a pivotal role in large-scale inspection and repair tasks, ensuring precision and scalability. The chapter explores the collation of AI and robotics, emphasizing data-driven approaches and real-world case studies to highlight their potential in achieving sustainable and inclusive urban development. It also discusses environmental, economic, and social benefits while addressing challenges like technical constraints, ethical concerns, and safety. Future directions envision smarter cities with innovative AI-driven solutions.

Robotics has revolutionized global construction, boosting productivity and safety while advancing sustainable cities. However, its adoption in Nigeria faces major obstacles. This study identifies key barriers hindering robotics adoption in the Nigerian built environment. A quantitative approach was employed, using a structured questionnaire based on a comprehensive literature review. Data analysis involved mean item score, exploratory factor analysis, the Mann-Whitney U test, and frequency distribution. Results highlight electricity supply limitations, high initial costs, and workforce displacement concerns as primary barriers. Four critical barrier groups emerged: investment in innovation, environmental impediments, technological gaps, and resource and risk concerns. The Mann-Whitney U test revealed significant differences in perceptions between indigenous and foreign firms. Addressing these

Foreword

The environmental crises confronting our planet today are unparalleled in both scale and complexity. Climate change, pollution, biodiversity loss, and disaster vulnerability are not isolated events; they are systemic challenges demanding equally systemic, intelligent responses. It is within this pressing context that the book *"Addressing Environmental Challenges with AI, Robotics, and Augmented Reality"* offers an invaluable contribution to academic, policy, and technological discourse. It is a pleasure and a privilege for me to write this foreword, not merely as a professor who has long studied the intersection of environment and technology, but as a firm believer in the transformative power of interdisciplinary innovation.

This volume is especially relevant and influential because it passionately examines how next-generation technologies like artificial intelligence (AI), robotics, and augmented reality (AR) are not merely theoretical possibilities but rather workable solutions that are already influencing real-world results. The chapters within this book are diverse in their geographic focus, rich in empirical grounding, and insightful in their forward-looking visions. The authors do not shy away from the challenges inherent in deploying these tools, especially in resource-constrained or politically complex environments. Instead, they offer grounded strategies, case studies, and policy frameworks that illuminate both the potential and the pitfalls of our technological moment.

From the use of machine learning in disaster prediction to the application of AR in environmental education, this book showcases how emerging technologies can catalyze behavioral change, improve resource allocation, and reimagine governance. Particularly commendable is the inclusion of research from both the Global North and South, reflecting the truly global nature of both environmental degradation and technological innovation. This diversity strengthens the book's relevance and makes it a vital reference for practitioners, scholars, and decision-makers alike. The attention paid to infrastructure, energy, and agriculture especially motivates me as fields where sustainability usually depends on the successful implementation of innovation into policy and practice.

I warmly commend this book to students, researchers, policymakers, and innovators who seek to engage deeply with the ways in which intelligent technologies can, and must, serve the planet. It is my hope that this work will not only inform but also encourage the kind of transformative thinking needed in this decisive decade for global sustainability.

Jamal Mabrouki
Mohammed V University, Rabat, Morocco

Preface

Emerging technologies and environmental science are now actively pursuing research, policy, and implementation at this intersection, which is no longer merely theoretical. *Addressing Environmental Challenges with AI, Robotics, and Augmented Reality* emerges at this critical juncture, offering a timely and multidisciplinary exploration of how cutting-edge technologies are being leveraged to combat some of the most urgent ecological crises of our time. This volume reflects a growing consensus that the fusion of artificial intelligence, robotics, and immersive digital technologies is not merely a futuristic proposition but a present-day necessity for sustainable development and resilience building.

In today's world, where climate change, urban pollution, energy inefficiencies, and disaster vulnerability converge with increasing frequency, traditional approaches have proven inadequate. Technology, particularly AI and robotics, is offering new paradigms that go beyond incremental change; they are catalyzing systemic transformation. From agricultural optimization and urban infrastructure maintenance to real-time disaster forecasting and governance reform, these tools are reshaping environmental management. Augmented reality further enhances this landscape by transforming how societies learn about and interact with their environments, cultivating awareness and sustainable behaviors from classrooms to community centers.

The motivation for this book derives from the realization that addressing environmental degradation requires more than isolated expertise. It demands interdisciplinary vision, global collaboration, and adaptive governance. This compilation brings together contributors from varied regions and fields, including engineering, policy, agriculture, education, and urban planning, to reflect this complexity. Their collective voices offer insight into real-world applications, strategic frameworks, and theoretical advancements that place technology at the heart of ecological preservation and restoration.

In Chapter 1, the book begins by exploring the integration of AI into waste management practices in smart cities. It highlights the role AI plays in optimizing waste collection, recycling, and resource recovery, making urban waste management more efficient and sustainable. This chapter sets the tone by focusing on how technological advancements are helping create smarter, greener cities.

Chapter 2 delves into the role of AI in monitoring air pollution, particularly in developing countries. This chapter discusses how AI systems can improve air quality monitoring by providing cost-effective, data-driven solutions. The challenges of implementing such systems in resource-limited environments are also explored, with a particular focus on the importance of policy reforms and infrastructure improvements to facilitate better pollution management.

In Chapter 3, the role of robotics in agriculture and waste management is explored in depth. The authors discuss how robotics, combined with AI, is revolutionizing farming practices by improving efficiency and reducing environmental impact. The use of robotics for waste management is also examined, particularly in how automated systems are improving recycling efforts and reducing waste in both urban and rural settings.

Chapter 4 shifts focus to the policy frameworks necessary to support AI applications in environmental monitoring and governance. The chapter emphasizes the need for adaptive governance systems that can manage the ethical, regulatory, and operational challenges associated with AI technologies. The authors propose strategies to ensure that AI-based environmental solutions are transparent, inclusive, and accountable.

Chapter 5 explores the potential of Augmented Reality (AR) to promote environmental awareness and sustainability education. AR, by creating interactive and immersive experiences, helps users understand complex environmental issues in a tangible and engaging way. This chapter discusses the role AR can play in shaping sustainable behaviors and encouraging environmental stewardship, especially among younger generations.

Chapter 6 focuses on the implementation of IoT in waste management in urban Bangladesh. It highlights how IoT technologies can optimize waste collection and improve recycling efforts in cities facing rapid urbanization. The challenges of implementing these technologies in developing countries are discussed, with solutions offered to overcome infrastructure and financial barriers.

Chapter 7 discusses how AI and robotics are transforming urban infrastructure maintenance. Predictive maintenance, real-time monitoring, and the use of robotics in large-scale inspections and repairs are showcased as solutions to improve the sustainability and efficiency of urban infrastructure. The chapter explores how these technologies can help cities manage resources more effectively while reducing environmental impacts.

In Chapter 8, the barriers to adopting robotics in the construction sector in Nigeria are explored. Despite the numerous advantages that robotics offers, including increased efficiency and safety, the chapter identifies key obstacles to adoption, such as high costs, workforce displacement concerns, and infrastructure limitations. The authors propose solutions to overcome these barriers and enable broader adoption of robotics in the Nigerian built environment.

Chapter 9 discusses how blockchain technology is revolutionizing renewable energy distribution and management. By enhancing transparency and enabling peer-to-peer energy trading, blockchain helps optimize energy use, ensure compliance with sustainability laws, and improve the resilience of energy markets. This chapter provides a forward-looking perspective on how blockchain can contribute to a more sustainable and resilient energy ecosystem.

In Chapter 10, the use of machine learning and IoT in real-time disaster management is explored. The authors present a framework for integrating these technologies to predict and manage natural disasters, such as floods, cyclones, and wildfires. The chapter highlights how these solutions surpass traditional disaster management methods in terms of efficiency, scalability, and eco-friendliness.

Chapter 11 delves into the integration of AI, robotics, and AR in disaster management. By examining case studies and technological advancements, the chapter demonstrates how these technologies improve disaster preparedness and response. The authors stress the need for ethical frameworks and inclusive deployment strategies to ensure that these technologies are effectively integrated into global disaster management systems.

Finally, Chapter 12 focuses on the transition to electric vehicles (EVs) in India. The chapter highlights the environmental benefits of EVs, examines the challenges of their widespread adoption, and outlines the policies needed to support this transition. It discusses the role of EVs in reducing greenhouse gas emissions and moving toward a more sustainable and greener transportation system.

This book is envisioned as a reference, a guide, and a catalyst. For researchers, it offers scholarly grounding and methodological diversity. For practitioners, it presents implementable frameworks and case studies. For policymakers, it provides critical evaluation and regulatory foresight. For educators, it opens up new pathways for interdisciplinary teaching and student engagement.

Above all, *Addressing Environmental Challenges with AI, Robotics, and Augmented Reality* contributes meaningfully to the global conversation on environmental sustainability by advocating for intelligent, inclusive, and integrative solutions. As we confront a future shaped by both ecological uncertainty and technological possibility, this book affirms that innovation, when guided by ethics and purpose, is our most powerful companion in securing a livable planet for generations to come.

Imran Hossain
Varendra University, Rajshahi, Bangladesh

Chapter 1
AI-Enabled Strategic Waste Management Practices for Environmental Sustainability in Smart Cities

Babasaheb Ramdas Jadhav
https://orcid.org/0000-0002-1933-0558
Dr. D.Y. Patil Vidyapeeth's Global Business School and Research Centre, India

Renuka Deshmukh
MIT World Peace University, India

Imran Hossain
https://orcid.org/0000-0002-7602-053X
Varendra University, Bangladesh

ABSTRACT

This study closes a significant knowledge gap on AI's unique offering to smart cities and provides useful information for those involved in information technology, social and sustainable development, urban planning, climate science, and intelligent urban planning ventures. It functions as a scholarly research output and strategic guideline for improving the layout of smart cities. It also emphasizes how AI may be used to build efficient, ecological, and dynamic metropolitan landscapes. The results showed how AI affects (i) outstanding health care, (ii) availability and effectiveness, (iv)

DOI: 10.4018/979-8-3373-1892-9.ch001

education, (v) security for the citizens, and (iii) public amenities and connectivity. Similar tactics were put forth to encourage the incorporation of AI in prospective upcoming smart cities, including capacity building, digital infrastructure, collaborative ecosystems, and explicit ethics and guidelines.

1. INTRODUCTION

AI is transforming the key components of smart cities, radically changing how cities function and interact with their citizens (Chauhan, C. et.al., 2022). AI contributes greatly to the sustainability of the world by optimizing resource allocation. Smart cities can utilize data intelligence for decision making through the analytics of data concerning waste production, energy utilization, and traffic patterns (Abdallah, M. et.al. 2020; Hossain et al., 2024a). Apart from that, AI assists with environmental surveillance by monitoring and managing the pollution of cities, thereby fostering better urban health. In addition, AI technology is critical for the management of water resources, predicting infrastructure upkeep, and automating recycling processes to help create sustainable urban ecosystems (Bianchini, A., et. al. 2019; Hossain et al., 2024b).

AI equally impacts the realms of public services and connectivity. Smart healthcare systems enable the running of hospitals, patient care is tailored to the individual, and even outbreaks can be forecasted (Balasubramanian, S., et. al. 2021). Educational optimizations are also done through artificial intelligence (AI), which uses predictive analytics to determine regions of focus for students and formulate unique learning pathways. This study attempts to bridge the gap around the impacts of AI on smart city development infrastructures. Also, there is a lack of ample comprehensive, post driven research meant to explore the anticipated trends and applications of AI in urban settings (Borrello, M., et al., 2017).

Given the speed at which AI technology is developing, these gaps are critical because they impede strategic planning and well-informed decision-making for the advancement of intelligent towns.

Aim of this research work is to investigate how AI may be integrated and how it can improve smart city drivers. The following were the study's aims in order to successfully actualize the stated aim: (i) Determine the main forces behind smart cities and look at how AI is currently enhancing these forces; (ii) Examine and pinpoint patterns in the use of Artificial Intelligence in intelligent towns, emphasizing on domains wherein AI has the greatest influence; (iii) Determine how AI may affect smart city drivers in the future; (iv) Determine ways to deal with obstacles and ethical issues in AI implementation. The objectives of this study were to be achieved by developing the below mentioned research questions:

a) What effect does AI currently have on the main elements influencing the growth of smart cities?
b) Considering the most relevant analytical information and studies, which are the key latest innovative upcoming trends for growth for Artificial intelligence significance in modern towns and cities?
c) Which moral dilemmas as well as challenges emerge during AI integration in smart cities, and by what means may they be addressed?

These goals and research questions helped to clarify the current and future functions of AI usages in intelligent towns and the wider significance of their applicability. This study particularly increases the conceptual knowledge of artificial intelligence as well as its real-world use in the emerging multidisciplinary fields of information technology, sociological studies, finance, climate economics, ecology, and urban design by solving the shortcomings. The study's findings will be useful to interdisciplinary stakeholders planning and carrying out smart city initiatives.

The limitations of systematic literature review (SLR) are acknowledged in this work, including the possibility of bias in choosing of publications as well as the likelihood of overlooking key major investigations that aren't included within the search engines of choice. Because AI technology is developing so quickly, it may not be sufficiently covered in the literature.

In the next part, the research approach that was employed is highlighted. Next are the investigations work's qualitative and quantitative study results. In line with the proposed conceptual framework, the research inquiries, as well as juxtaposition with pertinent prior studies, the investigation's findings are examined in the last section. The overview of the research as well as upcoming examination directions are offered in the concluding part.

2. METHODS

The Systematic Literature Review methodology was used in this study to investigate the influence and integration of Artificial Intelligence in intelligent towns. The Systematic Literature Review technique was selected due to its methodical, apparent, and repeatable tactic, which allowed for a thorough concoction of the body of knowledge about AI's function in developing intelligent towns. The suggested methodology's general architecture is shown in Figure 1.

According to the objectives of the study, this research methodology was used to present a range of perspectives and findings that contributed to the development of an extensive knowledge of AI regarding intelligent communities, particularly regarding identifying potential areas for growth in AI application. It is significant because it offers an extensive and in-depth awareness of Artificial Intelligence present state and its future applications in the creation of intelligent towns through the usage of this structured SLR approach.

In addition to providing useful information to stakeholders in sociological research, business development, artificial intelligence, metropolitan layout and design, intelligent and green and environmentally friendly city campaigns, and cerebral communities growth, this directly addresses the goals of the study. The research techniques employed in this study are listed in the next subsections.

Figure 1. Techniques for research outlining the steps and techniques applied

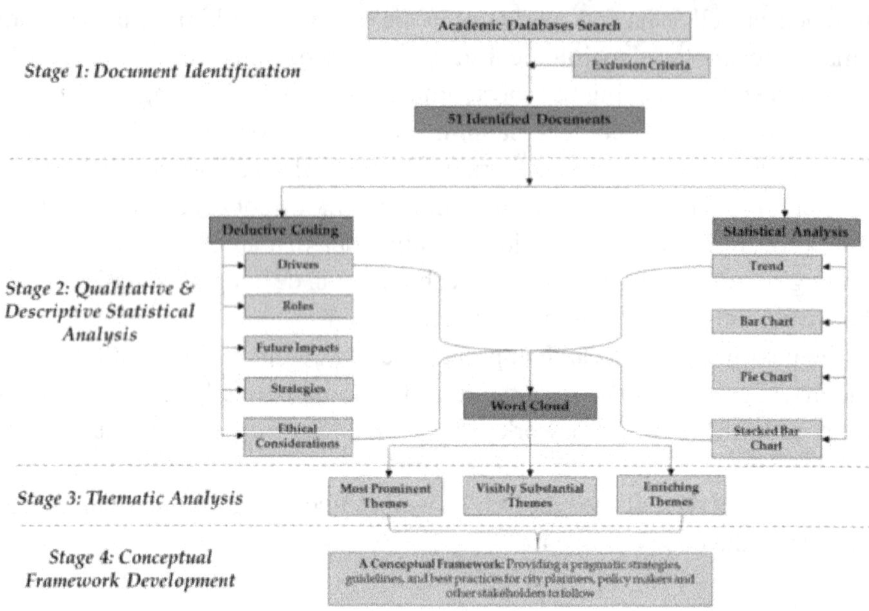

Figure 2. Procedure for choosing articles using search engines

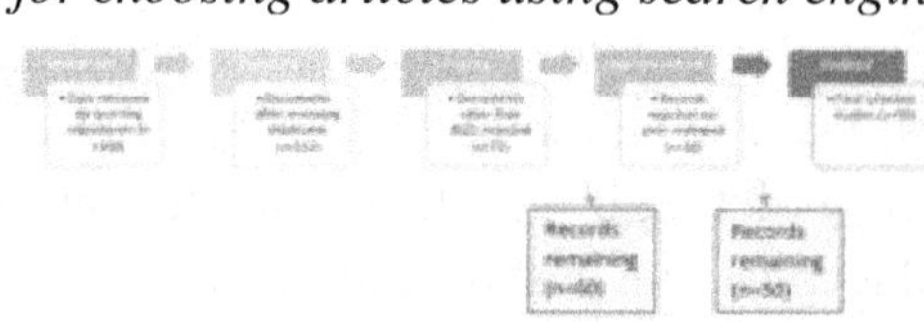

Table 1. Information is gathered and evaluated with inductive algorithms

Algorithms	Definition
Motivating elements	The main forces behind intelligent towns, or those that are crucial in determining how metropolitan areas grow
Functions	Knowing how AI is being used to power intelligent towns
Upcoming impacts	Perspectives on the possible effects of AI in the years ahead
Tactics	The methods for encouraging the incorporation of AI in prospective upcoming years
Moral considerations	The difficulties and moral issues surrounding the application of AI in intelligent towns

3. LIMITATIONS OF RESEARCH

Despite a thorough effort to investigate AI's function in smart cities, this study had several shortcomings that should be noted. First, it may overlook the continuity and growth of concepts from earlier studies, even while its sole emphasis on publications spanning 2022 and 2023 is useful for acquiring the most current findings. This time frame can make it more difficult to understand the development and context of including artificial intelligence in intelligent cities. Second, experts rigorously evaluate scientific articles and publish academic literature but can skip real-world experiences and practical knowledge found in industry reports, case studies, and literature.

This type of source may provide a more sophisticated perspective on the application of AI in metropolitan environments. Quantitative tests, presented primarily as figures and word clouds, provide a perspective on topic development. However, tackling the qualitative wealth of artificial intelligence and its purpose in intelligent cities is not short. These procedural options may not fully reflect the subtlety and delicacy of the impact of artificial intelligence on average life expectancy in a metropolitan city. Similarly, learning has linguistic bias and can overlook important research and development published in other languages by focusing on English sources.

Finally, this study recognizes the quick speed at which AI and smart city technologies are developing. Because technology is changing so quickly, the results and conclusions of this study could become outdated very quickly. It can be difficult to keep the study findings current and applicable over time due to the field's dynamic nature. Notwithstanding these drawbacks, the research offers a solid grasp of Artificial Intelligence's function in creating modern metropolitans as well as insightful information that may be built upon and expanded upon in other studies.

4. FINDINGS AND THE CREATION OF CONCEPTUAL MODELS

4.1 Annual Progression of Research on Artificial Intelligence Enabled Intelligent Towns

This study's secondary research goal is reflected in these findings. A thorough representation of the chronological distribution of research in the Systematic Literature Review concentrating on AI and intelligent towns before the inclusion criteria of chosen articles from 2022 and 2023 is provided by the line graph in diagram three. Setting a benchmark for early research activity, the chart shows 13 papers from the "pre-2005" era, which marks the beginning of the field's foundational phase. A notable upward trend was noted over the "2005–2010" timeframe, which saw a rise to 24 studies. This reflected growing scholarly interest as well as possible developments in AI and smart city technology. There are 13 studies in the "2011–2015" timeframe, indicating a stagnation in research and perhaps a maturation in the studies' topic focus.

Nonetheless, there is a minor drop to nine studies over the "2016–2019" era, suggesting a possible change in the field's research objectives or paradigms. This tendency reversals significantly in the "2020–2021" period, as the number of research rises to 21, coinciding with a global explosion in AI and smart technologies propelled by both increased public interest and technological developments. This upward trend is maintained in the most recent years, 2022 and 2023, with 30 and 24 investigations, correspondingly, highlighting the growing significance, applicability, and perseverance of this field of study. Given its recent upward trend, the line graph successfully captures the dynamic and changing character of academic work in AI and smart cities, emphasizing the growing focus on recent advancements and discoveries in the field. This graphic illustrates how research has developed over time and highlights the increasing scholarly interest in cutting-edge AI and urban development technology.

Figure 3. SLR's annual progression pattern

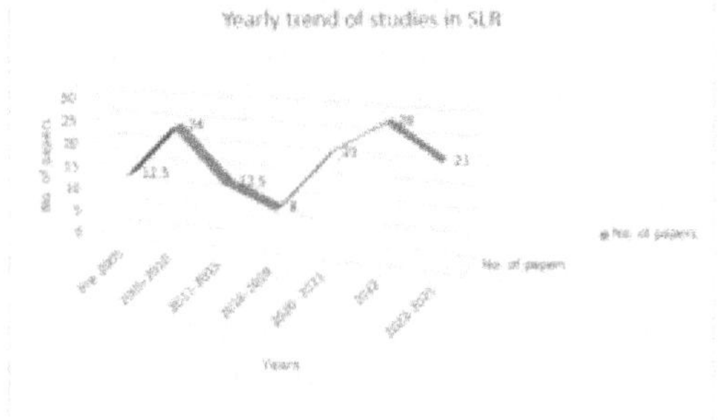

4.2. Evaluation of Content Retrieval in AI Enabled Modern Towns

A thorough selection of 150 publications was first taken into consideration for the review. But after a thorough screening procedure, some manuscripts were rejected based on standards: This review's focus on current and pertinent findings was ensured by excluding 70 papers because their publication dates were prior to 2022. The academic rigor of this review was also maintained by excluding 10 papers because they either lacked adequate peer review or were primarily grey literature. Additionally, to prevent repetition, 19 papers were found to be duplicates and eliminated. 51 publications were selected for examination after a rigorous screening procedure that considered factors like academic credibility, recentness, and relevance. This quantitative depiction highlights the careful and selective methodology used in this study, guaranteeing an emphasis on relevant and high-quality works on Artificial Intelligence and intelligent towns.

4.3. Theme Allocation Related with Artificial Intelligence in Intelligent Communities

The dissemination of Artificial Intelligence-associated research in the modern city arena in the years 2022 as well as 2023 is methodically displayed. The graphs. The chart's bars, each of which represents a different component and one of these years, each highlight a distinct area of research where AI integration is significant. With notable contributions in areas like artificial intelligence (AI) in sustainability, efficiency, connectivity, citizen engagement, public health, public utilities, controlling the surroundings, and designing cities, the graph for 2022 demonstrates a thorough engagement across a variety of smart city aspects.

A thorough examination of AI's role in advancing intelligent town projects is indicated by the height of the bar for 2022, which is the total number of research on each issue. Although the number of research per topic varies slightly, the distribution of themes for 2023 is comparable. The dynamic character of the study emphasis and the changing priorities within the discipline are reflected in this variety. In smart cities, efficiency optimally uses urban services and operations (Golbaz, S., et al. 2019)

Digital platforms that simplify government processes or intelligent traffic signals that ease congestion could be examples of this (Makarova, I., 2018). Establishing a connected city setting where information and data may be freely and efficiently shared is the aim of networking in smart cities (Lu, W. 2019; Hossain et al., 2025). This driver ensures that every part of the town is linked together, enabling seamless interaction and transfer of information for the efficient running of several municipal functions.

Figure 4. Word Stream

Using technology to enhance communication between the city and its citizens is known as citizen engagement in smart cities (Makarova, I., et.al. 2018). Technology must be used to facilitate interactions, involve residents in making choices, and ensure that the standards and preferences of the city's residents are met by its development. It refers to establishing a two-way dialogue in which inhabitants actively participate in the development of the city rather than merely being passive beneficiaries of its services (Modgil, S. et.al., 2021). Table 2 shows how AI is now improving the four major drivers that have been identified, along with examples. According to the findings, artificial intelligence is dramatically changing smart city drivers and improving urban life.

Table 2. AI's current contributions to strengthening the main forces behind smart cities

Sustainability	(i) Examine enormous information sets pertaining to resource allocation, garbage disposal, and usage of energy; (ii) Streamline power lines, forecast the need for electricity, and minimize utilization of resources and wasteful utilization of energy.	Among these are (i) Microsoft's for energy efficiency; (ii) Google's acquisition of Nest Thermostat Predix for predictive maintenance.
Efficacy	Digitalizing and streamlining municipal processes; forecasting city facilities upkeep	AI-powered smart traffic management systems are a good illustration. Real-world examples include (i) Waymo's driverless taxi service in Phoenix, Arizona; (ii) the predictive analytics used for urban planning by Singapore's Urban Redevelopment Authority; and (iii) Barcelona's intelligent trash cans.
Interconnectedness	Handle massive amounts of data from different IoT gadgets, enhancing the incorporation and accessibility of many urban amenities.	Autonomous drones with artificial intelligence are being utilized in Dubai, United Arab Emirates, to monitor traffic jams and react rapidly to crises.
Participation of residents	Enabling intelligent technologies that facilitate communication	One actual example is the MyLA311

AI is a revolutionary force that is changing how cities function and provide for their citizens, not just a tool for development in each of these areas. More responsive, effective, and sustainable urban settings are being made possible by integrating AI into smart city drivers. In these environments, technology and human intuition collaborate to improve living conditions.

5. ARTIFICIAL INTELLIGENCE'S POSSIBLE FUTURE EFFECT ON INTELLIGENT TOWNS

The aim of the study is connected to these findings. With context to modern cities, artificial intelligence (AI) holds promise for significantly impacting several important factors that have not yet been completely utilized. Four sectors that could be impacted by AI in the future were identified in this study.

Initially, public health. Initially public health. Public health is the first potential topic identified in this study. Because KI can help with aggressive healthcare management to improve health delivery and monitor and predict public health risks. AI's predictive capabilities help cities allocate better resources for healthcare. This saves lives and reduces health costs (Mourtzis, 2022). Public pension companies are managed in second place. Management of public suppliers such as energy and water are another area that is promising in an AI-intelligent city. By hoping for tighter usage trends and distributions, AI can improve the efficiency of these suppliers. For example, AI systems can predict high power requirements and dynamically change offers, leading to a more sustainable and balanced pattern of energy consumption. City planners can recognize the possible impact of decision-making by using AI to simulate a variety of urban development scenarios that can lead to strategy and healthy urban development (Owojori, O. M. and Okoro, C., 2022). Finally, environmental protection and surveillance. AI interventions are welcome in these areas. Supporting residents' health and well-being and sustainability of the urban environment, AI applications ensure that intelligent cities remain livable and comfortable for future generations (Paes, M.X., et al., 2019).

The fourth goal of methods to promote AI integration in possible future incentives for intelligent cities is supported by these results to improve the importance of AI in less good research components of intelligent cities. Regulations should be introduced to promote integration (Pagoropoulos, A. et.al., 2017). The following paragraphs provide the identified strategies. First, we bring together the cooperative ecosystems, authorities, IT companies, universities, and the public. The tech infrastructure money that AI needs to flourish is second in. The data architecture, which enables effective data processing and analysis, and the physical infrastructure, which comprises sensors and connectivity networks, are included in this. From environmental monitoring to traffic control, cities will be able to use AI for a variety of purposes with the help of a strong digital infrastructure (Perey, R., et.al. 2019).

Third, enhancing the ability of local government and citizens. By increasing the general public's and local leaders' level of AI literacy, training programs help elucidate Artificial Intelligence besides embolden its espousal.

Developing unblemished strategies as well as moral norms for the use of AI in intelligent towns is the final phase. By adopting this, concerns about security and privacy will be addressed, and all populations will equitably benefit from AI (Prieto-Sandoval, V. et.al., 2018).

5.1. The Challenges and Moral Concerns of Using AI

Like the results in the previously mentioned the preceding section, the results in this section also support the fourth objective of the study. There are certain ethical questions and problems raised by the incorporation artificial intelligence into intelligent towns, which necessitate careful consideration. The ensuing paragraphs address the highlighted difficulties and moral dilemmas.

First, making sure that the enormous volumes of data that smart cities produce, and gather are secure and private. Because AI systems depend so heavily on data to function properly, there is a possibility that sensitive and personal data may be compromised (Rajamanikam, A. and Solihin, M.I., 2019). The challenging element is putting robust security protocols for data into place while allowing the unfettered exchange of data necessary for artificial neural networks to function competently. This delicate complementary performance necessitates progressive technology elucidations besides stringent supremacy rules to prevent illicit use or exploitation of knowledge.

The possibility of bias in AI schemas is second place. In sensitive areas such as social services, employment, and law enforcement, this can lead to unfair or biased outcomes. AI systems must be properly designed and monitored regularly to ensure that they function fairly and fairly to meet this challenge. Additionally, there is a need for an integrated and diverse data record that is highly representative of the people in the city (Sadraei, R. et al., 2023). Third, the ability of AI to create social inequality worries about the possibility that AI can produce socioeconomic inequality. Although AI can increase comfort and efficiency, their benefits could not be shared equally across all social segments (Raworth, K., 2017). For example, AI-controlled services can primarily serve people with access to the internet to digitally formed people, leaving people who are not technical. As this digital gap could widen the socioeconomic gap, it is essential that the Smart City project includes measures to promote fair access to AI technology.

Fourth, it is morally required to consider how AI will affect human autonomy and decision-making in the long run. There is a risk of becoming overly dependent on AI systems as they become more complex and essential to city operations, which could undermine human abilities and judgment (Secinaro, S. et.al., 2021). This calls into doubt the appropriate amount of decision-making authority for AI, particularly in crucial areas of public welfare. The difficulty lies in finding a balance so that AI enhances human judgment rather than takes the place of it.

6. CONCEPTUAL MODEL

The recognized obstacles and moral issues that are the crucial chauffeurs of intelligent towns, the best ways to promote AI, the practical impact and contributions of AI, and the functions of Artificial intelligence in augmenting the vital motorists of intelligent towns were all taken into consideration when proposing a theoretical structure as an organizing principle or recommendations for effective planning and implementation of smart cities. Figure 5 displays the framework. If approved, the theoretical framework will act as a blueprint for applying AI to maximize smart city performance. For the benefit of municipal planners, legislators, and other interested parties, it provides useful tactics, standards, and best practices. Although the conceptual model is based on theoretical ideas, it retains practical application insights.

Figure 5. A conceptual strategy for optimizing smart city drivers that strategically integrates AI

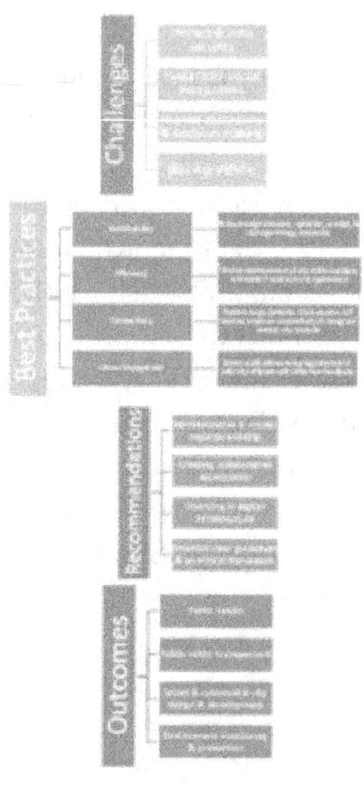

7. DISCUSSION

Based on existing research from both theoretical and practical viewpoints, this essay suggested a conceptual framework for promoting AI As an operational instrument for meritoriously and profitably objectifying Artificial Intelligence bearings on the architecture of impending modern towns, the theoretical structure was released to address potential challenges and ethical issues.

Through a search of the Scopus database, more than fifty works on intelligent towns with a theoretical structure or blueprint between 2015 and 2023 were discovered. However, from the standpoint of the literature assessment, only two existing research considered the importance of creating conceptual frameworks (Singh, D. and Satija, A., 2018). One of the existing studies focused on the governance of sustainable mobility, while the other study examined the difficulties in developing modern towns.

This revision presented a theoretical outline that engrossed on the ethical concerns and issues of AI to safeguard the hurdle less integration of Artificial Intelligence duties in developing significant forces in modern towns. Similarly, the conceptual framework suggested in this study covers the wide range of possible future effects of Artificial Intelligence on the development of intelligent towns, including efficiency, connection, climate science (ESG), besides public participation.

The selection/document identification procedure as well as the search terms and phrases employed in the databases that were chosen could be explained by this aberration. There are many more AI obstacles, even though the current research concentrated on intelligent towns as well as AI issues in the healthcare industry.

Based on a careful analysis of previous studies, experts within the area of artificial intelligence throughout the framework of ecological architecture have looked at the wide implications of artificial intelligence within the context of cities. A study that thoroughly examined the role of urbanism in the increase of AI use was carried out in 2023 by Cugurullo and associates. A study like this was carried out by the primary author, who examined the contrast between AI and city planning from both a theoretical and practical standpoint (Tang, M. and Liao, H., 2021). In separate research, the European Commission examined the connection amongst artificial intelligence and the growth of cities (Upadhyay, A. et.al., 2021). The current study has significantly added to the body of information on AI and intelligent towns by examining how AI can be utilized to develop and enhance smart city drivers.

Similarly, compared to this study, which specifically looked at the ethical issues and encounters encompassing the use artificial intelligence in the setting of intelligent towns, the literature currently in publication has not sufficiently tackled the challenges and moral quandaries exclusive to applying AI in various intelligent town expansion domains. The next section addresses the findings while being true to the research questions. The findings of the SLR, which comprise qualitative as well as quantitative evaluation, provide insightful answers to the problems surrounding artificial intelligence research in smart cities. These findings bridge the knowledge disparity exists amongst AI theory and practical city design applications.

The topic of the study is “In what ways is artificial intelligence presently impacting the key pillars of intelligent town growth, including as long-term viability effectiveness, and interaction?” According to the qualitative investigation, AI plays a revolutionary role in enhancing the main forces behind modern towns. What sets AI apart as a sustainability tool is its ability to reduce energy waste, enhance resource management, and facilitate environmental monitoring. AI has greatly increased the efficiency of city operations, especially in the areas of infrastructure predictive maintenance and congestion control.

With phrases like “Sustainability,” “Efficiency,” and “Connectivity” showing up frequently and highlighting their significance in the literature, as depicted in the word cloud, confirms the prominence of the above-mentioned issues.

About the study's topic, “What are the possible growth areas and emerging trends for artificial intelligence deployments in intelligent towns depending on existing analytical information as well as evaluation?” The quantitative study revealed new patterns and intriguing possibilities for artificial intelligence's expansion in intelligent towns, with a particular focus on public utilities management, urban planning, and public health. AI has shown a lot of potential, especially in the realm of public health, where it may be used to improve healthcare delivery and monitor and predict health issues. The trend analysis also highlighted the increasing significance of AI in managing public utilities like power and water and in urban planning. These expanding areas of interest were underlined by the distribution of research topics in 2022 and 2023, as the stacked bar chart illustrates.

The results of qualitative research have resolved many competition and moral concerns regarding the inclusion of artificial intelligence in intelligent cities. The main challenges include addressing potential distortions in AI systems, ensuring data security and private spheres, and limiting the possibility of an increase in social inequality. The sustainable impact of artificial intelligence on autonomy and people's judgments is also morally astounding. The results of qualitative and quantitative research fully understand the problem and moral dilemmas, and the current and future impacts of AI on intelligent cities in this field.

8. CONCLUSIONS AND FUTURE RESEARCH TRENDS

Research shows that AI can fundamentally change the urban environment. However, this study includes limitations that can limit the breadth of knowledge obtained, such as This study provides a strong basis for further research despite these limitations. These limitations recommend approaches such as expanding historical domains, integrating various sources of literature, and further inclusion of the qualitative effects of AI. It helps us to better understand the possibilities of artificial intelligence and uses these powerful innovations to raise all living standards, not only modern, but also equitable, ecologically robust, and beneficial, in metropolitan and socioeconomic environments.

Prospective research in the field of urban artificial intelligence can take in many directions. Another promising approach is to include grey literature from case studies, industry reports, and research frameworks. These materials can provide a more applied viewpoint to the academic one by offering concrete examples, practical ideas, as well as insights of professionals in the field.

REFERENCES

Abbasi, M., Abduli, M. A., Omidvar, B., & Baghvand, A. (2014). Results uncertainty of support vector machine and hybrid of wavelet transform-support vector machine models for solid waste generation forecasting. *Environmental Progress & Sustainable Energy*, *33*(1), 220–228. DOI: 10.1002/ep.11747

Abdallah, M., Talib, M. A., Feroz, S., Nasir, Q., Abdalla, H., & Mahfood, B. (2020). Artificial intelligence applications in solid waste management: A systematic research review. *Waste Management (New York, N.Y.)*, *109*, 231–246. DOI: 10.1016/j.wasman.2020.04.057

Agrawal, A., Gans, J., & Goldfarb, A. (2019). *The Economics of Artificial Intelligence: An Agenda*. University of Chicago Press. DOI: 10.7208/chicago/9780226613475.001.0001

Bag, S., & Pretorius, J. H. C. (2020). Relationships between industry 4.0, sustainable manufacturing and circular economy: Proposal of a research framework. *The International Journal of Organizational Analysis*, *30*(4), 864–898. DOI: 10.1108/IJOA-04-2020-2120

Balasubramanian, S., Shukla, V., Sethi, J. S., Islam, N., & Saloum, R. (2021). A readiness assessment framework for Blockchain adoption: A healthcare case study. *Technological Forecasting and Social Change*, *165*, 120536. DOI: 10.1016/j.techfore.2020.120536

Bianchini, A., Rossi, J., & Pellegrini, M. (2019). Overcoming the main barriers of circular economy implementation through a new visualization tool for circular business models. *Sustainability (Basel)*, *11*(23), 6614. DOI: 10.3390/su11236614

Borrello, M., Caracciolo, F., Lombardi, A., Pascucci, S., & Cembalo, L. (2017). Consumers' perspective on circular economy strategy for reducing food waste. *Sustainability (Basel)*, *9*(1), 141. DOI: 10.3390/su9010141

Cascante, M.Q., Garcıa, N.A. and Fold, N. (2022), "The role of external forces in the adoption of aquaculture innovations: an ex-ante case study of fish farming in Colombia's southern Amazonian region", Technological Forecasting and Social Change, Vol. 174, 121185, doi: . techfore.2021.121185.DOI: 10.1016/ j

Cha, G.-W., Kim, Y.-C., Moon, H.J. and Hong, W.-H. (2017), "New approach for forecasting demolition waste generation using chi-squared automatic interaction detection (CHAID) method", Journal of Cleaner Production, Vol. 168, pp. 375-385, doi: . 09.025.DOI: 10.1016/j.jclepro.2017

Chauhan, C., Parida, V. and Dhir, A. (2022), "Linking circular economy and digitalisation technologies: a systematic literature review of past achievements and future promises", Technological Forecasting and Social Change, Vol. 177, 121508, doi: . 121508.DOI: 10.1016/j.techfore.2022

European Commission (2019), Accelerating the Transition to the Circular Economy: Improving Access to Finance for Circular Economy Projects, Publications Office, LU.

European Commission. (2020). *Categorisation System for the Circular Economy*. European Commission.

European Parliament (2022), "Towards carbon neutrality through ambitious transformation of the EU energy system".

Golbaz, S., Nabizadeh, R., & Sajadi, H. S. (2019). Comparative study of predicting hospital solid waste generation using multiple linear regression and artificial intelligence. *Journal of Environmental Health Science & Engineering*, *17*(1), 41–51. DOI: 10.1007/s40201-018-00324-z

Haleem, A., Khan, S., Luthra, S., Varshney, H., Alam, M., & Khan, M. I. (2021). Supplier evaluation in the context of circular economy: A forward step for resilient business and environment concern. *Business Strategy and the Environment*, *30*(4), 2119–2146. DOI: 10.1002/bse.2736

Hannan, M. A., Arebey, M., Begum, R. A., Basri, H., & Al Mamun, M. A. (2016). Content-based image retrieval system for solid waste bin level detection and performance evaluation. *Waste Management (New York, N.Y.)*, *50*, 10–19. DOI: 10.1016/j.wasman.2016.01.046

Hossain, I., Haque, A. K. M., Ullah, S. M., & Salehin, A. (2025). Assessing the Role of Urban Governance in Promoting Sustainable Cities and Communities: Evidence from Four City Corporations in Bangladesh. *International Journal of Community Well-being*, •••, 1–28. DOI: 10.1007/s42413-025-00247-2

Hossain, I., Haque, A. M., Kılıç, Z., Ullah, S. A., Azrour, M., & Mabrouki, J. (2024b). Exploring Household Waste Management Practices and IoT Adoption in Barisal City Corporation. In *Smart Internet of Things for Environment and Healthcare* (pp. 27–45). Springer Nature Switzerland. DOI: 10.1007/978-3-031-70102-3_2

Hossain, I., Haque, A. M., & Ullah, S. A. (2024a). Assessing Sustainable Waste Management Practices in Rajshahi City Corporation: An Analysis for Local Government Enhancement using IoT and AI Technology. *Environmental Science and Pollution Research International*. Advance online publication. DOI: 10.1007/s11356-024-33171-7

Kouhizadeh, M., Sarkis, J., & Zhu, Q. (2019). At the nexus of blockchain technology, the circular economy, and product deletion. *Applied Sciences (Basel, Switzerland), 9*(8), 1712. DOI: 10.3390/app9081712

Lanzalonga, F., Oppioli, M., Dal Mas, F., & Secinaro, S. (2023). Drones in Venice: Exploring business model applications for disruptive mobility and stakeholders' value proposition. *Journal of Cleaner Production, 423*, 138764. DOI: 10.1016/j.jclepro.2023.138764

Lehtokunnas, T., Mattila, M., & Mesiranta, N. (2020). Towards a circular economy in food consumption: Food waste reduction practices as ethical work. *Journal of Consumer Culture, 22*(1), 227–245. DOI: 10.1177/1469540520926252

Lu, W. (2019), "Big data analytics to identify illegal construction waste dumping: a Hong Kong study", Resources, Conservation and Recycling, Vol. 141, pp. 264-272, doi: . 10.039.DOI: 10.1016/j.resconrec.2018

Lu, W., & Chen, J. (2022). Computer vision for solid waste sorting: A critical review of academic research. *Waste Management (New York, N.Y.), 142*, 29–43. DOI: 10.1016/j.wasman.2022.02.009

Ma, S., Zhang, Y., Liu, Y., Yang, H., Lv, J., & Ren, S. (2020). Data-driven sustainable intelligent manufacturing based on demand response for energy-intensive industries. *Journal of Cleaner Production, 274*, 123155. DOI: 10.1016/j.jclepro.2020.123155

Makarova, I., Shubenkova, K., & Pashkevich, A. (2018), "The Concept of the decision support system to plan the reverse logistics in automotive industry", 2018 26th International Conference on Software, Telecommunications and Computer Networks (SoftCOM), IEEE, pp. 1-6. DOI: 10.23919/SOFTCOM.2018.8555760

Marrucci, L., Daddi, T., & Iraldo, F. (2021). The contribution of green human resource management to the circular economy and performance of environmental certified organisations. *Journal of Cleaner Production, 319*, 128859. DOI: 10.1016/j.jclepro.2021.128859

Modgil, S., Gupta, S., Sivarajah, U., & Bhushan, B. (2021). Big data-enabled large-scale group decision making for circular economy: An emerging market context. *Technological Forecasting and Social Change, 166*, 120607. DOI: 10.1016/j.techfore.2021.120607

Mourtzis, D., Angelopoulos, J., & Panopoulos, N. (2022). A literature review of the challenges and opportunities of the transition from industry 4.0 to society 5.0. *Energies, 15*(17), 6276. DOI: 10.3390/en15176276

Owojori, O. M., & Okoro, C. (2022). The private sector role as a key supporting stakeholder towards circular economy in the built environment: A scientometric and content analysis. *Buildings (Basel, Switzerland)*, *12*(5), 695. DOI: 10.3390/buildings12050695

Paes, M. X., de Medeiros, G. A., Mancini, S. D., Ribeiro, F. M., & Puppim de Oliveira, J. A. (2019). Transition to circular economy in Brazil: A look at the municipal solid waste management in the state of S~ ao Paulo. *Management Decision*, *59*(8), 1827–1840. DOI: 10.1108/MD-09-2018-1053

Pagoropoulos, A., Pigosso, D.C. and McAloone, T.C. (2017), "The emergent role of digital technologies in the Circular Economy: a review", Procedia CIRP, Vol. 64, pp. 19-24, doi: . 02.047.DOI: 10.1016/j.procir.2017

Perey, R., Benn, S., Agarwal, R., & Edwards, M. (2018). The place of waste: Changing business value for the circular economy. *Business Strategy and the Environment*, *27*(5), 631–642. DOI: 10.1002/bse.2068

Prieto-Sandoval, V., Jaca, C. and Ormazabal, M. (2018), "Towards a consensus on the circular economy", Journal of Cleaner Production, Vol. 179, pp. 605-615, doi: . 12.224.DOI: 10.1016/j.jclepro.2017

Rajamanikam, A., & Solihin, M. I. (2019). Solid waste bin classification using Gabor wavelet transform. *Innovative Technology and Exploring Engineering*, *8*, 114–117.

Raworth, K. (2017). *Doughnut Economics: Seven Ways to Think like a 21st-Century Economist*. Chelsea Green Publishing.

Sadraei, R., Biancone, P., Lanzalonga, F., Jafari-Sadeghi, V., & Chmet, F. (2023). How to increase sustainable production in the food sector? Mapping industrial and business strategies and providing future research agenda. *Business Strategy and the Environment*, *32*(4), 2209–2228. DOI: 10.1002/bse.3244

Secinaro, S., Brescia, V., Calandra, D., & Biancone, P. (2021). Towards a hybrid model for the management of smart city initiatives. *Cities (London, England)*, *116*, 103278. DOI: 10.1016/j.cities.2021.103278

Singh, D., & Satija, A. (2018). Prediction of municipal solid waste generation for optimum planning and management with artificial neural network—case study: Faridabad in Haryana State (India). *International Journal of System Assurance Engineering and Management*, *9*(1), 91–97. DOI: 10.1007/s13198-016-0484-5

Stahel, W. R. (2016). The circular economy. *Nature*, *531*(7595), 435–438. DOI: 10.1038/531435a

Tang, M., & Liao, H. (2021). multi-attribute large-scale group decision making with data mining and subgroup leaders: An application to the development of the circular economy. *Technological Forecasting and Social Change*, *167*, 120719. DOI: 10.1016/j.techfore.2021.120719

United Nations. (2015). *SDGs- Sustainable Development Goals*. United Nations.

United Nations. (2023). *The Sustainable Development Goals Report 2023*. United Nations.

Upadhyay, A., Mukhuty, S., Kumar, V., & Kazancoglu, Y. (2021). Blockchain technology and the circular economy: Implications for sustainability and social responsibility. *Journal of Cleaner Production*, *293*, 126130. DOI: 10.1016/j.jclepro.2021.126130

Yetilmezsoy, K., Ozkaya, B., & Cakmakci, M. (2011). Artificial intelligence-based prediction models for environmental engineering. *Neural Network World*, *21*(3), 193–218. DOI: 10.14311/NNW.2011.21.012

Zhang, A., Venkatesh, V. G., Liu, Y., Wan, M., Qu, T., & Huisingh, D. (2019). Barriers to smart waste management for a circular economy in China. *Journal of Cleaner Production*, *240*, 118198. DOI: 10.1016/j.jclepro.2019.118198

Chapter 2
AI in Air Pollution Monitoring and Its Implementation in Developing Countries

Shiblee Noman
https://orcid.org/0000-0001-5800-7844
University of Dhaka, Bangladesh

Taskia Haq Lyric
http://orcid.org/0009-0003-4505-044X
University of Dhaka, Bangladesh

ABSTRACT

This study looks into air pollution in developing countries and how Artificial Intelligence (AI) can contribute to air quality monitoring and data-driven interventions in those nations. Using a systematic literature review, it identifies successful AI implementations, like that from South Africa, which demonstrates cost-effective monitoring, and one from India that emphasizes the need for enabling legislation and policy. The major challenges are data integration, infrastructure constraints, and the need for energy sector reform because fossil fuel combustion remains one of the major sources of pollution. The research also suggests the potential of smart city projects to improve air pollution control. Overall, the research highlights the use of AI in addressing environmental and health issues towards the realization of Sustainable Development Goals (SDG) 3—Good Health and Well-Being, and SDG 13—Climate Action.

DOI: 10.4018/979-8-3373-1892-9.ch002

Addressing Environmental Challenges With AI, Robotics, and Augmented Reality

1. INTRODUCTION

Air pollution is a serious problem for the environment and public health, especially in developing nations where urbanization, industrialization, and lax regulations all contribute to air quality decline. Conventional air pollution monitoring techniques frequently have drawbacks such as exorbitant prices, a dearth of real-time data, and insufficient coverage. On the other hand, by improving data-driven policymaking, predictive analytics, and monitoring effectiveness, artificial intelligence (AI) provides a game-changing solution.

Poor air quality remains a significant issue in many developing countries, increasing the risk of respiratory illnesses, cardiovascular complications, and early mortality. People in areas with high air pollution levels have been found to have decreased lung capacity; one study in Delhi indicated that around 40% of its population had reduced lung capacity (Rautela & Goyal, 2024). Current monitoring systems are often outdated, unavailable, or incapable of providing comprehensive and up-to-date data. AI-driven models can analyze large datasets, identify pollution trends, and make precise predictions about future air quality levels. To improve monitoring coverage and forecasting, machine learning and deep learning algorithms can utilize satellite data, Internet of Things sensors, and meteorological variables. Traditional air quality stations are expensive to build and maintain, which is not a viable solution for developing countries where people are at highest risk due to air pollution (Mckenzie et al., 2024). Here, AI-powered systems can substantially reduce monitoring costs, making them more feasible for resource-limited nations. By using cloud-based AI models to remotely access real-time air quality data, governments and regulators can make more informed decisions. Developing countries often lack the infrastructure and trained personnel necessary for extensive environmental monitoring. With minimal human involvement, AI can automate data processing and provide valuable insights. International collaboration and public-private partnerships can enhance capacity building and facilitate AI deployment across various sectors. AI-powered air pollution monitoring can equip decision-makers with evidence-based information to support the enactment of sound environmental policies. This research can help pinpoint the ethical issues, regulatory challenges, and policy frameworks needed for AI-driven environmental governance.

Research Objectives

1) To find out how AI can be used to mitigate the challenges of air pollution in developing countries where traditional air quality monitoring stations are expensive to build and maintain.
2) To find the barriers to implementing AI-based solutions for air pollution in developing countries in government policy, infrastructure, and public awareness.

Research Questions

1) How can AI be adapted to the unique challenges of air pollution in developing countries?
2) What are the barriers to implementing AI-based solutions for air pollution in these regions?

2. LITERATURE REVIEW

2.1 Air Pollution

The environment and human health are seriously threatened by air pollution, which is the release of toxic compounds into the atmosphere (Rani et al., 2011). Deforestation, automobile pollution, and industrialization are some of the causes of air pollution (Soni et al., 2020). Air pollution is toxic and causes serious respiratory and cardiovascular problems (Lawan et al., 2023). Children, the elderly, and people with pre-existing diseases are especially susceptible to the acute and long-term health impacts of both indoor and outdoor air pollution (Yamaoka, 1986). According to Rani et al. (2011), respiratory conditions, allergies, and even cancer are common health problems linked to air pollution. Primary pollutants are those that are released directly into the atmosphere, while secondary pollutants are created by chemical processes in the atmosphere (Yamaoka, 1986). Numerous control measures, including technology advancements, regulatory reforms, and the usage of cleaner energy sources, have been put into place to solve this worldwide issue (Kumar & Katoria, 2013). According to Kumar and Katoria (2013), these include enhancing fuel quality, switching to compressed natural gas for automobiles, and putting scrubbing procedures in place for industrial emissions.

2.2 Artificial Intelligence

The replication of human intellect in machines, known as artificial intelligence (AI), allows them to carry out tasks that normally demand human cognitive abilities (Nagendraswamy & Salis, 2021; Morandín-Ahuerma, 2022). Artificial intelligence can create machines which can learn, adapt and complete work that would otherwise need human intelligence (Saxena et al., 2023). Computer science, biology, psychology, and engineering are just a few of the fields that are included in artificial intelligence (Sonone & Dharme, 2019). It can be classified according to autonomy (reactive, deliberative, cognitive, or entirely autonomous) and cognitive capacity (weak, general, or superlative) (Morandín-Ahuerma, 2022). Performance in manufacturing, service systems, and expert systems has been greatly enhanced by AI applications (Pannu, 2015). These applications cover a wide range of industries, including weather forecasting, business, engineering, research, and medicine (Pannu, 2015). AI is particularly useful in the fields of computer gaming, medical picture categorization, network intrusion prevention, and power system oscillation damping (Pannu, 2015). It is anticipated that as AI technology develops, it will either complement or completely replace human abilities in a variety of domains, resulting in greater accuracy and efficiency when doing complicated tasks (Morandín-Ahuerma, 2022).

2.3 Artificial Intelligence and Environmental Management

The potential of artificial intelligence (AI) to address environmental issues and promote sustainability is increasingly recognized. Applications of AI are present in various fields, including water resource management, transportation, energy management, and biodiversity conservation (Raihan et al., 2024). These technologies enable advanced prediction models for renewable energy production while enhancing power grid management and resource allocation (Raihan et al., 2024). Furthermore, AI facilitates sustainable resource management, climate change modelling, and environmental monitoring (Shuford, 2024). Biodiversity researchers are utilizing machine learning and natural language processing techniques to forecast ecological services (Raihan et al., 2024). However, the rapid advancement of AI requires a solid legal understanding and oversight to uphold safety and ethical standards (Parveen M.A. Ansari et al., 2021). The ethical implications and potential socioeconomic impacts of AI-driven environmental solutions must be considered, even as AI offers innovative solutions for resource optimization and climate change mitigation (Lohani, 2024).

2.4 Artificial Intelligence and Air Pollution

A potent technique for predicting and forecasting air pollution is artificial intelligence (AI). With a spike in research since 2017, recent studies demonstrate the growing use of AI techniques in this field (Li et al., n.d.). Among the most popular AI techniques for predicting air quality include machine learning, deep learning, and fuzzy logic (A. Masood & K. Ahmad, 2021). Because these strategies can handle complex, non-linear data, they have demonstrated higher performance when compared to traditional statistical methods (S. Subbiah & S. Paramasivan, 2024). To anticipate the air quality index (AQI) and lower pollution on city roadways, artificial intelligence (AI)-based algorithms have been created (Rohan Gupta et al., 2024). In terms of research production, China and the US are at the top, with the Chinese Academy of Sciences holding a commanding position (Li et al., n.d.). A promising area of study for air pollution is the use of AI with environmental science techniques and inexpensive air quality sensors (Li et al., n.d.).

Research Gap

The research gap exists in the challenges of implementing AI-based solutions for air pollution mitigation in developing countries such as Bangladesh in South Asia. Thus, this research aims to find how AI can be applied to mitigate air pollution and what the challenges are in adopting AI applications.

3. METHODOLOGY

A systematic literature review methodology has been used in this work by examining academic papers, reports, and case studies of artificial intelligence (AI) in air pollution monitoring and its use in developing nations. A structured search for relevant academic literature has also been conducted across multiple databases and sources, such as academic databases, government and institutional reports, and industry and technical reports. Academic databases like Google Scholar and Semantic Scholar were used to search for resources. Most papers studied have been published in the last ten years, so the current research is thoroughly analyzed. Peer-reviewed papers from recent years were studied to ensure the authenticity of the research. Journals and case studies on AI applications for air pollution in developing countries have been analyzed to find out the different AI systems that were used, their affordability, and policy implications. Existing policies and infrastructure have also been analyzed to determine their readiness to implement AI solutions, and the direction for future research will be identified. It was studied whether other research supported the findings from the resources used for credibility.

4. FINDINGS AND DISCUSSION

4.1 AI in Air Pollution Monitoring

4.1.1 Risks of Air Pollution

The presence of pollutants in the air from fossil fuel combustion and other industrial processes has harmful effects on human health. Exposure over a long period may cause cardiovascular disease and respiratory diseases such as asthma and bronchitis. Children and the elderly are especially prone to such risks. Cardiovascular diseases like heart attack and stroke are also caused by exposure to harmful pollutants. It can also harm mental health. Harmful pollutants also cause hypertension, headaches and lack of attention. The WHO estimates that about 4.2 million people die every year from air pollution. The risk of air pollution is particularly high in areas with high population density. In addition to human health hazards, pollutants also harm plant and animal lives, causing biodiversity loss. It also causes a lack of productivity and strain on healthcare facilities (Mckenzie et al., 2024; Rautela & Goyal, 2024).

4.1.2 Challenges with Traditional Air Pollution Monitoring

Traditional air quality and weather monitoring stations are expensive to build and maintain. Since they cannot be built easily and, in many places, there is limited coverage by traditional monitoring stations. There are also resource constraints, and the stations cannot monitor data at high frequency and cannot always do real-time monitoring (Mckenzie et al., 2024).

For these challenges, it is necessary to build cost-effective sensors that can be easily made in large quantities and provide extensive coverage for better detection of pollution sources and main hotspots. They can also provide real-time data so that more effective public health interventions can be taken and better policy decisions can be made (Mckenzie et al., 2024).

The introduction of AI and IoT has transformed air pollution monitoring by enabling real-time, data-driven decision-making. IoT-enabled sensors collect data on pollutants like PM2.5, NOx, CO, and VOCs, which AI models analyze to find patterns, identify pollution sources, and predict air quality. Unlike traditional fixed-location systems, this AI can offer greater accuracy and responsiveness. AI can generate instant pollution alerts through apps and software, allowing authorities to take prompt action, such as issuing health advisories or rerouting traffic. In developing regions, where traditional systems are limited and expensive, low-cost IoT sensors combined with AI provide a scalable and affordable solution. This integration also supports long-term policymaking by offering insights into pollution trends and intervention outcomes, greatly enhancing public health and environmental sustainability (Oko & Rweyemamu, 2024).

Internet of Things (IoT) devices can be used to collect air pollution data, as studied by Mani et al. (2021) in their research. A prototype IOT device for collecting air pollutants was developed, which would collect air pollutants like carbon monoxide, ozone and ammonia. The device used sensor modules to register the data. Different statistical models were used to see which would give the best results in forecasting data. It was seen that the Auto-Regression Integrated Moving Average Model (ARIMA) was the better option after comparing the forecast data and test data. Though the device was used in one location, it was cost-effective to use it in different places across India for accurate real-time air pollution tracking. So, even in developing countries like India, AI can be used to monitor air pollution.

Another cost-effective solution is the use of virtual sensors for air quality monitoring, as studied by Zaidan et al. (2023). Virtual sensors are machine learning models integrated into low-cost sensors (LCS) to collect measurements of different pollutants. Such virtual sensors can be used to increase the amount of air quality monitoring data by having more monitoring stations which are small in scale. Virtual sensors can be used to monitor pollutants that are costly or difficult to measure. They do so by analyzing the correlation between pollutants and environmental variables. Thus, this allows for low-cost sensors to be deployed for wide-scale monitoring and offers a cost-effective solution for air quality monitoring.

4.1.3 Case Study SACAQM and AI_r

The South African Consortium of Air Quality Monitoring (SACAQM) was established with the aim of bringing together public, private and research institutions to create cost-effective air quality monitoring systems. They mainly created a system that will complement the existing air quality monitoring stations and not replace them. These will help to enhance the readings of existing stations (Mckenzie et al., 2024).

SACAQM developed the AI_r system to complement the existing monitoring stations. The AI_r consisted of a printed circuit board, which a battery or external power supply can power through a USB cable. Both ways were designed so that the system could work with very little power consumption. A Sensirion SEN55 sensor was used to monitor Particulate Matter (PM), temperature, humidity, Volatile Organic compounds (VOC), and nitrogen oxides. IoT devices were used to connect separate sensors. The data gathered from the AI_r device and traditional monitoring system were calibrated, and it was seen that both have similar consistency and accuracy, so the AI_r system was also reliable. Calibration is used to compare the AI_r device readings and data from traditional stations and see if the AI_r data is accurate. In this regard, the comparison revealed that the system's data is reliable. The data were presented on a website where people can access it easily, understand it and make informed decisions based on the data. From the data of the nodes, Artificial Intelligence and Machine Learning were used to create predictions and forecasting for air quality data (Mckenzie et al., 2024).

The AI_r devices were deployed in school buildings in several cities in South Africa. This would raise awareness among people about air pollution and its monitoring. Thus, it would also help to see how air pollution can affect vulnerable and poor areas. Also, using real-time data, teachers can make informed decisions, like whether to continue outdoor activities for the children (Mckenzie et al., 2024).

Figure 1. Process outline of SACAQM AI_r

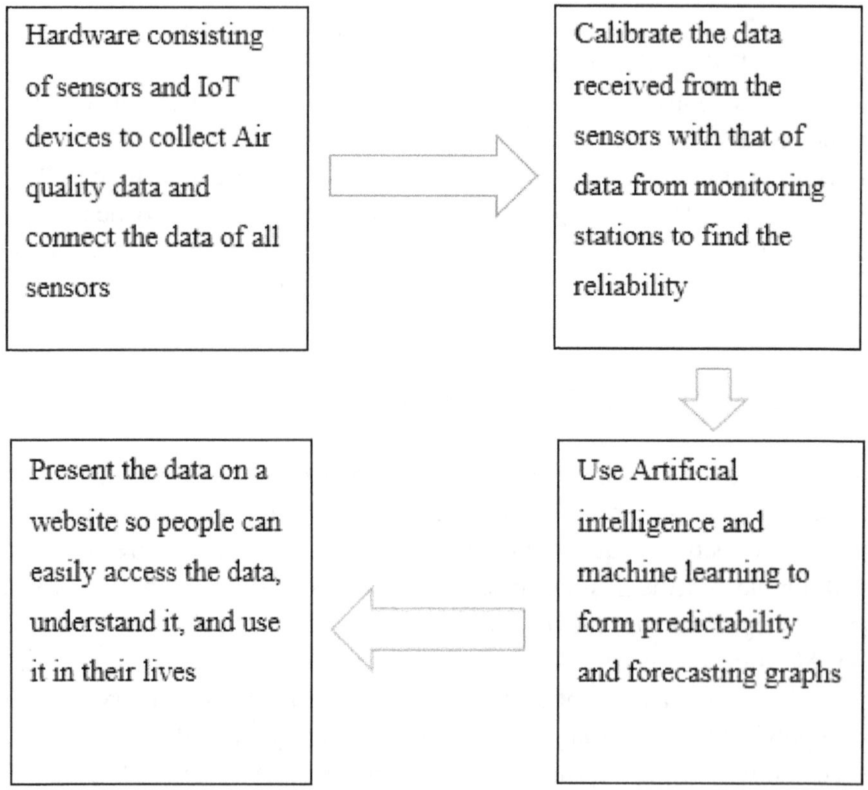

Note: Developed by the authors' comprehension of AI r

4.1.4 Case Study India's Encounter with Air Pollution

4.1.4.1 Pollutants

The main pollutants in India are particulate matter (PM), organic aerosols, black carbon, water-soluble brown carbon and volatile organic compounds. These are released from traffic emissions and residential cooking and heating. Generally, fuel combustion is the dominant source of these pollutants, so finding cleaner energy sources is very important. However, there are also atmospheric interventions, such as UV rays, that create some pollutants. Therefore, it is also necessary to gain more knowledge on air pollution (Rautela & Goyal, 2024).

4.1.4.2 Remedy

India has been enacting legislation aimed at tackling air pollution since 1905, and the process is still ongoing in the 21st century in the internet era. Key legislation is the Air (Prevention and Control of Air Pollution) Act of 1981, which was amended multiple times. Then, there is the Environment Protection Act 1986 and the Motor Vehicles Act. There have also been many judicial interventions and executive actions to control air pollution. Judicial interventions in the form of petitions filed by M. C. Mehta were quite influential in forming guidelines and policies. There were also executive actions, such as the Auto Fuel Policy in 2003, which was revised later to take clean energy initiatives (Rautela & Goyal, 2024).

India has been trying to monitor air quality since the 1980s when it formulated the National Ambient Air Quality Standards (NAAQS), and the Central Pollution Control Board (CPBP) has been monitoring it and providing guidelines for the public. There are Continuous Ambient Air Quality Monitoring System (CAAQMS) stations in 1082 locations for real-time data and decision-making (Rautela & Goyal, 2024).

Many studies have been conducted in India to predict air pollutants and different air quality variables using different artificial intelligence and machine learning models. The pollutants included sulfur dioxide, ozone, carbon monoxide and PM 2.5. Most studies were successful in predicting the levels of pollutants so it shows that AI models can be used for air quality monitoring in India (Rautela & Goyal, 2024).

4.1.4.3 Challenges in India

In the first place, there is a technological barrier as highly advanced technology is expensive, and the rural areas lack adequate infrastructure for new technological solutions. Then, there is a problem in regulation, though there are established laws, frameworks and policies that are not implemented accordingly, and thus better mechanisms to punish non-compliance are required. Thirdly, there is a large gap in public awareness about the causes and effects of air pollution, which necessitates people to be made more aware of the hazards of air pollution so that they can take proactive measures to reduce their carbon footprints. India has many harmful agricultural practices, such as burning crop residues, which release large quantities of pollutants into the air. Farmers should be provided with proper alternatives to dispose of crop residue rather than burning it. Urbanization and infrastructure development also release pollutants such as dust and cement into the air. Sustainable and green building technologies need to be used that release fewer pollutants and prioritize air quality. Also, there is the issue of cross-border pollution, for which India needs to collaborate with its neighbours to address air quality concerns. Furthermore, since air pollution is also related to climate change, long-term climate change goals are required to cope with it. Lastly, there is also the challenge of social equity, so no group is affected more by reforms (Rautela & Goyal, 2024).

4.1.4.4 Potential for India

Advancements in Artificial Intelligence and Machine Learning models can be used with weather forecasting numerical models to predict and manage air pollution. These technologies can offer a deeper understanding of air pollution, the effect of atmospheric conditions on air pollution and the dispersal patterns of pollutants. Moreover, these technologies can create breakthroughs in sustainable energy sources and can lead the transformation from traditional fossil fuels to cleaner alternatives, thus reducing the overall pollution in India. However, India also needs to collaborate with other countries to share knowledge, find the best practices, and conduct joint research initiatives. In this way, India can contribute to global knowledge and understanding of air pollution and benefit from it. India needs access to global expertise, technologies and resources for which collaboration is essential (Rautela & Goyal, 2024).

4.2 Controlling Air Pollution by AI

Odirichukwu et al. (2024) have analyzed how AI can be used to develop early warning systems that will help people with respiratory illnesses stay inside. They used an IoT device to monitor air quality in Nigeria and find data that showed hazardous levels of air pollutants. Many IoT devices are being manufactured for use, such as the Canazree AI, but they are not available in developing countries like Nigeria, and that is an issue. Therefore, the government should put resources into building monitoring infrastructures to measure air quality. Such initiatives are required for both short-term and long-term environmental protection and achieving SDG goals 3,

6, 11, 13 and 17, which include good health, clean water and sanitation, sustainable cities, and climate action partnership for goals.

Monitoring can be done with AI, and it can be used to control air pollution in many ways. By monitoring the level of pollutants, effective policies to mitigate them can be put in place, such as industrial regulations. With AI, industrial processes can be optimized so that waste and emissions can be reduced and stopped. Further monitoring can reveal whether the industries are following the standards or not.

AI can be used in making smart grids and energy management that will help reduce loss in various systems by making them more efficient Cheng & Yu (2019). This will reduce energy waste and the demand for energy, and it will help reduce pollution caused by energy production. AI is also used to integrate various renewable energy sources into the main grid, optimize the system, and reduce costs (Boza & Evgeniou, 2021). Renewable energy sources do not pollute the environment by emitting greenhouse gases, so it can be beneficial to reduce air pollution.

AI can help in smart traffic control using real-time sensor data and forecasting to have a positive impact on mobility and the environment. AI can be integrated with GPS, surveillance cameras, and vehicles to monitor where there is a high level of traffic congestion and help people avoid those areas, thus reducing the idle time spent in traffic congestion. Besides saving time, this also means that less fuel is burnt by cars, and this means less greenhouse emissions and a reduction in air pollution (Nallaperuma, 2019).

4.3 Policy Implications

Ramadan et al. (2024) demonstrate that real-time air quality can be monitored in industrial areas for pollutants such as ammonia, carbon monoxide, nitrogen dioxide, methane, carbon dioxide, sulfur dioxide, and ozone. This research carries significant policy implications, as real-time monitoring systems facilitate continuous data collection and prompt action to address pollution levels. Additionally, such monitoring will help establish comprehensive standards for the maintenance of air quality monitoring systems, including data accuracy and system calibration. The data generated from this system will enable industries to implement measures to control emissions and develop safety protocols. Moreover, it will foster public-private partnerships, allowing collaboration to identify optimal air pollution control methods, while the government can provide assistance or subsidies to small firms addressing air pollution. Lastly, this research has implications for promoting worker safety and regulatory compliance.

4.3.1 Governance and Air Pollution Monitoring

In a study, a dataset was created on 10,000 urban centres in the Global South. The dataset examined the Air Quality Monitoring (AQM) practices in these low to middle-income cities. It was found that approximately 90% of these cities did not engage in any AQM activities, while the remaining 10% exhibited varying degrees of such activities. AQM activities were found to depend on the income levels of the cities and the characteristics of their institutions, particularly the extent of democracy present. Urban areas with higher income levels and more democratic institutions are more likely to implement AQM activities in response to high levels of air pollution. In democracies, public authorities are more responsive to the demands of citizens and facilitate the involvement of various sectors in the development and distribution of low-cost sensors. In contrast, non-democracies have fewer monitoring stations, and democracies experience a higher proliferation of low-cost sensors. Additionally, democracies conduct monitoring in both high and low-pollution areas, whereas non-democracies primarily focus on low-pollution areas (Schoch et al., 2025).

4.3.2 Implication for public health and urban planning

When compared to conventional statistical and deterministic models, AI-driven models—especially those driven by machine learning and deep learning algorithms—have shown impressive accuracy and efficiency in predicting air pollution levels. To produce accurate air quality forecasts, these sophisticated models make use of enormous volumes of historical and real-time data, such as traffic patterns, industrial emissions, satellite imaging, and meteorological variables. The ability of AI-based models to combine spatial and temporal data is one of its main advantages; this enables them to capture intricate correlations between various environmental parameters and pollution sources. Artificial intelligence (AI) systems continuously learn from data, enhancing their predicting abilities over time, in contrast to traditional models that could rely on preset equations and assumptions. Given how quickly weather patterns and human activity may alter air quality, this flexibility is especially useful for short-term air pollution forecasting (Neo et al., 2022).

Since they allow policymakers and medical practitioners to evaluate the possible health hazards associated with poor air quality, accurate air pollution forecasts are crucial for public health management. Predictive models powered by artificial intelligence (AI) can assist in predicting pollution spikes and provide early warnings, enabling communities and governments to put preventive measures in place, like limiting industrial emissions, modifying traffic patterns, or informing susceptible groups to take precautions. Healthcare providers can also use these forecasts to plan for a rise in respiratory and cardiovascular illnesses associated with exposure to air pollution. Additionally, by incorporating air quality forecasts into environmental policy and urban planning, AI models aid in the creation of smart city projects (Neo et al., 2022).

AI-powered monitoring systems, for example, can assist in identifying the ideal sites for green areas, maximizing the use of public transit to cut emissions, and facilitating the transition to clean energy. All things considered, the use of AI in air pollution forecasting is a revolutionary step toward proactive environmental management and the defence of public health. AI-driven algorithms help to create healthier, more sustainable urban settings by consistently enhancing the precision and dependability of air quality predictions (Neo et al., 2022).

4.4 International and Regional Policies on AI in Environmental Management

4.4.1 EU

The European Union (EU) is making major efforts to incorporate sustainability considerations into its AI legislation to guarantee that AI technologies support environmental objectives while respecting fundamental rights and ethical norms. The EU is investigating methods to reinterpret current legislative frameworks, like the General Data Protection Regulation (GDPR), in a way that is consistent with sustainability goals as part of this endeavour. For instance, the GDPR's data-sharing rules might be reexamined to allow AI-powered emissions tracking and environmental monitoring while upholding stringent privacy safeguards. The EU is proposing new laws, most notably the EU AI Act, which intends to create comprehensive guidelines for the creation, application, and use of AI systems, in addition to amending existing ones. The AI Act places more stringent rules on high-risk AI systems and classifies AI applications according to risk levels. Additionally, the law promotes accountability, transparency, and moral AI practices, with sustainability emerging as a crucial factor. To ensure that AI systems help lower energy usage and carbon footprints, the EU is anticipated to incorporate environmental impact evaluations into AI regulatory compliance (Perucica & Andjelkovic, 2022; Hacker, 2022).

The EU is adopting co-regulation—a governance model that blends industry-driven standards and best practices with regulatory oversight—to achieve a balance between individual rights and larger environmental concerns. This strategy makes it possible for AI development to be flexible while maintaining environmental sustainability as a top concern. The EU wants to establish an AI ecosystem that links technical advancement with climate action and sustainable resource management by working with companies, researchers, and civil society. Sustainability by design, which stresses incorporating environmental factors into AI development from the beginning, is another essential element of the EU's approach. These imply that resource optimization, carbon reduction, and energy efficiency should all be considered when designing AI systems. For example, AI-driven data centres could have to run on renewable energy, and AI models might be tuned to use the least amount of computing power possible.

With these diverse initiatives, the EU hopes to establish itself as a world leader in sustainable AI governance and establish a standard for how AI can be governed to benefit the environment and society. It is anticipated that incorporating sustainability concepts into AI regulations will have an impact on other areas and further the worldwide goal of developing AI in an ecologically conscious manner (Perucica & Andjelkovic, 2022; Hacker, 2022).

4.4.2 China

Thanks to robust government support and technology breakthroughs, Chinese businesses are increasingly using artificial intelligence (AI) to improve environmental management. The Chinese government has acknowledged AI as a crucial instrument for accomplishing its lofty sustainability objectives, which include becoming carbon neutral by 2060. China is aggressively pushing companies to include AI-driven solutions for energy efficiency, pollution control, and sustainable resource management through funding, legislation, and regulatory frameworks (Sun et al., 2024).

Energy optimization is one of the main uses of AI in environmental management. Smart grids driven by AI assist in balancing the supply and demand for electricity, cutting down on energy waste and improving power distribution efficiency. AI algorithms are being used by many Chinese businesses, especially in sectors like manufacturing, logistics, and construction, to forecast patterns in energy usage, modify operations accordingly, and lower their carbon footprints. AI-driven predictive maintenance reduces energy loss and equipment downtime in industrial facilities by ensuring that machinery runs as efficiently as possible (Sun et al., 2024).

AI is also being used to check the quality of the air and water, which is important considering China's history of pollution-related issues. AI-powered sensor networks are being deployed by businesses to continually monitor air pollution, identify dangerous pollutants, and give regulatory agencies real-time information. AI is being utilized in the water industry to optimize wastewater treatment, guaranteeing that contaminants are effectively eliminated before water is released back into the environment.

Furthermore, AI is revolutionizing the creation of smart cities. To cut emissions and increase sustainability, cities like Beijing, Shanghai, and Shenzhen are incorporating AI into trash management, transportation, and urban planning. Artificial intelligence (AI)-powered traffic management systems maximize road utilization by minimizing traffic and fuel consumption, which immediately improves air quality (Sun et al., 2024).

The success of these AI-powered environmental projects emphasizes the importance of public-private partnerships in China. To encourage companies to work with academic institutions and government organizations to create AI-based green solutions, the government has started several programs, such as the AI Innovation Action Plan for Ecological and Environmental Protection. Furthermore, big private companies like Alibaba, Baidu, and Tencent, as well as state-owned enterprises (SOEs), are aggressively funding AI-driven sustainability initiatives. China is establishing itself as a global leader in AI-driven environmental management by encouraging cooperation between the public and commercial sectors, which proves that AI can be a potent force for industrial sustainability and ecological preservation. Chinese businesses will be vital in determining how AI-powered environmental governance develops in the future as these technologies advance (Sun et al., 2024).

4.5 Global Energy Sector

Artificial intelligence (AI) is changing the global oil and gas sector by transforming environmental management, promoting sustainability, and supporting the transition to cleaner energy. Countries like the USA, UK, Sweden, and the UAE are using AI to optimize drilling and extraction, waste reduction, and boost energy efficiency through tools like machine learning, predictive maintenance, and digital twins. AI also assists in detecting and mitigating environmental hazards such as carbon emissions, methane leaks, and oil spills via drones, IoT sensors, and satellites. Plus, AI supports the integration of renewable energy in the energy grids by balancing outputs from solar and wind with traditional oil and gas infrastructure. It also assists in regulatory compliance, helping companies monitor emissions and forecast policy shifts. These innovations are advancing sustainability goals while modernizing energy systems globally (Chutcheva et al., 2022).

4.6 AI, Smart Cities and Air Pollution Control

United Nations Sustainable Development Goal 11 is to create sustainable cities and communities that place importance on economic opportunities, safety, resilience, and sustainability. It also includes robust infrastructure with public transportation and green spaces, plus less resource consumption and less waste and emission. Smart cities focus on using technology to improve efficiency and provide intelligent solutions for energy efficiency, waste management and other criteria. So, smart cities and sustainable cities complement each other; that is, smart cities can become sustainable. Global South has been seeing a rapid increase in urbanization, with many city centres aiming to become smart cities (Das, 2025).

AI can be used in many different sectors in a smart city, where the main sectors are planning and management and the economy include finance and optimum resource allocation. Then, there is infrastructure and mobility that consists of transportation and mobility, energy management and waste management. AI can also help in environmental monitoring and reducing emissions that will help reduce air pollution. It will also help in governance, participation, responsiveness, equity, quality of life, healthcare and education (Das, 2025).

4.6.1 Barriers to Smart Cities

There are many barriers to creating smart, sustainable cities; first of all, there is an infrastructure deficit as many cities do not have good broadband internet networks, reliable energy grids, and efficient transport systems. Also, many cities have unequal access to healthcare and education. Then, effective policies for digital technology and artificial intelligence are not in place in many cities. Plus, many cities have limited access to funding and investment.

4.6.2 Smart Cities in the Global South

Despite the barriers, many cities are implementing smart city initiatives. First of all, India has created smart financial solutions in digital payment scenarios, which increase transparency and accountability and reduce corruption. Kenya is also implementing smart city initiatives in main city centres like Nairobi, which have smart grid technologies enabled by AI that monitor real-time energy consumption and help reduce waste. Rwanda also has a digital payment system and has implemented an IoT-enabled car parking system to reduce congestion (Das, 2025). Thus, smart city initiatives can be used to monitor and enhance energy usage and traffic so that there will be less wastage and more efficient energy consumption, leading to better air quality as less fuel will be needed for energy.

4.6.3 Policy Recommendations for Implementing Smart Cities

- Cities should identify key areas that require technological innovation, such as transportation, healthcare, waste management and energy infrastructure and prioritize these sectors first.
- City authorities should encourage partnerships between the public and private sectors so that the private sector has funds to research, implement changes

and involve international experts and other stakeholders, which will increase efficiency and transparency.

- Governments should develop their institutional capacities through training in AI and collaboration with universities and research institutes.
- Encourage people's participation so that marginalized communities will not be left behind, and create inclusive policies so that all people will benefit equally and the digital divide will be reduced.

4.7 Key Challenges in AI Adoption and Solutions

Artificial Intelligence (AI) adoption is a game-changer, especially in fields like public services, healthcare, and environmental management. However, a number of significant obstacles stand in the way of its optimal application and broad adoption. To guarantee that the advantages of AI are optimized, these issues need to be resolved through cooperation, efficient governance, and ethical concerns.

4.7.1 Data Integration Issues

Data integration is one of the biggest obstacles to the adoption of AI. For AI systems to make wise conclusions, high-quality, varied, and structured data is essential. Integrating data from various sources, including government databases, Internet of Things sensors, satellite imaging, and public records, can be challenging for both public and private companies. Additionally, the data could be inaccurate, fractured, or compartmentalized, which makes training and using AI models more difficult (Uwaoma et al., 2024). Establishing interoperable data systems that can easily communicate and process information across several platforms and stakeholders is necessary to overcome data integration challenges. Businesses and governments must invest in data processing and cleaning technology, establish open data rules, and standardize data formats.

4.7.2 Lack of AI Literacy

The lack of AI literacy among important stakeholders, such as legislators, corporate executives, and the broader labour force, is another significant obstacle. To make wise decisions in the complicated realm of artificial intelligence, one must have a firm grasp of algorithms, data analysis, and machine learning concepts. There are skill gaps that restrict people's capacity to use and profit from AI technologies in many developing nations and even in some developed economies (Uwaoma et al., 2024). Governments and educational establishments should establish AI education and training initiatives at all levels, from professional development centres to universities, to address this. Decision-makers should be the focus of AI literacy programs to make sure they can oversee AI initiatives and comprehend their ramifications.

4.7.3 Resistance to Technological Change

Many facets of society, including both persons and institutions, frequently oppose technological advancement. People can worry that AI would require large expenditures in new infrastructure, cause data privacy issues, or replace jobs. A societal hesitancy to embrace AI is frequently also a result of doubts about its efficacy or its unforeseen repercussions (Uwaoma et al., 2024). Clear communication about the advantages of AI and its potential long-term benefits—such as improved productivity, cost savings, and the creation of jobs in new industries—is necessary to overcome reluctance. Governments can take a proactive approach to change management by offering incentives for the adoption of AI and developing frameworks that facilitate workforce changes, such as retraining initiatives. AI solutions should also be made easy to use and integrated into current systems to lessen resistance and friction. As was seen from the case study in India, people are still clinging to old practices of burning crop residue, which greatly increases the level of pollutants in the air. People are less aware of the causes and consequences of air pollution, and thus, they need to be educated about such practices which damage the environment. Governments should develop other sustainable methods of waste disposal rather than burning, which are less harmful to the environment, and AI can be used to make the new systems efficient.

4.7.4 Ethical Considerations

Several ethical questions, including those about privacy, bias, accountability, and transparency, may be brought up by AI technologies. For example, if the data used to train AI algorithms reflects past disparities, the algorithms may unintentionally reinforce biases. Furthermore, because AI systems frequently depend on vast amounts of sensitive or personal data to operate efficiently, worries about data privacy are intensified. It is also challenging for people and organizations to comprehend how and why decisions are made when AI decision-making systems are opaque (Uwaoma et al., 2024). Strong regulatory frameworks must be created to guarantee that AI systems are developed and implemented ethically and responsibly to lessen these ethical dilemmas. Organizations and businesses must also embrace ethical AI standards that support the eradication of prejudice and protect individuals' privacy.

4.7.5 Collaborative Efforts and Robust Governance

Lastly, cooperation amongst various stakeholders—including governments, businesses in the private sector, academic institutions, and civil society—is necessary to overcome the obstacles to AI adoption. Addressing the cross-cutting problems related to AI deployment requires a concerted effort. To ensure that AI advances the public benefit without violating people's rights or escalating social injustices, strong governance frameworks must be established to supervise its development and application. Public forums should be formed for people's participation in AI and air quality monitoring issues (Uwaoma et al., 2024). As was seen from the study by Schoch et al. (2025), it is important for democratic participation to ensure that there is adequate air quality monitoring. So, general people also need to be more active in voicing their problems due to air pollution and pressure the authorities to act in the public's interests.

4.8 Future Research Implication

This study paves the way for future research that can address more local challenges, such as the consequences of limited infrastructure, scarcity of data for AI language models, and policy gaps in local governance. Future studies can work on solutions that can be implemented in a more local and decentralized way that are affordable and scalable and cater to local needs. Low-cost sensors and AI models should be given more importance for air quality monitoring and the use of data for public health and environmental policies. Additionally, inclusive and sustainable policy frameworks that encourage community understanding and participation need to be given due importance.

5. CONCLUSION

AI has the potential to transform air pollution monitoring, particularly in underdeveloped nations where conventional methods often fail due to a lack of infrastructure, resources, and data. Developing countries can enhance their ability to predict pollution levels, monitor air quality in real time, and implement prompt interventions by leveraging AI technologies such as machine learning, predictive modelling, and IoT integration. In addition to offering a more economical and effective way to evaluate and reduce air pollution, AI-driven solutions can be tailored to address specific local challenges, thereby enhancing environmental preservation, public health, and policymaking. As demonstrated in the earlier discussion, developing countries like India and Nigeria have successfully experimented with AI data monitoring. Therefore, it can be applied globally to mitigate air pollution.

However, several challenges must be addressed for AI to be effectively utilized in air pollution monitoring in underdeveloped nations, including poor data integration, inadequate technical infrastructure, and low levels of AI literacy. Strong governance structures, adequate training for local stakeholders, and ethical considerations are also essential for the implementation of AI technology to ensure equitable access and prevent increasing social inequality.

Governments, international organizations, and the business sector must collaborate and invest in infrastructure development, capacity building, and the establishment of strong regulations to ensure the long-term effectiveness of AI in air pollution monitoring. With proper support, AI has the potential to significantly improve air quality control, enhance citizen well-being, and assist developing nations in achieving their environmental and public health goals. Undoubtedly, digital technology will play a crucial role in the future of air pollution monitoring, with AI serving as a key enabler of this transformation.

Figure 2. Summary of findings compiled by the authors

Traditional monitoring system

- Expensive
- Less coverage
- Less real-time data

AI enhanced monitoring system

- Less expensive
- Can be deployed in greater quantities so there will be more coverage
- Since, more can be made and distributed, it will enable better real-time data

Barriers

- Technological gap
- Regulatory and policy gap
- Lack of public awareness

Policy recommendations

- More global approach for information sharing and expertise sharing
- Encourage people to take initiatives and participate in decision making process for air quality monitoring and pollution control

REFERENCES

Ansari, S. P., Odongo, J. O., Nomani, M. Z., Salahuddin, G., Faez Hasan, M., & Pallathadka, L. K. (2021). Evaluating the role of environment management system based on artificial intelligence. *Materials Today: Proceedings*.

Boza, P., & Evgeniou, T. (2021). Artificial intelligence to support the integration of variable renewable energy sources to the power system. *Applied Energy*, *290*, 116754. DOI: 10.1016/j.apenergy.2021.116754

Cheng, L., & Yu, T. (2019). A new generation of AI: A review and perspective on machine learning technologies applied to smart energy and electric power systems. *International Journal of Energy Research*, *43*(6), 1928–1973. DOI: 10.1002/er.4333

Chutcheva, Y., Kuprianova, L., Seregina, A., & Kukushkin, S. (2022). Environmental management of companies in the oil and gas markets based on AI for sustainable development. *Frontiers in Environmental Science*, *10*, 952102. Advance online publication. DOI: 10.3389/fenvs.2022.952102

Das, D. K. (2025). Digital Technology and AI for Smart Sustainable Cities in the Global South: A Critical Review of Literature and Case Studies. *Urban Science (Basel, Switzerland)*, *9*(3), 72. DOI: 10.3390/urbansci9030072

Gupta, R., Manna, A., Sharma, H., Pandit, P., Pan, S., & Singh, Y. (2024). Air Pollution Prediction Systems in City Highways using Artificial Intelligence. *2024 International Conference on Electrical Electronics and Computing Technologies (ICEECT)*, *1*, 1-6. DOI: 10.1109/ICEECT61758.2024.10739234

Hacker, P. (2024). Article: Sustainable AI Regulation. *Common Market Law Review*, *61*(2), 345–386. Advance online publication. DOI: 10.54648/COLA2024025

Kumar, S., & Katoria, D. (2013). Air Pollution and its Control Measures.

Lawan, R., Uumar, H. S., & Auwalu, A. (2023). EFFECTS OF AIR POLLUTION ON HUMAN HEALTH: A REVIEW. *International Research Journal of Modernization in Engineering Technology and Science*. Advance online publication. DOI: 10.56726/IRJMETS33055

Li, X., Jahani, A., Kokol, P., cnYaolongChen, Guo, Q., Ren, M., Wu, S., Sun, Y., Wang, J., Wang, Q., Ma, Y., & Song, X. Applications of artificial intelligence in the field of air pollution: A bibliometric analysis.

Lohani, N. (2024). *AI-based Environmental Sustainability: Transforming Conservation Efforts*. International Journal For Multidisciplinary Research.

Mani, G., Viswanadhapalli, J. K., & Sriramalakshmi, P. (2021, November). AI powered IoT based real-time air pollution monitoring and forecasting. []. IOP Publishing.]. *Journal of Physics: Conference Series, 2115*(1), 012016. DOI: 10.1088/1742-6596/2115/1/012016

Masood, A., & Ahmad, K. (2021). A review on emerging artificial intelligence (AI) techniques for air pollution forecasting: Fundamentals, application and performance. *Journal of Cleaner Production, 322*, 129072. DOI: 10.1016/j.jclepro.2021.129072

Mckenzie, R., Cervello, A., Chabalala, V., Chiraira, I., Mathaha, T., Mayana, L., Njara, N., Tshuma, M. T., Kong, J., Cheng, L., Smith, B., & Mellado, B. (2024, November). A Cost- Effective Air Quality Monitoring System for the Global South. In *2024 IEEE International Humanitarian Technologies Conference (IHTC)* (pp. 1-7). IEEE. DOI: 10.1109/IHTC61819.2024.10855074

Morandín-Ahuerma, F. (2022). What is Artificial Intelligence? *International Journal of Research Publication and Reviews.*

Nagendraswamy, C., & Salis, A. (2021). A review article on artificial intelligence. *Annals of Biomedical Science and Engineering.*

Nallaperuma, D., Nawaratne, R., Bandaragoda, T., Adikari, A., Nguyen, S., Kempitiya, T., De Silva, D., Alahakoon, D., & Pothuhera, D. (2019). Online Incremental Machine Learning Platform for Big Data-Driven Smart Traffic Management. *IEEE Transactions on Intelligent Transportation Systems, 20*(12), 4679–4690. DOI: 10.1109/TITS.2019.2924883

Neo, E., Hasikin, K., Mokhtar, M., Lai, K., Azizan, M., Razak, S., & Hizaddin, H. (2022). Towards Integrated Air Pollution Monitoring and Health Impact Assessment Using Federated Learning: A Systematic Review. *Frontiers in Public Health, 10*, 851553. Advance online publication. DOI: 10.3389/fpubh.2022.851553 PMID: 35664109

Odirichukwu, J. C., Odirichukwu, S. P. C., Okolie, S. A., Njoku, O. A., Dimoji, C., Ndigwe, C., Ekuma, G.O., Onuoha, G., Igwe, F.A., Chika-Ugada, P.K. & Chima, T. Leveraging AI- Driven IoT Systems for Enhanced Air Quality Management in Nigeria: An Impact Examination towards Sustainable Development Goal.

Ooko, S., & Rweyemamu, E. (2024). *Monitoring and Predicting African Rural Household Air Pollution Using Internet of Things and Artificial Intelligence.* Pan-African Journal of Health and Environmental Science., DOI: 10.56893/ajhes2024v03i01.06

Pannu, A. (2015). Artificial Intelligence and its Application in Different Areas.

Perucica, N., & Andjelkovic, K. (2022). *Is the future of AI sustainable? A case study of the European Union. Transforming Government: People*. Process and Policy., DOI: 10.1108/TG-06-2021-0106

Raihan, A., Paul, A., Rahman, M.S., Islam, S., Paul, P., & Karmakar, S. (2024). Artificial Intelligence (AI) for Environmental Sustainability: A Concise Review of Technology Innovations in Energy, Transportation, Biodiversity, and Water Management. *Journal of Technology Innovations and Energy*.

Ramadan, M. N., Ali, M. A., Khoo, S. Y., Alkhedher, M., & Alherbawi, M. (2024). Real-time IoT- powered AI system for monitoring and forecasting of air pollution in industrial environment. *Ecotoxicology and Environmental Safety*, *283*, 116856. DOI: 10.1016/j.ecoenv.2024.116856 PMID: 39151373

Rani, B., Singh, U., Maheshwari, R.K., & Garh, L. (2011). Menace of Air Pollution Worldwide.

Rautela, K. S., & Goyal, M. K. (2024). Transforming air pollution management in India with AI and machine learning technologies. *Scientific Reports*, *14*(1), 20412. DOI: 10.1038/s41598-024-71269-7 PMID: 39223178

Saxena, P., Saxena, V., Pandey, A., Flato, U. a. P., & Shukla, K. (2023). Multiple aspects of artificial intelligence. https://doi.org/DOI: 10.60148/muasartificialintelligence

Schoch, M., Fournier De Lauriere, C., & Bernauer, T. (2025). Monitoring Urban Air Pollution in the Global South: Large Gaps Associated with Economic Conditions and Political Institutions. bioRxiv, 2025-03.

Shuford, J. (2024). Interdisciplinary Perspectives: Fusing Artificial Intelligence with Environmental Science for Sustainable Solutions. *Journal of Artificial Intelligence General science (JAIGS) ISSN:3006-4023*.

Soni, S., Kumar, C. K., Rawat, S. S., & Prajapat, Y. K. (2020). A REVIEW ON AIR POLUTION. *International Journal of Advance Research and Innovative Ideas in Education*, *6*, 460–464.

Sonone, S., & Dharme, A. (2019). A review paper on simulated intellect. *International Journal of Physics and Mathematics*.

Subbiah, S. S., & Paramasivan, S. K. (2024). Prediction of air quality pollutants using artificial intelligence techniques: A review. *WOMEN IN PHYSICS:7th IUPAP International Conference on Women in Physics*. DOI: 10.1063/5.0183240

Sun, J., Guan, X., Zeng, Y., Zhang, J., Chen, X., & Zhan, X. (2024). Addressing pollution challenges for enterprises under diverse extreme climate conditions: Artificial intelligence- driven experience and policy support of top Chinese enterprises. *Frontiers in Public Health*, *12*, 1436304. Advance online publication. DOI: 10.3389/fpubh.2024.1436304 PMID: 39301513

Uwaoma, P., Eleogu, T., Okonkwo, F., Farayola, O., Kaggwa, S., & Akinoso, A. (2024). AI's Role in Sustainable Business Practices and Environmental Management. *International Journal of Research and Scientific Innovation*. https://doi.org/.DOI: 10.51244/IJRSI.2023.1012029

Yamaoka, S. (1986). HEALTH EFFECTS OF AIR POLLUTION. *Chemistry & Industry*, 827–830.

Zaidan, M. A., Motlagh, N. H., Boor, B. E., Lu, D., Nurmi, P., Petäjä, T., Ding, A., Kulmala, M., Tarkoma, S., & Hussein, T. (2023). Virtual Sensors: Toward High-resolution Air Pollution Monitoring Using AI and IoT. *IEEE Internet of Things Magazine*, *6*(1), 76–81. DOI: 10.1109/IOTM.001.2200103

Chapter 3
The Role of Robotics in Sustainable Agriculture and Waste Management

Kalirajan Murugasridevi
https://orcid.org/0000-0001-8955-9000
Amrita School of Agricultural Sciences, Amrita Vishwa Vidyapeetham, India

V. R. Mageshen
https://orcid.org/0000-0001-6611-6223
Amrita School of Agricultural Sciences, Amrita Vishwa Vidyapeetham, India

Ramesh Poornima
Vanavarayar Institute of Agriculture, India

Ambikapathi Ramya
University of Eastern Finland, Finland

ABSTRACT

Robotics in agriculture revolutionizes conventional farming and improvises its efficiency, precision, and sustainability. This chapter explores advancements in UAV-based pesticide and herbicide applications, automated pest and weed control, planting to harvesting tactics emphasizing efficacy, spray deposition, and environmental impact compared to traditional methods. On the other hand, smart waste management strategies, including sensor-based monitoring and AI-driven analytics, facilitate the efficient recycling of agricultural waste and prevent environmental degradation. Despite these advancements, challenges persist, including regulatory constraints, high initial costs, and the need for field-specific calibration of robotics for optimized performance. Overall, UAV technology and robotics, coupled with AI

DOI: 10.4018/979-8-3373-1892-9.ch003

and precision agriculture, present a transformative solution for modern agriculture, balancing productivity with environmental stewardship.

INTRODUCTION

As global issues such as climate change, population growth, and resource depletion escalate, robotic technologies are emerging as vital tools for promoting sustainability in various domains (Rame et al. 2024). The emergence of agricultural robots signifies a revolutionary period in traditional farming, indicating a paradigm shift in cultivation practices. These advanced machines are functioning as accelerators for innovation, altering traditional methods by effortlessly incorporating automation into the agricultural landscape. These robots aim to tackle the numerous issues facing the agriculture sector, serving as technological sentinels that introduce unparalleled efficiency, precision, and sustainability. Automation, a major aspect of this agricultural revolution, is crucial in addressing the sector's diverse issues. Automation addresses labor shortages and resource inefficiency, serving as a fundamental answer that promises to enhance agricultural methods. The incorporation of robotics in agriculture has significantly improved precision farming methods, enabling autonomous machines equipped with Artificial Intelligence (AI), machine learning algorithms, and advanced sensor systems to perform essential tasks like planting, weeding, and harvesting with exceptional accuracy (Padhiary et al., 2024; Wakchaure et al., 2023). Recent years, especially the 2020s, have witnessed significant developments, including the development of robotic devices for precise weed detection and eradication, alongside real-time crop monitoring and disease identification and also for harvesting economic produces. Moreover, robots are utilized in vertical farming systems for effective indoor growing. Anticipation surrounds the prospective combination of genetic engineering for tailored crop development and the emergence of fully automated agricultural systems, managed and operated by robots and artificial intelligence. Robotic applications in agriculture, particularly in precision farming, are classified into planting, inspection, spraying, and harvesting operations. Technologies such as infrared sensors and computer vision improve efficiency and sustainability. Industry 4.0 integration with IoT facilitates real-time monitoring. Selective spraying methods and vision-based harvesting enhance pesticide utilization and precision. The development of lightweight robotic systems mitigates soil compaction, a major issue associated with traditional heavy machinery, hence fostering soil conservation and long-term agricultural sustainability. The electrification of robotic farm tractors advances environmental goals by reducing dependence on fossil fuels and lowering greenhouse gas emissions. Pearson et al. (2022) shown that electrified robotic vehicles can significantly reduce the greenhouse gas footprint of agricultural techniques. Besides agriculture, robotics markedly improves the productivity of waste management systems by enhanced gathering of waste, dividing, and disposal (Olawade et al., 2024). Even though agricultural robotics has advanced considerably, some studies still raise questions that impede its broad use in sustainable farming. Many current robotic systems lack adaptability for varied and dynamic real-world agricultural settings since they are designed and tested under controlled circumstances. Moreover, the use of robotics especially for irrigation optimization and water conservation has not been fully investigated. Especially for small and medium-scale farmers, high costs and system complexity also present significant obstacles. These technologies' practical use is also diminished by a lack of interoperability with current farm infrastructure and inadequate integration of AI-driven real-time decision-making. Importantly, there is little region-specific study, particularly in developing nations, that customizes robotic applications to particular agro-climatic situations. Filling in these holes will help to fully exploit robotics in promoting sustainable agriculture. The primary objective of this work is to investigate ways that robotics might support sustainable waste management and agricultural practices. This covers assessing the efficacy of present robotic interventions in overcoming significant agricultural obstacles, considering how automation addresses labor shortages, and determining the particular ecological impacts these innovations reduce.

PRECISION FARMING AND RESOURCE OPTIMIZATION

Precision farming has revolutionized modern agriculture by integrating advanced technologies to optimize resource management and productivity. The application of GIS and GPS enables precise mapping and site-specific interventions, reducing wastage of inputs and ensuring efficient resource distribution. Remote sensing, through satellite imagery and UAVs, facilitates continuous crop monitoring, allowing early detection of deficiencies and pest infestations, which aids in timely interventions. IoT-driven environmental sensors enhance real-time monitoring, particularly in irrigation management, ensuring optimal water use and improved crop health (Kumar et al.,2024; Sharma & Shivandu,2024). Big data analytics and predictive modeling contribute to decision-making, allowing farmers to adopt proactive strategies to enhance sustainability (Mana et al.,2024). While these technologies significantly enhance efficiency and reduce environmental impact, barriers such as high costs, data complexity, and the need for technical expertise remain. Addressing these challenges through policy support, financial incentives, and training programs can accelerate the widespread adoption of precision agriculture, ultimately fostering sustainable and eco-friendly farming practices. The key aspects of precision farming has been represented in Table 1.

Table 1. Significant advantages of precision farming

Technology	Function	Impact	References
GIS & GPS	Accurate field mapping, site-specific management	Reduces input waste, improves efficiency	Pedersen & Lind (2017)
Variable Rate Technology (VRT)	Precise application of fertilizers and pesticides	Minimizes environmental impact, optimizes resources	He et al. (2023)
Remote Sensing (Satellite & UAVs)	Monitors crop health, soil conditions, and weather	Enables real-time decision-making, early issue detection	Inoue et al. (2020)
IoT Sensors	Monitors environmental parameters (moisture, temperature)	Enhances precision irrigation and resource efficiency	Narayana et al. (2024)
Big Data Analytics	Processes agricultural data for decision-making	Supports predictive analytics and farm optimization	Chen et al. (2024)
Decision Support Systems (DSS)	Provides data-driven recommendations	Enhances productivity and sustainability	Tonle et al. (2024)

Automated Weed and Pest Control

Automated technologies have revolutionized pest and weed management in modern agriculture by integrating UAVs, AI, and precision spraying techniques. The adoption of UAVs, particularly drone-mounted sprayers, has become essential due to their superior operational efficiency, high speed, extensive area coverage, minimal spray drift, and the added advantage of eliminating human exposure to pesticides. UAVs (Unmanned Aerial Vehicles) equipped with high-resolution sensors enable real-time field surveillance, capturing detailed imagery that machine learning algorithms process to distinguish crops from invasive weeds. This accurate identification enables the creation of focused weed dispersion maps, permitting site-specific herbicide treatment, thus decreasing chemical usage and mitigating environmental impact. Azghadi et al. (2025) emphasizes the creation and field testing of a custom-built robotic spot applying device for sugarcane fields. Aiming at weeds with computer vision and deep learning, this technology greatly decreases herbicide utilization. While cutting herbicide use by 35% on average and up to 65% in low weed pressure areas, field tests showed that robotic spot herbicide application is 97% as effective as conventional broadcast spraying. This method also resulted in 39% lower herbicide levels in irrigation runoff and 54% lower herbicide loads, suggesting significant environmental advantages. Huang et al. (2025) investigated how solar-powered robotic mowers could be integrated into no-tillage (NT) and intercropping systems for organic cherry tomato production. The findings indicate that NT together with intercropping increases crop yield, energy efficiency, and economic profitability while lowering the carbon footprint. By lowering fossil fuel use and increasing carbon sequestration, solar mowers help to promote sustainability even more. This strategy offers a hopeful framework for sustainable, energy-efficient methods in organic agriculture. Zhang et al. (2025) presented a self-driving weeding robot controlled by corn row that combines seedling and weed identification, navigation, and exact weeding. The robot attains more than 79.8% weeding accuracy with little crop damage by means of an improved YOLOv5 model and smart control systems. It tackles important shortcomings in present systems, including inadequate weed detection and absence of actuator integration in dry areas. This development shows how robots could increase crop field management accuracy and efficiency. Gerhards et al. (2025) showed that robotics—especially camera-guided and AI-based weeding systems—greatly enhance weed control and crop yield in vegetable production. Compared to traditional techniques, robotic weeding had the best weed control effectiveness—up to 87% and lower crop damage. In vegetables, where herbicide use is limited and manual weeding is expensive, these technologies are particularly useful. The results show that robots are a viable, efficient substitute for precise weed control in high-value crops.

Robotics in agriculture enables timely and accurate pest detection, minimizing crop losses and environmental harm. AI-based robotic traps, like SpyFly, monitor insects in real time and forecast population growth using deep learning models. These systems integrate environmental data to improve prediction accuracy and support early intervention. The result is reduced pesticide use and more sustainable, data-driven farming practices (Kapetas et al.,2025). By precisely identifying Red Imported Fire Ant nests with over 90% accuracy, Su et al. (2024) showed how AI-powered robotic dogs can transform pest control. With less mistakes, the CyberDog-YOLOX system found more nests than human inspectors, proving its efficiency and scalability. This strategy improves environmental monitoring and lowers labor needs. All things considered, it shows a significant progress in applying robots for effective and sustainable management of agricultural pests. Romano & Stefanini (2025) investigated how biomimetic robots in animal–robot interaction (ARI) might affect the behavior of the Mediterranean fruit fly, a significant agricultural pest. Robotic agents drew flies by imitating the appearance and movement of the fly, therefore proving possible for behavior-based pest control. The results show ARI as a viable, sustainable way to control insect populations without using chemicals. In precision agriculture and integrated pest control, this invention creates fresh opportunities for robotics.Song et al. (2020) found that utilizing the combination of chlorfenapyr, chlorantraniliprole, and lufenuron through drone technology attained a control efficacy of 94.94%, significantly diminishing the population of the Fall armyworm, *Spodoptera frugiperda* J.E. Smith, by roughly 94.86%. The bioefficacy and toxicity of Dinotefuran 8% + Diafenthiuron 48% WG over sap-sucking pests in Bt cotton utilizing a drone-mounted sprayer was evaluated by Sreenivas et al. (2024). They have also demonstrated the consistent droplet deposition, improved application efficiency, and reduced spray drift when administered at a height of 1.2 m above the crop canopy. The findings suggest that drone technology may reduce the application of pesticides while maintaining successful pest control, preserving efficiency, and avoiding phytotoxicity. Using deep learning-based detection, Meshram et al. (2024) introduced a smart autonomous pesticide spraying robot improving precision and safety in pest control. With 97% accuracy, the robot finds pests and only applies pesticide where required, therefore reducing human exposure and environmental effect. Its adaptive spraying system guarantees efficient pesticide use while navigating fields autonomously. By means of robotics, this invention represents a major development in safe and sustainable agricultural practices.

Intelligent spraying technologies improve this method by employing sophisticated machine vision and AI-based recognition to identify and selectively address weeds. These strategies significantly improve the utilization of resources by employing herbicides solely in necessary areas, thus minimizing chemical misuse, lowering costs, and preventing herbicide resistance. The integration of deep learning models, particularly convolutional neural networks (CNNs), improves UAV-based spraying by increasing accuracy in detecting along with precision, leading to more efficient and environmentally sustainable weed control techniques. Paul et al. (2024) noted that UAV application at 30 L ha^{-1} resulted in superior weed management through efficient particle distribution. The findings underscore UAV spraying as a feasible alternative to manual knapsack spraying, facilitating efficient herbicide application with reduced carrier volume.

Table 2. Uses and advantages of environmental technologies

Technology	Application	Advantages	References
UAVs (Drones)	Capturing high-resolution images for weed identification	Enables site-specific management, reduces pesticide use	Raj et al. (2024)
Machine Learning	Analyzing field images to distinguish healthier and unhealthy crops	Improves accuracy in pest detection and control	Kalirajan et al. (2024)
Smart Spraying Systems	Real-time pest and disease detection and targeted pesticide application	Minimizes chemical usage, lowers costs	Verma et al. (2024)
Deep Learning & CNNs	Enhancing UAV-based spraying precision	Increases efficiency and reduces environmental impact	Taseer & Han (2024)

Efficient planting and harvesting

Robotic farming and harvesting are promoting the progress of intelligent, sustainable agriculture (Botta et al., 2022). Notwithstanding limitations like expense and infrastructure, the potential for enhanced yields, reduced labor dependence, and precision agriculture makes robotic technologies revolutionary in modern agriculture (Adewusi et al., 2024). Figure 1 provides an overview of robotic operation to elucidate the procedure.

Figure 1. Overview of robotic operation in agricultural field

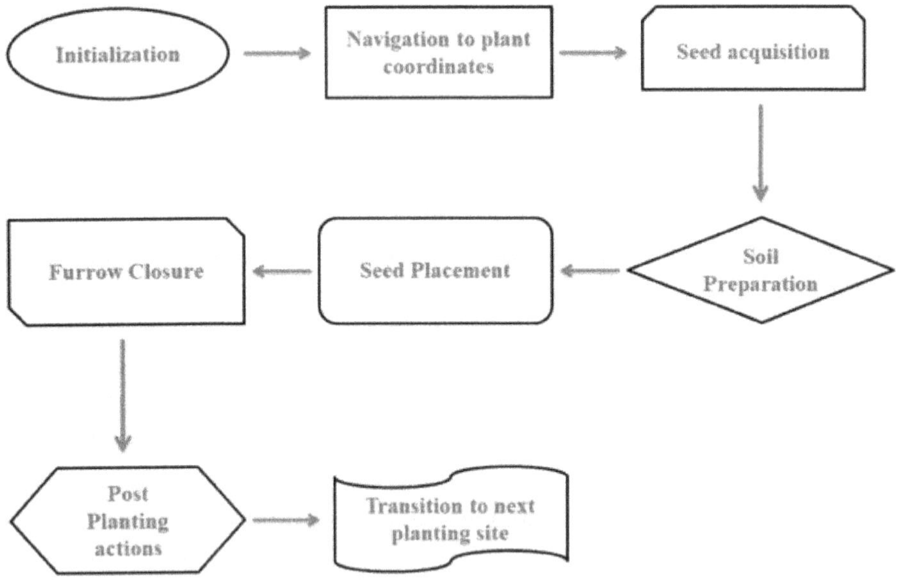

Automated Precise Sowing

Robots utilizing GPS and geo-mapping technologies generate comprehensive images of the terrain to determine optimal planting locations. The robotic device utilizes this data to sow seeds at precise depths and intervals, enhancing the likelihood of germination (Vinod Chandra et al., 2024).

Seed Distribution Techniques

Robots employ crank-slider mechanisms or alternative mechanical technologies to effectively implant seeds in the soil (De baerdemaeker, 2013). Certain robots utilize sucking vacuums or hydraulic methods to accurately position seeds.

Automating and Sensors Connectivity

Sensors assess soil water, temperature, and levels of nutrients to guarantee appropriate seed positioning.AI-driven algorithms modify seeding techniques according to actual time environmental variables (Getahun et al., 2024).

Speed and Accuracy

Robots sow seeds more rapidly than human labourers while ensuring excellent precision. They can acclimate to diverse environments and circumstances, rendering them suitable for both local and big agricultural operations (Laveglia et al., 2024). By means of artificial intelligence, IoT, and computer vision technologies, Cossio-Montefinale et al. (2025) emphasizes how important robotics is for sweet cherry farming. The system precisely evaluates cherry maturity using agroclimatic wireless sensor networks and video-based fruit detection, therefore attaining a low error rate of 5% in maturity estimation. This strategy improves smart, sustainable production in cherry farming and solves labor shortages. Bhattarai et al. (2025) shows how robotic systems might be used for precise apple pollination, so helping to solve the loss of natural pollinators. With encouraging fruit quality results, the robotic system achieved moderate pollination success in Honeycrisp and Fuji apple orchards using machine vision and motion planning. For more general use in various settings and types, though, more development is still required.

Automated Picking of Fruit

Robots equipped with AI-driven optical systems identify ripe fruits by assessing size, hue, and texture. The robot's higher quality sensors and machine learning algorithms enable them to track mature and immature fruits efficiently and carefully. Zhao et al. (2025) offers a new self-supervised approach for apple ripeness identification that solves problems with in-field occlusion and sparse labeled data. The system exceeds several other techniques in accuracy by means of a continuous ripeness score prediction and reconstruction of hidden apple components. The approach improves precision agriculture by using few parameters and providing flexibility for selective harvesting. This development increases the possibilities of robots in ripeness evaluation and orchard management.

Crop Classification and Categorization

Robots can assess the state of maturity of crops utilizing sensors and deep learning techniques. The crops that were harvested are instantly sorted according to size and grade. This precise categorization increases sales value, lowers post-harvest expenditures and optimizes supply chain logistics in contemporary agriculture

Exact Cutting and Grabbing

Robotic arms equipped with grips, sucking cups, or knives delicately harvest or sever crops without inflicting damage. Suction process gently elevate and segregate tender produce whereas precision blades accomplish crisp cuts which avoid scratches or decay in fruits.

Selective Harvesting Approach

Robots can exclusively identify and gather ripe food, deferring the collection of unripe crops for subsequent picking. This mitigates food waste and enhances agricultural productivity.

Rapid Harvesting

Robots operate more swiftly than human workers, traversing extensive areas in reduced time. In relevance to human labour susceptible to tiredness and time limitations, robots can perform task relentlessly offering more efficient and consistent harvesting process.

ROBOTS IN CROP HEALTH AND MONITORING

The progression of robots in agriculture has transformed the monitoring of soil and crop health (Karunathilake et al., 2023). Conventional techniques for monitoring soil and crop health are difficult, time-consuming, and susceptible to human error (Sharma and Shivandu, 2024). Autonomous robots present a viable answer by delivering continuous, high-resolution data to enhance resource utilization and increase production (Mahlein et al., 2024). Developing robots for soil and crop health monitoring beyond mere automation; it involves the creation of more intelligent and sustainable agricultural systems (Table 2). Robotic agriculture tracking technologies enhance sustainable agriculture elevating crop yield, lowering chemical usage and preserving supplies (Singh et al., 2021). Ospina & Noguchi (2025) presents a GIS-based remote surveillance tool for controlling several farming robots, so improving operational efficiency by means of work advances estimation and real-time robot velocity adjustment. Even with up to 40 robots, the system guarantees low latency and reduces task time variations by up to 87%. Results from experiments confirm its capacity to cut postpones and maximize completion of tasks. Its capacity to change robot speeds to fit terrain conditions is the innovation, therefore enhancing the general efficiency of agricultural activities.

Table 3. Things to be considered for designing a robot for monitoring soil and crop health

S. No	Particulars	Specifications
1	Sensor Suite	Inclusion of sensors for measuring soil properties and also high resolution cameras to monitor crop health
2	Movement and Direction	A four wheel drive base rover along with robotic gripper can augment the robots capacity to engage with surroundings
3	Data Handling and transmission	Provide real time data processing and analysis process for faster decision making. Collected data is send to cloud based platform via Wi-Fi or LTE
4	Power supply	Rechargeable batteries are preferred for longer field work. Inclusion of solar panels also add energy and enhance working hours
5	Adaptation and environmental endurance	To withstand tough agricultural situation build the robot through dust, moisture and temperature resistant elements.
6	Guarding Attributes	Install sensors and algorithms towards identifying and avoiding barriers surrounding crops. For avoiding accidents and damage of equipment employ emergency pause and network diagnostics
7	Data Linkage and user experience	Determine the robot's data sources work with farming operations and decision making system. Farmers can constantly track and make relevant information With the help of smartphone application or digital panels, farmers

Livestock and Farm Management

Robots have altered livestock care by streamlining monotonous activities, boosting animal wellness and accelerating efficacy (Monteiro *et al.*, 2021). The deployment of robotic tolls into livestock activities aids resilient farming via lowering labour dependability, circumventing substance waste while promoting animal health (Figure 2)

Figure 2. Function of robots towards efficient animal husbandry

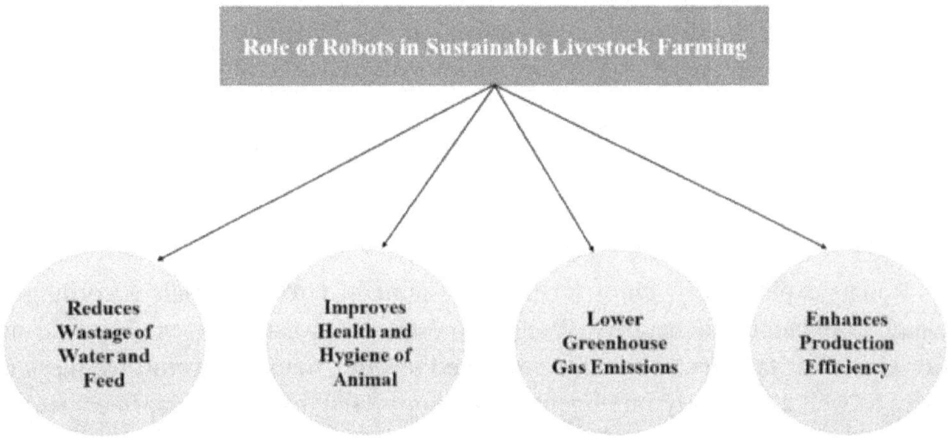

1. Automated nourishment mechanism

Robots considerably assist in the oversight of livestock grazing by streamlined allocation of food over consistent period. Provision of harmonious food in compliance with animal nutritional requirements. Alleviating overindulgence and inhibiting feed wastefulness

Example: Juno Feed Pusher- Manually delivers feed onto livestock over predetermined durations

2. Robotic Milking Devices

Autonomous milking devoid of individual engagement. Helps in recognition of specific animals through sensors

Example: Lely Astronaut- Develops operational efficiency as well as raises milk flow by means of superior sanitary practices

3. Robots for health surveillance

Robots deploy IoT devices alongside AI algorithms to recognize strange actions suggestive of disease or sickness.

Example: Ida Welfare tracking technology- Records cattle wellness and notify producers

4. Robots assisted waste handling

Robots aid in the disinfection of animal habitats by streamlined elimination of waste from animal. Preserving hygiene in agricultural farms. Optimizing laborious work and welfare hazards.

Example: Lely discovery- Retractable barn sanitation robot

5. Robots for screening and tracing animal

Robots exploit RFID chips for locating animals. Group animals according to weight, wellbeing and maturity. Track the presence of livestock on extensive farms. For Cattle ear tagging, robots are employed towards tracking overall wellness of cows as well as mobility Intelligent collars supporting ovine and caprine species- Assessing grazing habits along with animal posture

Farm Mechanization and Automation

Robots in agriculture are revolutionizing conventional farming practices by delivering expedited, precise, and sustainable solutions (Eissa, 2024). They enhance production, mitigate labor shortages, and preserve crop quality (Rehman et al., 2024). Innovations in AI, IoT, and precision agriculture are rendering robot-assisted harvesting an essential component of smart farming systems (Jin and Han, 2024). Farm mechanization along with automation through robotics are now vital for environmentally friendly farming system via boosting material efficacy, decreasing rely on manpower while increasing production (Table 3). The merging of robots, IoT alongside AI promotes the progression of bright agricultural devices which comply towards worldwide environmental objectives (Bucci et al., 2018). The list of robots employed for particular crops and associated technologies is listed in Table 4.

Table 4. Robots employed for particular crops and associated technologies

Crop	Robot name	Technology	Function
Rice	Agribit transplanter	GPS + AI	Automatic transplanting
Wheat	Robotti	GPS + Autonomous navigation	Weeding and seeding
Maize	Rowbot	GPS + IoT sensors	Fertilizer application and control of weeds
Cotton	Cotton harvester robot	Machine vision	Picking of cotton
Sugarcane	Cane harvester robot	GPS + IoT sensors	Harvesting sugarcane
Tomato	Sweeper robot	Computer vision	Harvesting tomato
Grapes	Vinerobot	Hyperspectral imaging	Detection and tracking of diseases
Potato	Diggerbot	GPS + Machine vision	Harvesting potatoes
Apple	Agrobot SW6010	AI + Robotic arms	Harvesting apples
Strawberries	Octinion Rubion	Soft grippers + AI vision	Strawberry picking

Perks of Robots Towards Agricultural Mechanization

- ✓ Elevated accuracy
- ✓ Dwindled labour dependability
- ✓ Expedited operations
- ✓ Resource Maintenance
- ✓ Augmented crop productivity
- ✓ Optimized carbon footprint

Drawbacks of Robots Towards Agricultural Mechanization

- ✓ Massive initial investment
- ✓ Inadequate practical competence
- ✓ Inspection and servicing problems
- ✓ Cybersecurity hazards
- ✓ Minor agricultural properties in emerging countries

Robotics for Societal Benefit

Especially in terms of labor shortages and environmental effects, robotics in agriculture has the ability to solve a great variety of social issues. Here is a broader perspective on how these technologies address these problems:

1. Labor scarcity

Labor scarcity in agriculture, particularly in areas with an aging population or migration of workers to cities, have been a mounting worry. By automating processes historically needing human labor, robotics can offer answers. Among these duties are crop monitoring, weeding, planting, and harvesting. Farmers can guarantee constant production by using autonomous machines, therefore lowering their reliance on seasonal or temporary labor. Many of the tedious chores in agriculture, including weed control and irrigation management, can be automated, therefore lowering the need for manual labor. Robots like automated tractors, harvesters, and drones can carry out these tasks at scale and with more accuracy. Advancements in robotics and artificial intelligence allow agricultural equipment to be controlled remotely or even autonomously. This is especially useful for areas lacking competent labor. By remotely operating or monitoring the robots, farmers can maximize time and reduce labor expenses.

2. Effects on the Environment

Excessive pesticide use, soil degradation, water waste, and carbon emissions from conventional farming practices all contribute to the environmental sustainability of agriculture, which is a major issue. Robotics provides various answers that directly tackle these ecological concerns. Robotic systems can do precision farming by aiming at particular areas needing intervention, such as applying fertilizers, pesticides, or water only where needed. This lowers the total input use and lessens environmental pollution. For instance, drones with sensors and autonomous tractors can identify plant stress, therefore enabling precise pesticide or nutrient application. Robots made for weeding, such as those employing machine vision, can pick out weeds without using chemical herbicides. This helps to lower herbicide and pesticide use, therefore lessening the effect on non-target organisms and lowering chemical runoff into adjacent ecosystems. Powered by renewable energy sources, autonomous electric tractors and harvesters can help lower the carbon footprint of farming activities. These machines lower emissions connected to conventional agriculture by replacing fossil fuel-powered equipment. Robotic irrigation systems such as automated drip irrigation or moisture-monitoring drones help in the exact application of water depending on real-time data. This guarantees that crops get the correct amount of water at the appropriate time and helps to solve the increasing problem of water scarcity in agriculture by preventing waste of water.

3. Other Social Advantages

Robots can help raise the overall yield per acre by optimizing the use of resources—labor, water, pesticides, etc.—thereby preserving environmental sustainability. Producing more food with less resources helps to enhance world food security. Automation of hazardous activities like pesticide spraying or operating heavy machinery can enhance the health and safety of workers by lowering their exposure to toxic chemicals and the physical strain of manual labor.

ROLE OF ROBOTICS IN WASTE MANAGEMENT

AI applications in waste management are gaining significant influence and promise, with AI-powered systems improving waste collection routes, optimizing waste sorting, and optimizing waste monitoring (Di Vaio et al., 2022; Pallathadka et al., 2023). Traditional waste collection methods often struggle with inefficiency due to inadequate routing and unpredictable timetables. AI-powered systems can monitor waste levels in real-time, allowing for more effective collection and resource allocation (Bamakan et al., 2022; Hoque and Krishna and Sharma, 2023, Yadav and Karmakar, 2020; Yu et al., 2021). AI-powered route optimization algorithms assess historical data, traffic patterns, and waste creation rates to identify the most efficient routes for waste collection vehicles. Dynamic scheduling algorithms optimize collection schedules based on real-time data, ensuring timely garbage removal while lowering operational expenses (Hussain et al., 2020; Ghahramani et al., 2022; Lin et al., 2022). AI technologies, including image recognition and machine vision, are increasingly being used to automate and increase garbage sorting accuracy. Advanced image recognition algorithms can analyze visual data and accurately identify various waste materials, allowing for efficient separation of organic, recyclables, and non-recyclables waste (Das et al., 2019; Guo et al., 2021; Lu et al., 2022; Zhang et al., 2021). Robotic sorting systems combined with AI algorithms can efficiently separate and sort waste items, speeding the sorting process and increasing recycling rates (Dong et al., 2022; Kumar et al., 2021; Majchrowska et al., 2022; Sundaralingam and Ramanathan, 2023).

AI is also playing an important role in garbage recycling by optimizing several phases of the process (Wu et al., 2023). Material identification and sorting methods powered by AI allow for accurate identification and separation of recyclable materials, maximizing resource recovery and minimizing waste contamination (Onoda, 2020; Sharma et al., 2021, 2022). AI-powered process optimization algorithms examine data on recycling parameters to determine the best operating conditions and increase efficiency. Waste monitoring is another area where AI is making progress (Ghanbari et al., 2023; Zhu et al., 2023). Real-time monitoring systems interact with IoT devices and sensor networks, collecting and analyzing data on waste generation, collection, and disposal operations. AI algorithms analyze this data, offering useful insights for waste management operations. The types of robots in waste management, key technologies used, its application, benefits, challenges and future trends are represented in Figure 3. Predictive analytics models use historical and real-time data to estimate trash generation trends, improve resource allocation, and allow for data-driven waste management decisions (de Sousa et al., 2019). Collaboration among waste management agencies, technology suppliers, researchers, and policymakers are critical for resolving these issues and encouraging ethical and sustainable usage of AI in trash management (Oyedotun and Moonsammy, 2021).

Figure 3. Types, application, benefits, challenges and future trends of robotics in waste management

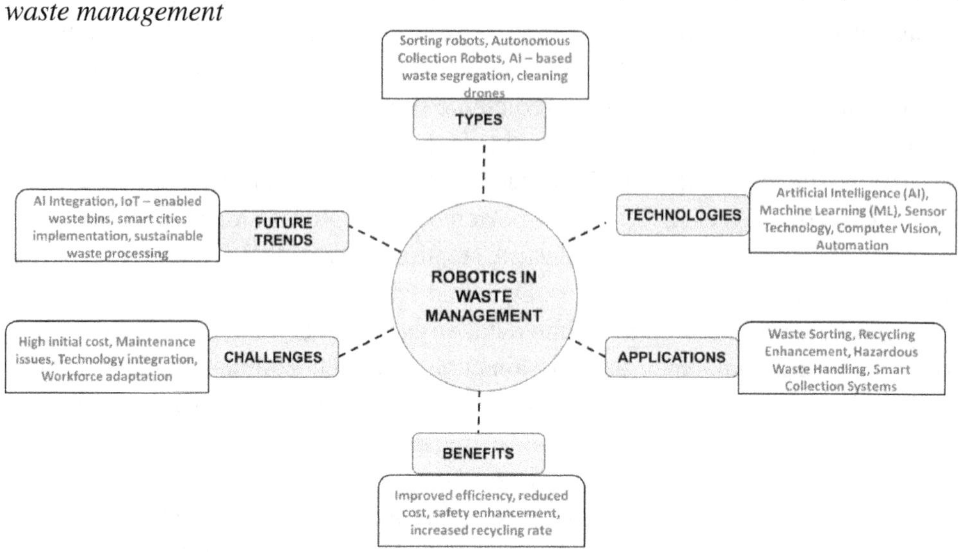

Waste Sorting

Robotic sorting systems utilize robotic arms equipped with sensors, cameras, and AI algorithms to automate waste sorting (Subramanian et al., 2021). Advancements in these systems include new grasping technologies, greater sensor capabilities, and AI algorithms. Robots can handle various trash products with varied forms, sizes, and textures using grippers with changeable grabbing mechanisms and touch sensors (Sarc et al., 2019; Ahmed and Asadullah, 2020; Shreyas Madhav et al., 2022). AI algorithms, such as reinforcement learning, allow robots to learn the best picking strategies and adapt to changing trash composition. Collaborative robots, known as cobots, are being used to supplement human operators in waste sorting operations, improving productivity and safety. However, the use of AI-powered robotics raises ethical concerns, particularly regarding employment displacement and environmental justice. As AI-powered robots automate tasks previously performed by people, there is a risk of job loss and economic disruption, particularly for workers in the waste management business. Additionally, AI deployment may exacerbate environmental justice challenges by disproportionately impacting communities burdened by pollution and waste facilities (Bag, 2023; Nwokediegwu and Ugwuanyi, 2024). Therefore, it is crucial to evaluate the social and environmental implications of AI technology and adopt policies that enable equitable access to opportunities while mitigating negative consequences on workers and communities (Comninos et al., 2019).

Waste Recycling

Robotics and automation technologies are revolutionizing the trash recycling sector by automating labor-intensive operations and increasing production (Brintha et al., 2020; Bharti et al., 2022; Ejimofor et al., 2022; Nafiz et al., 2023). AI-powered robots with sensors, cameras, and sophisticated algorithms can sort, dismantle, and process recyclable materials (Mao et al., 2021). Current advancements in robotics and automation include the integration of AI algorithms with robotic systems to provide precise and efficient waste processing and sorting (Rutqvist et al., 2020). These robots can adapt to various shapes, sizes, and weights of rubbish, using AI algorithms to recognize and sort recyclable elements. Collaborative robots are being used alongside human operators to improve recycling operations' safety and efficiency. The latest trends in AI-driven trash recycling focus on improving material identification and sorting accuracy, streamlining recycling operations, improving quality control and inspection, and utilizing robotics (Bui et al., 2023; K. Wang et al., 2021). These improvements help the trash recycling industry become more sustainable and efficient. However, ongoing research and development continue to focus on challenges such as integrating AI technologies with existing recycling infrastructure, standardizing data and processes, and ensuring the scalability and cost-effectiveness of AI solutions (Monzambe et al., 2019). Collaboration among researchers, waste management firms, technology suppliers, and regulatory agencies is critical in addressing these problems and driving widespread adoption of AI in garbage recycling.

Data management and analytics are also important trends in AI-powered garbage recycling. The gathering, storage, and analysis of huge amounts of data generated by sensors, cameras, and other monitoring equipment allows for ongoing development and optimization of recycling processes (Gopalakrishnan, 2019). Advanced data analytics tools, such as machine learning and artificial intelligence, are used to extract important insights from the data, allowing for better decision-making and further improving recycling operations. The incorporation of AI technology with Internet of Things (IoT) devices in garbage recycling is another noteworthy trend. IoT devices, such as smart sensors and connected machinery, can provide real-time information about waste composition, equipment performance, and operating parameters (Nahaei and Naziri-Oskuei, 2021; Salem et al., 2023). By merging AI and IoT, waste recycling systems can become more networked and responsive, allowing for preventative maintenance, real-time monitoring, and improved overall process control (Rakhio, 2024). AI-powered trash recycling platforms allow for seamless integration and collaboration among various stakeholders along the recycling value chain. These platforms allow for the sharing of data, information, and best practices, resulting in a more transparent and efficient recycling environment. AI-powered technologies open up new revenue streams via data-driven insights and predictive analytics, allowing recycling companies to provide tailored solutions and value-added services to their clients (Licardo et al., 2024).

In conclusion, integrating AI technologies with existing recycling infrastructure presents various issues that must be addressed for successful adoption (Andeobu et al., 2022). One problem is the compatibility of AI systems with legacy infrastructure and processes, which may necessitate significant changes or upgrades to accommodate new technologies (Wan et al., 2020). Another difficulty is ensuring interoperability and data compatibility between various AI systems and existing recycling equipment and databases. Communication protocols, standardized data formats, and integration interfaces are required to allow for easy data sharing and collaboration throughout the recycling value chain. Stakeholders such as facility operators, workers, and regulators may be skeptical about AI-driven technologies' reliability, accuracy, and safety. Addressing these issues requires a collaborative effort among technology developers, recycling facility operators, legislators, and other stakeholders (Lu et al., 2013). Strategies include offering financial incentives and subsidies to offset implementation costs, providing training and support programs to educate workers on AI technology, and developing industry-wide standards and best practices for AI integration in recycling operations (Niˇzeti'c et al., 2019).

Waste Collection and Transportation

The integration of IoT and AI technologies has revolutionized municipal solid waste (MSW) management, enabling more efficient, sustainable, and environmentally friendly solutions (Bibri et al., 2024; Zyoud and Zyoud, 2024). Traditional methods, such as collection schedules and manual sorting, are being replaced by real-time data and intelligent decision-making procedures (Zoumpoulis et al., 2024). AI and IoT in waste management systems enable cities to address the complex challenges of collection, sorting, and recycling with remarkable efficiency and accuracy. Smart bins with sensors, AI-driven route optimization algorithms, and predictive analytic models are among the most significant advances in trash management in urbanized areas. These bins are equipped with sensors that measure waste levels, identify odors, and monitor bin status, providing real-time data to MSW management systems. This data enters into a centralized system that monitors waste levels around the city, allowing for more efficient resource scheduling and allocation. Smart bins can detect scents, preventing instances where a bin is nearly empty but the waste inside it has begun to decompose or emit unpleasant odors in public places. This capability is especially useful in high-traffic areas like city centers, parks, and tourist attractions, where cleanliness is critical for both the local environment and public satisfaction.

IoT-enabled bins provide dynamic collection schedules that adjust to actual garbage amounts, reducing the number of unneeded pickups and allowing waste management agencies to better deploy their resources. They can help target collection services to high-usage zones while limiting attention to low-waste areas in towns where certain areas generate more rubbish than others. This optimization is especially useful for towns with tight budgets, as it saves fuel usage, vehicle wear and tear, and personnel costs. IoT-based smart bins also help reduce greenhouse gas emissions by reducing the need for frequent excursions and improving waste collection routes based on real-time data. For example, improving waste collection routes based on real-time data from smart bins can cut truck fuel use by up to 30%, reducing carbon dioxide emissions and allowing towns to get closer to their environmental sustainability goals. Route optimization is another key improvement enabled by the use of AI and IoT in MSW management (Olawade et al., 2024). AI algorithms can use various data inputs, such as traffic patterns, bin fill levels, energy consumption, and vehicle capacity, to calculate the most effective routes for garbage collection vehicles. Traditional waste collection routes often use predetermined paths that may not reflect current road conditions or waste levels, resulting in inefficient routes that waste fuel and time. AI-driven route optimization allows waste management systems to update routes in real time, taking into consideration factors like traffic congestion, road closures, and high-waste regions that require rapid care.

AI and IoT have the potential to revolutionize urban waste management by enhancing efficiency, reducing costs, and improving waste collection processes (Olawade et al., 2024). By adjusting routes and optimizing waste collection, cities can reduce their carbon footprint and improve environmental health. This not only reduces fuel consumption and emissions but also increases the lifespan of collecting vehicles, reducing wear and tear and minimizing vehicle repair and replacement expenses (Bibri et al., 2024; Wolniak and Stecula, 2024). Predictive analysis is another tool in AI's growing toolset, allowing cities to accurately predict garbage generation patterns based on historical data and variables such as seasonal trends, event schedules, and neighborhood demographics. This allows cities to adapt their collection schedules ahead of time to account for transient surges in waste generation, ensuring that waste bins are not overfilled and public places remain clean. This is particularly useful in cities with high tourism rates, where variable population numbers can make garbage management difficult and dynamic. AI-based predictive analysis helps avoid unexpected increases in waste, increasing service dependability and public satisfaction. It also accounts for long-term trash production trends, such as seasonal variations in garbage generation. Understanding these trends allows MSW management systems to plan ahead, deploying more resources during peak times and scaling back during low-waste periods (Jogarao et al., 2024; Oyejobi et al., 2024). AI algorithms can also account for changes in community behaviors and legislation, such as higher recycling rates or waste reduction programs, allowing for continuous adjustments to collection techniques.

The integration of AI and IoT into trash management creates opportunities for improved recycling practices, contributing to the creation of a circular economy. Smart bins equipped with sensors can help separate recyclables from general rubbish, speeding up the recycling process and reducing the amount of garbage transported to landfills. AI-powered sorting systems in trash facilities automate the categorization and identification of recyclables. Blockchain technology combined with IoT sensors can provide a record of garbage from collection to processing, ensuring that recyclable materials are not diverted to landfills (Olawade et al., 2024). This level of traceability ensures that recycling targets are met while providing stakeholders with assurance that recyclable materials are appropriately managed. AI-powered systems can encourage recycling by rewarding residents who properly sort their waste, promoting a culture of responsible disposal. In conclusion, AI and IoT-based waste management systems offer a realistic solution to urban carbon emissions by reducing waste collection frequency, streamlining routes, and minimizing landfill usage through increased recycling. By diverting garbage from landfills and improving recycling procedures, smart waste management promotes the transition to a circular economy, where materials are reused and repurposed rather than discarded after a single use.

Hazardous Waste Handling and E – Waste Recycling

Robotics technologies offer a novel solution to hazardous waste management by reducing human exposure to toxic compounds. These machines can navigate complex areas and perform delicate jobs with accuracy, making them ideal for handling hazardous materials in urban settings. Sensors on these modern machines can detect hazardous gases, radiation, and other harmful substances, allowing them to work securely in contaminated environments. Programming robotic systems to quickly sort, separate, and package hazardous trash speeds up the disposal process and reduces the danger of environmental contamination. Many robots have been successfully utilized in e-waste sorting, dismantling, recovering, and disposal. E-waste can be classified using various criteria, such as product type, material composition, size, and more (Sutikno et al., 2024). The WEEE ID project, funded by VINNOVA, collaborated with ReFind Technologies to create an intelligent, automated sorting system that uses sensors and intelligent data processing to detect whether used electronic products are good for refurbishment, reuse, or recycling, and sorts them accordingly (Harper et al., 2019; Chen and Shen, 2017).

The automated disassembly of WEEE is a challenging problem with no universal answer. However, certain EEE manufacturers have been able to construct product-specific disassembly robots, such as Apple's Liam and Daisy robots. The Liam system consists of 29 robots capable of disassembling an iPhone 6 phone into 8 discrete components in 11 seconds (Rujanavech et al., 2016), while the Daisy robot is claimed to be capable of disassembling nine iPhone models at a rate of 200 devices per hour with greater efficiency than existing techniques such as shredding. Reconfigurable recycling systems (RRSs) are systems with the built-in ability to rearrange or modify their recycling processes to adapt to the specific characteristics of a waste stream. Barwood et al. (2015) demonstrated an RRS based on the Staübli RX160 robotic arm capable of semi-destructive disassembly and separation of electric vehicle components. Robotics are widely used in the metallurgical industry, as they can shape recovered metals during the forging process (Bennettsz et al., 2012; Madhav et al., 2022). However, around 8% of the global e-waste is discarded in waste bins and subsequently landfilled or incinerated. Researchers are exploring how robots can help with garbage collection by separating e-waste from household waste and how robots can be used to locate gas emission sources once e-waste has reached the landfill.

Landfill Management and Waste Reduction

Emerging waste management technologies, notably automated landfill management, hold promise for addressing urban trash disposal concerns. One important part of this innovation is the use of deep learning algorithms and robotics engineering to speed garbage separation and landfill operations. The automated garbage separation system uses sensor technologies and deep learning models to precisely categorize various forms of waste, increasing the efficiency and accuracy of waste management systems. Furthermore, the use of robotic arms to sort and handle waste demonstrates the potential for improved safety and sustainability in landfill operations. As concerns about solid waste management (SWM) in cities develop, smart technology such as low-speed shredder compactors and automated pneumatic waste collecting systems may assist limit environmental damage while also making trash disposal easier. Municipalities can use modern technology like these to move toward more efficient and ecologically responsible landfill management methods, eventually contributing to the advancement of urban smart waste management programs (Sutikno et al., 2024).

Furthermore, waste-to-energy robotics are critical to upgrading urban smart waste management systems. Sensors and AI algorithms enable these robots to efficiently sort through different sorts of rubbish, reliably distinguishing recyclables from non-recyclables. Cities may use robotics to expedite waste sorting and disposal, maximizing energy recovery while minimizing environmental effect. Research has demonstrated that waste-to-energy robotics can greatly improve the overall efficiency of waste management operations, resulting in cost savings and lower greenhouse gas emissions. Furthermore, robotic technologies in this industry are constantly advancing, with ongoing research aimed at improving robot capabilities and broadening their uses in waste-to-energy plants. Waste-to-energy robotics can help metropolitan regions achieve sustainable and environmentally acceptable waste management techniques (Sutikno et al., 2024).

When determining the cost-effectiveness of robotic waste disposal systems in metropolitan areas, it is critical to account both the initial investment and long-term operations costs. While the initial expenses of introducing robotic technology may be high, studies indicate that, over time, these systems can result in significant savings by increasing operational efficiency and lowering labor costs. Robotic waste collectors, for example, can work nonstop without taking breaks or receiving extra pay, enhancing productivity and lowering overall costs. Furthermore, the use of artificial intelligence and predictive analytics in waste management can assist optimize routes, cut fuel consumption, and minimize environmental effect, resulting in cost savings. Decision-makers can better grasp the economic benefits of incorporating robotics into garbage disposal operations in cities by carefully examining the total cost of ownership and potential return on investment (Sutikno et al., 2024).

Ocean and Environmental Cleanup

The use of underwater robot technology in ship cleaning not only frees divers from heavy labor, but it also results in safe and efficient industrial goods. By improving the underwater cleaning robot's operating procedure, damage to the ship's underlying coating can be reduced. Cavi-Jet International, for example, offers a wide range of hull cleaning solutions, including handheld equipment and diver-operated vehicles. Divers utilize Cavi-Jet pistols to clean complex surfaces with various shapes and locations that large cleaning equipment cannot reach. The water and sandblasting Cavi-Jet pistol is specifically developed to remove hard maritime fouling from the hull. Similarly, Yan et al. (2019) created an underwater cleaning robot that uses cavitation cleaning technology. Furthermore, Kostenko et al. (2019) created a new underwater cleaning system consisting of an underwater robot and laser cleaning equipment that may be utilized to clean the hull. While considering permanent magnetic adhesion technology, the first-generation ARMUS robot is a three-axis tracked system that attaches to the hull using neodymium magnets. The robot tracks produce an attraction force of 336 kg to the surface, which is sufficient to maintain the robot glued to the hull surface even when the ship is moving. The second-generation ARMUS robot is meant to remain underwater for an indefinite period of time and work on both sides (external and interior, such as cargo holders) of the ship's hull. The permanent magnet adsorption method has the advantage of not requiring external energy for magnetic force maintenance, which means that the system's adsorption capacity does not increase the capacity of the power system. Simultaneously, as the thickness of the contaminated layer increases, the magnetic attraction force weakens, making it impossible to enhance the electromagnetic force by consuming electricity. The application of robotics in waste management is summarized in Table 5.

Table 5. Application of robotics in waste management

Study Area	Robot	Technology Used	Outcome
Finland	ZenRobotics Recycler	Computer Vision, Artificial Intelligence	Waste sorting with 98% accuracy
USA	AMP Robotics	AI – driven robotic arms	Improved effectiveness in recycling centres
Singapore	Smart Bin Project	Internet of Things (IoT) – enabled waste bins	30% cost reduction in waste collection
Netherlands	Marine Drone Cleanup	Autonomous drones	The plastic pollution in ocean has drastically reduced

CONCLUSION

Robotics, therefore, provide great possibility in sustainable agriculture. But more multidisciplinary research is still required, particularly in the fields of long-term ecological evaluations, artificial intelligence-driven decision-making, and smallholder affordability. Exploitation of robotics to achieve sustainable farming and smart waste management compared to conventional practices may result in sustainability, precision and potential methods to enhance yield. As robotics have made agriculture elated from sowing to harvest including automatic planting, harvesting and AI based crop monitoring have in turn decreased labour shortage issues and remains environmental friendly by effectively utilizing the resources. Smart farming strategies leads to precise operational tactics, decreases synthetic pesticides and fertilizers usages and eliminates the exposure of human to detrimental chemicals. On the other hand, incorporation of robotics in waste management sector by using automatic garbage grading, smart dust bins and smart arms for gathering, sorting and recycling has made revolution in this sector compared with traditional system. In addition, assistance of AI and IoT in these sectors have given option for real time monitoring thereby resulting preciseness in every instance. Eventhough, they are highly helpful,smart enough to deal with agricultural problems, the hindering factor to afford robot is high cost,lack of skill to operate. Hence, policy makers, government and stakeholders may aid the growers with subsides to afford the robots in a community basis through regulatory frameworks. Proper training is necessary for the growers to acquire adequate knowledge to handle the robots in farming. Recently emerging cobots (collaborative robots) is an another better option to implement it in agriculture to undertake multitasking in the real field conditions. Holistically, indulging robotics in combination with AI, IoT and sensors may be a better strategy to endup in smart sustainable agriculture and waste management.

REFERENCES

Ahmed, A.A.A., Asadullah, A.B.M., (2020). Artificial intelligence and machine learning in waste management and recycling. Eng. Int. 8, 43–52. https://doi.org/. v8i1.498DOI: 10.18034/ei

Al Duhayyim, M., Mohamed, H. G., Aljebreen, M., Nour, M. K., Mohamed, A., Abdelmageed, A. A., Yaseen, I., & Mohammed, G. P. (2022). Artificial ecosystem-based optimization with an improved deep learning model for IoT-assisted sustainable waste management. *Sustainability (Basel)*, *14*(18), 11704. DOI: 10.3390/su141811704

Andeobu, L., Wibowo, S., & Grandhi, S. (2022). Artificial intelligence applications for sustainable solid waste management practices in Australia: A systematic review. *The Science of the Total Environment*, *834*, 155389. DOI: 10.1016/j.scitotenv.2022.155389 PMID: 35460765

Azghadi, M. R., Olsen, A., Wood, J., Saleh, A., Calvert, B., Granshaw, T., & Philippa, B. (2025). Precision robotic spot-spraying: Reducing herbicide use and enhancing environmental outcomes in sugarcane. *Computers and Electronics in Agriculture*, *235*, 110365. DOI: 10.1016/j.compag.2025.110365

Bag, T. 2023. Socio-economic impacts of scientific-technological advancements. Int. J. Multidiscip. Educ. Res. 12. https://ijmer.s3.amazonaws. com/pdf/volume12/volume12-issue8(4)/13.pdf

Bamakan, S. M. H., Malekinejad, P., & Ziaeian, M. (2022). Towards blockchain-based hospital waste management systems; applications and future trends. *Journal of Cleaner Production*, *349*, 131440. DOI: 10.1016/j.jclepro.2022.131440

Barwood, M., Li, J., Pringle, T., & Rahimifard, S. (2015). Utilisation of Reconfigurable Recycling Systems for Improved Material Recovery from E-Waste. *Procedia CIRP*, *29*, 746–751. DOI: 10.1016/j.procir.2015.02.071

Bharti, S., Fatma, S., & Kumar, V. (2022). AI in Waste Management: The Savage of Environment. In Paul, P. K., Choudhury, A., Biswas, A., & Singh, B. K. (Eds.), *Environmental Informatics: Challenges and Solutions* (pp. 97–123)., DOI: 10.1007/978-981-19-2083-7_6

Bhattarai, U., Sapkota, R., Kshetri, S., Mo, C., Whiting, M. D., Zhang, Q., & Karkee, M. (2025). A vision-based robotic system for precision pollination of apples. *Computers and Electronics in Agriculture*, *234*, 110158. DOI: 10.1016/j.compag.2025.110158

Bibri, S. E., Krogstie, J., Kaboli, A., & Alahi, A. (2024). Smarter eco-cities and their leading-edge artificial intelligence of things solutions for environmental sustainability: A comprehensive systematic review. *Environmental Science and Ecotechnology*, *19*, 100330. DOI: 10.1016/j.ese.2023.100330 PMID: 38021367

Bibri, S. E., Krogstie, J., Kaboli, A., & Alahi, A. (2024). Smarter eco-cities and their leading-edge artificial intelligence of things solutions for environmental sustainability: A comprehensive systematic review. *Environmental Science and Ecotechnology*, *19*, 100330. DOI: 10.1016/j.ese.2023.100330 PMID: 38021367

Brintha, V. P., Rekha, R., Nandhini, J., Sreekaarthick, N., Ishwaryaa, B., & Rahul, R. (2020). Automatic Classification of Solid Waste Using Deep Learning, in: Kumar, L.A., Jayashree, L.S., & Manimegalai, R. (Eds.), *Proceedings of International Conference on Artificial Intelligence, Smart Grid and Smart City Applications*. Springer International Publishing, Cham, pp. 881–889. DOI: DOI: 10.1007/978-3-030-24051-6_83

Bui, T.-D., Tseng, J.-W., Tseng, M.-L., Wu, K.-J., Lim, M.K., (2023). Municipal solid waste management technological barriers: A hierarchical structure approach in Taiwan. Resour. Conserv. Recycl. 190, 106842 https://doi.org/. resconrec.2022. 106842DOI: 10.1016/j

Chen, H., & Shen, J. (2017, October). A degradation-based sorting method for lithium- ion battery reuse. *PLoS One*, *12*(10), e0185922. DOI: 10.1371/journal. pone.0185922 PMID: 29023485

Chen, J., Begicheva, S., & Nazarov, D. (2024). The role of big data in advancing precision agriculture and ensuring food security. In BIO Web of Conferences (Vol. 121, p. 02013). EDP Sciences. DOI: 10.1051/bioconf/202412102013

Comninos, A., Muller, E. S., & Mutung'u, G. (2019). Artificial intelligence for sustainable human development. Association for Progressive Communications (APC), Article, 19. https://tarbiatbadani.farafile.ir/content/demo/202009/eea99b6b -c4 8b-46e7-8703-bce6845e64e1.pdf

Cossio-Montefinale, L., Quiñinao, C., & Verschae, R. (2025). Orchard sweet cherry color distribution estimation from wireless sensor networks and video-based fruit detection. *Computers and Electronics in Agriculture*, *235*, 110334. DOI: 10.1016/j. compag.2025.110334

Das, S., Lee, S.-H., Kumar, P., Kim, K.-H., Lee, S. S., & Bhattacharya, S. S. (2019). Solid waste management: Scope and the challenge of sustainability. *Journal of Cleaner Production*, *228*, 658–678. DOI: 10.1016/j.jclepro.2019.04.323

de Sousa, W.G., de Melo, E.R.P., Bermejo, P.H.D.S., Farias, R.A.S., Gomes, A.O., (2019). How and where is artificial intelligence in the public sector going? A literature review and research agenda. Gov. Inf. Q. 36, 101392 https://doi.org/. giq.2019.07.004DOI: 10.1016/j

Di Vaio, A., Hassan, R., D'Amore, G., & Strologo, A. D. (2022). Digital technologies for sustainable waste management on-board ships: An analysis of best practices from the cruise industry. *IEEE Transactions on Engineering Management.*

Dong, Z., Chen, J., & Lu, W. (2022). Computer vision to recognize construction waste compositions: A novel boundary-aware transformer (BAT) model. *Journal of Environmental Management, 305*, 114405. DOI: 10.1016/j.jenvman.2021.114405 PMID: 34995944

Ejimofor, M. I., Aniagor, C. O., Oba, S. N., Menkiti, M. C., & Ugonabo, V. I. (2022). Artificial intelligence in the reduction and management of land pollution. In *Current Trends and Advances in Computer-Aided Intelligent Environmental Data Engineering* (pp. 319–333). Elsevier., DOI: 10.1016/B978-0-323-85597-6.00009-4

Fang, B., Yu, J., Chen, Z., Osman, A. I., Farghali, M., Ihara, I., Hamza, E. H., Rooney, D. W., & Yap, P. S. (2023). Artificial intelligence for waste management in smart cities: A review. *Environmental Chemistry Letters, 21*(4), 1959–1989. DOI: 10.1007/s10311-023-01604-3 PMID: 37362015

Gal, I.-A., Ciocîrlan, A.-C., & M˘arg˘aritescu, M. (2021). State machine-based hybrid position/ force control architecture for a waste management mobile robot with 5DOF manipulator. *Applied Sciences (Basel, Switzerland), 11*(9), 4222. DOI: 10.3390/app11094222

Gerhards, R., Spaeth, M., Heyn, A., & Saile, M. (2025). Sensor-Guided Mechanical Weed Control in Transplanted Lettuce and Cabbage. *Journal of Crop Health, 77*(2), 67. DOI: 10.1007/s10343-025-01139-3

Ghahramani, M., Zhou, M., Molter, A., & Pilla, F. (2022). IoT-based route recommendation for an intelligent waste management system. *IEEE Internet of Things Journal, 9*(14), 11883–11892. DOI: 10.1109/JIOT.2021.3132126

Ghanbari, F., Kamalan, H., & Sarraf, A. (2023). Predicting solid waste generation based on the ensemble artificial intelligence models under uncertainty analysis. *Journal of Material Cycles and Waste Management, 25*(2), 920–930. DOI: 10.1007/s10163-023-01589-9

Gopalakrishnan, P., (2019). Blockchain Based Waste Management 8.

Gundupalli, S. P., Hait, S., & Thakur, A. (2018, August). Classification of metallic and non-metallic fractions of e-waste using thermal imaging-based technique. *Process Safety and Environmental Protection, 118,* 32–39. DOI: 10.1016/j.psep.2018.06.022

Guo, H., Wu, S., Tian, Y., Zhang, J., Liu, H., (2021). Application of machine learning methods for the prediction of organic solid waste treatment and recycling processes: A review. Bioresour. Technol. 319, 124114 https://doi.org/. biortech.2020.124114. DOI: 10.1016/j

Harper, G., Sommerville, R., Kendrick, E., Driscoll, L., Slater, P., Stolkin, R., ... & Anderson, P. (2019). Recycling lithium-ion batteries from electric vehicles. *nature, 575*(7781), 75-86.

He, L. (2023). Variable rate technologies for precision agriculture. In *Encyclopedia of Digital Agricultural Technologies* (pp. 1533–1542). Springer International Publishing. DOI: 10.1007/978-3-031-24861-0_34

Hernandez Bennetts, V., Lilienthal, A. J., Neumann, P. P., & Trincavelli, M. (2012). Mobile Robots for Localizing Gas Emission Sources on Landfill Sites: Is Bio-Inspiration the Way to Go? *Frontiers in Neuroengineering, 4*. Advance online publication. DOI: 10.3389/fneng.2011.00020 PMID: 22319493

Hoque, M. M., & Rahman, M. T. U. (2020). Landfill area estimation based on solid waste collection prediction using ANN model and final waste disposal options. *Journal of Cleaner Production, 256*, 120387. DOI: 10.1016/j.jclepro.2020.120387

Hsu, E., Barmak, K., West, A. C., & Park, A.-H. A. (2019). Advancements in the treatment and processing of electronic waste with sustainability: A review of metal extraction and recovery technologies. *Green Chemistry, 21*(5), 919–936. DOI: 10.1039/C8GC03688H

Huang, Q., Someya, K., Dewi, R. K., & Komatsuzaki, M. (2025). No-tillage intercropping with a robotic mower: Advancing a high productivity, low-carbon and energy-efficient organic farming system. *Energy, 324*, 135851. DOI: 10.1016/j.energy.2025.135851

Hussain, A., Draz, U., Ali, T., Tariq, S., Irfan, M., Glowacz, A., Antonino Daviu, J. A., Yasin, S., & Rahman, S. (2020). Waste management and prediction of air pollutants using IoT and machine learning approach. *Energies, 13*(15), 3930. DOI: 10.3390/en13153930

Iliev, D. (2022). Electronic waste management: A review of the limiting factors and robotic solutions. *Engineering 4.0 and the Internet of Everything*, 75.

Inoue, Y. (2020). Satellite-and drone-based remote sensing of crops and soils for smart farming–a review. *Soil Science and Plant Nutrition, 66*(6), 798–810. DOI: 10.1080/00380768.2020.1738899

Jogarao, M., Lakshmanna, B., & Naidu, S. (2024). *AI-Enabled circular economy management for sustainable smart cities: integrating artificial intelligence in resource optimization and waste reduction. Smart Cities and Circular Economy.* Emerald Publishing Limited.

Kalirajan, M., Mageshen, V. R., Aswitha, K., & Saranya, M. (2024). AI-Based Regulation of Water Supply and Pest Management in Farming. In *Artificial Intelligence Techniques in Smart Agriculture* (pp. 195–215). Springer Nature Singapore. DOI: 10.1007/978-981-97-5878-4_12

Kapetas, D., Christakakis, P., Goglia, V., & Pechlivani, E. M. (2025, January). AI-based robotic trap for real-time insect detection, monitoring and population prediction. In 2025 IEEE 6th International Conference on Image Processing, Applications and Systems (IPAS) (pp. 1-7). IEEE.

Khanal, S.K., Tarafdar, A., You, S., (2023). Artificial intelligence and machine learning for smart bioprocesses. Bioresour. Technol. 375, 128826 https://doi.org/. biortech.2023.128826.DOI: 10.1016/j

Kostenko, V. V., Bykanova, A. Y., & Tolstonogov, A. Y. (2019). Underwater robotics complex for inspection and laser cleaning of ships from biofouling. *IOP Conference Series. Earth and Environmental Science, 272*(2), 1–7. DOI: 10.1088/1755-1315/272/2/022103

Kumar, N. M., Mohammed, M. A., Abdulkareem, K. H., Damasevicius, R., Mostafa, S. A., Maashi, M. S., & Chopra, S. S. (2021). Artificial intelligence-based solution for sorting COVID related medical waste streams and supporting data-driven decisions for smart circular economy practice. *Process Safety and Environmental Protection, 152*, 482–494. DOI: 10.1016/j.psep.2021.06.026

Kumar, V., Sharma, K. V., Kedam, N., Patel, A., Kate, T. R., & Rathnayake, U. (2024). A comprehensive review on smart and sustainable agriculture using IoT technologies. *Smart Agricultural Technology, 8*, 100487. DOI: 10.1016/j.atech.2024.100487

Licardo, J. T., Domjan, M., & Orehovački, T. (2024). Intelligent robotics—A systematic review of emerging technologies and trends. *Electronics (Basel), 13*(3), 542. DOI: 10.3390/electronics13030542

Lin, K., Zhao, Y., Kuo, J.-H., Deng, H., Cui, F., Zhang, Z., Zhang, M., Zhao, C., Gao, X., Zhou, T., & Wang, T. (2022). Toward smarter management and recovery of municipal solid waste: A critical review on deep learning approaches. *Journal of Cleaner Production*, *346*, 130943. DOI: 10.1016/j.jclepro.2022.130943

Lu, J. W., Chang, N. B., & Liao, L. (2013). Environmental informatics for solid and hazardous waste management: Advances, challenges, and perspectives. *Critical Reviews in Environmental Science and Technology*, *43*(15), 1557–1656. DOI: 10.1080/10643389.2012.671097

Lu, W., Chen, J., & Xue, F. (2022). Using computer vision to recognize composition of construction waste mixtures: A semantic segmentation approach. *Resources, Conservation and Recycling*, *178*, 106022. DOI: 10.1016/j.resconrec.2021.106022

Majchrowska, S., Mikołajczyk, A., Ferlin, M., Klawikowska, Z., Plantykow, M.A., Kwasigroch, A., Majek, K., (2022). Deep learning-based waste detection in natural and urban environments. Waste Manag. 138, 274–284. https://doi.org/. wasman.2021.12.001.DOI: 10.1016/j

Mana, A. A., Allouhi, A., Hamrani, A., Rehman, S., El Jamaoui, I., & Jayachandran, K. (2024). Sustainable AI-based production agriculture: Exploring AI applications and implications in agricultural practices. *Smart Agricultural Technology*, *7*, 100416. DOI: 10.1016/j.atech.2024.100416

Mao, W.-L., Chen, W.-C., Wang, C.-T., & Lin, Y.-H. (2021). Recycling waste classification using optimized convolutional neural network. *Resources, Conservation and Recycling*, *164*, 105132. DOI: 10.1016/j.resconrec.2020.105132

Marconi, M., Palmieri, G., Callegari, M., & Germani, M. (2019, February). Feasibility Study and Design of an Automatic System for Electronic Components Disassembly. *Journal of Manufacturing Science and Engineering*, *141*(2), 021011. DOI: 10.1115/1.4042006

Meshram, A., Vanalkar, A., Meshram, K., Badar, A., Mehta, G., & Kaushik, V. (2024). Enhancing crop protection through smart autonomous pesticide spray bot in sustainable agriculture. *Engineering Research Express*, *6*(4), 045563. DOI: 10.1088/2631-8695/ad9545

Monzambe, G. M., Mpofu, K., & Daniyan, I. A. (2019). Statistical analysis of determinant factors and framework development for the optimal and sustainable design of municipal solid waste management systems in the context of industry 4.0. *Procedia CIRP*, *84*, 245–250. DOI: 10.1016/j.procir.2019.04.182

Nafiz, M. S., Das, S. S., Morol, M. K., Al Juabir, A., & Nandi, D. (2023, January). Convowaste: An automatic waste segregation machine using deep learning. In 2023 3rd International conference on robotics, electrical and signal processing techniques (ICREST) (pp. 181-186). IEEE.

Nahaei, V. S., & Naziri-Oskuei, F. (2021). Fuzzy clustering approach for marketing recycled products of tabriz municipality waste management organization. Int. J. Innov. Market. *Elem.*, *1*(1), 25–36. DOI: 10.59615/ijime.1.1.25

Narayana, T. L., Venkatesh, C., Kiran, A., Kumar, A., Khan, S. B., Almusharraf, A., & Quasim, M. T. (2024). Advances in real time smart monitoring of environmental parameters using IoT and sensors. *Heliyon*, *10*(7), e28195. DOI: 10.1016/j.heliyon.2024.e28195 PMID: 38571667

Nižetić, S., Djilali, N., Papadopoulos, A., & Rodrigues, J. J. (2019). Smart technologies for promotion of energy efficiency, utilization of sustainable resources and waste management. *Journal of Cleaner Production*, *231*, 565–591.

Nwokediegwu, Z. Q. S., & Ugwuanyi, E. D. (2024). Implementing ai-driven waste management systems in underserved communities in the USA. *Eng. Sci. Technol. J.*, *5*(3), 794–802. DOI: 10.51594/estj.v5i3.903

Olawade, D. B., Fapohunda, O., Wada, O. Z., Usman, S. O., Ige, A. O., Ajisafe, O., & Oladapo, B. I. (2024). *Smart waste management: a paradigm shift enabled by artificial intelligence, Waste Manage*. Bull.

Olawade, D. B., Fapohunda, O., Wada, O. Z., Usman, S. O., Ige, A. O., Ajisafe, O., & Oladapo, B. I. (2024). *Smart waste management: A paradigm shift enabled by artificial intelligence*. Waste Management Bulletin.

Onoda, H. (2020). Smart approaches to waste management for post-COVID-19 smart cities in Japan. *IET Smart Cities*, *2*(2), 89–94. DOI: 10.1049/iet-smc.2020.0051

Ospina, R., & Noguchi, N. (2025). Real-time work progress estimation based on GIS remote monitoring system for agricultural robot vehicles. *Computers and Electronics in Agriculture*, *234*, 110313. DOI: 10.1016/j.compag.2025.110313

Oyedotun, T. D. T., & Moonsammy, S. (2021). Linking national policies to beneficiaries: Geospatial and statistical focus to waste and sanitation planning. *Environmental Challenges*, *4*, 100142. DOI: 10.1016/j.envc.2021.100142

Oyejobi, D. O., Firoozi, A. A., Fern'andez, D. B., & Avudaiappan, S. (2024). Integrating circular economy principles into concrete technology: Enhancing sustainability through industrial waste utilization. *Results in Engineering*, *24*, 102846. DOI: 10.1016/j.rineng.2024.102846

Padhiary, M., Saha, D., Kumar, R., Sethi, L. N., & Kumar, A. (2024). Enhancing precision agriculture: A comprehensive review of machine learning and AI vision applications in all-terrain vehicle for farm automation. *Smart Agricultural Technology*, *8*, 100483. DOI: 10.1016/j.atech.2024.100483

Pallathadka, H., Mustafa, M., Sanchez, D. T., Sajja, G. S., Gour, S., & Naved, M. (2023). Impact of machine learning on management, healthcare and agriculture. *Materials Today: Proceedings*, *80*, 2803–2806. DOI: 10.1016/j.matpr.2021.07.042

Paul, R. A. I., Palanisamy, M. A., Peramaiyan, P., Kumar, V., Bagavathiannan, M., Gurjar, B., Vijayakumar, S., Djanaguiraman, M., Pazhanivelan, S., & Ramasamy, K. (2024). Spray volume optimization with UAV-based herbicide application for effective droplet deposition and weed control in direct-seeded rice. *Frontiers in Agronomy*, *6*, 1491842. DOI: 10.3389/fagro.2024.1491842

Pearson, S., Camacho-Villa, T. C., Valluru, R., Gaju, O., Rai, M. C., Gould, I., Brewer, S., & Sklar, E. (2022). Robotics and autonomous systems for net zero agriculture. *Current Robotics Reports*, *3*(2), 57–64. DOI: 10.1007/s43154-022-00077-6

Pedersen, S. M., & Lind, K. M. (2017). Precision agriculture–from mapping to site-specific application. *Precision agriculture: Technology and economic perspectives*, 1-20.

Raj, M., Harshini, N. B., Gupta, S., Atiquzzaman, M., Rawlley, O., & Goel, L. (2024). Leveraging precision agriculture techniques using UAVs and emerging disruptive technologies. *Energy Nexus*, *14*, 100300. DOI: 10.1016/j.nexus.2024.100300

Rakhio, A. (2024). Research advancements in recycling: YOLOV4 and darknet-powered object detection of hazardous items. *ParadigmPlus*, *5*(1), 1–11. DOI: 10.55969/paradigmplus.v5n1a1

Rame, R., Purwanto, P., & Sudarno, S. (2024). Industry 5.0 and sustainability: An overview of emerging trends and challenges for a green future. *Innovation and Green Development*, *3*(4), 100173. DOI: 10.1016/j.igd.2024.100173

Romano, D., & Stefanini, C. (2025). Animal–robot interaction induces local enhancement in the Mediterranean fruit fly Ceratitis capitata Wiedemann. *Bioinspiration & Biomimetics*, *20*(3), 036009. DOI: 10.1088/1748-3190/adbb42 PMID: 40014926

Rujanavech, C., Lessard, J., Chandler, S., Shannon, S., Dahmus, J., & Guzzo, R. "*Liam - An Innovation Story*," p. 8, 2016.

Rutqvist, D., Kleyko, D., & Blomstedt, F. (2020). An automated machine learning approach for smart waste management systems. *IEEE Transactions on Industrial Informatics*, *16*(1), 384–392. DOI: 10.1109/TII.2019.2915572

Said, Z., Sharma, P., Thi Bich Nhuong, Q., Bora, B. J., Lichtfouse, E., Khalid, H. M., Luque, R., Nguyen, X. P., & Hoang, A. T. (2023). Intelligent approaches for sustainable management and valorisation of food waste. *Bioresource Technology*, *377*, 128952. DOI: 10.1016/j.biortech.2023.128952 PMID: 36965587

Salem, K. S., Clayson, K., Salas, M., Haque, N., Rao, R., Agate, S., Singh, A., Levis, J. W., Mittal, A., Yarbrough, J. M., Venditti, R., Jameel, H., Lucia, L., & Pal, L. (2023). A critical review of existing and emerging technologies and systems to optimize solid waste management for feedstocks and energy conversion. *Matter*, *6*(10), 3113–p3684. DOI: 10.1016/j.matt.2023.08.003

Sarc, R., Curtis, A., Kandlbauer, L., Khodier, K., Lorber, K. E., & Pomberger, R. (2019). Digitalisation and intelligent robotics in value chain of circular economy oriented waste management – A review. *Waste Management (New York, N.Y.)*, *95*, 476–492. DOI: 10.1016/j.wasman.2019.06.035 PMID: 31351634

Sharma, H., Haque, A., & Blaabjerg, F. (2021). Machine learning in wireless sensor networks for smart cities: A survey. *Electronics (Basel)*, *10*(9), 1012. DOI: 10.3390/electronics10091012

Sharma, K., & Shivandu, S. K. (2024). Integrating artificial intelligence and internet of things (IoT) for enhanced crop monitoring and management in precision agriculture. *Sensors International*, *5*, 100292. DOI: 10.1016/j.sintl.2024.100292

Sharma, P., Vimal, A., Vishvakarma, R., Kumar, P., De Souza Vandenberghe, L. P., Kumar Gaur, V., & Varjani, S. (2022). Deciphering the blackbox of omics approaches and artificial intelligence in food waste transformation and mitigation. *International Journal of Food Microbiology*, *372*, 109691. DOI: 10.1016/j.ijfoodmicro.2022.109691 PMID: 35509146

Shreyas Madhav, A., Rajaraman, R., Harini, S., & Kiliroor, C. C. (2022, July). Application of artificial intelligence to enhance collection of E-waste: A potential solution for household WEEE collection and segregation in India. *Waste Management & Research*, *40*(7), 1047–1053. DOI: 10.1177/0734242X211052846 PMID: 34726090

Shreyas Madhav, A., Rajaraman, R., Harini, S., & Kiliroor, C. C. (2022). Application of artificial intelligence to enhance collection of E-waste: A potential solution for household WEEE collection and segregation in India. *Waste Management & Research*, *40*(7), 1047–1053. DOI: 10.1177/0734242X211052846 PMID: 34726090

Sláčala, J., Špačková, H., Menšík, M., Svobodová, P., & Barčák, T. "Use of robotic manipulation systems in automation of FORGING processes," presented at the METAL 2020, 2020, pp. 1200–1204. DOI: 10.37904/metal.2020.3662

Song, C., & Cui, W. (2020). Review of underwater ship hull cleaning technologies. *Journal of marine science and application, 19*(3), 415-429.

Song, X. P., Liang, Y. J., Zhang, X. Q., Qin, Z. Q., Wei, J. J., Li, Y. R., & Wu, J. M. (2020). Intrusion of fall armyworm (*Spodoptera frugiperda*) in sugarcane and its control by drone in China. *Sugar Tech, 22*(4), 734–737. DOI: 10.1007/s12355-020-00799-x

Sreenivas, A. G., Ranganath, , Wazid, , Vijayalakshmi, , & Shivayogiyappa, . (2024). Unmanned aerial vehicle (UAVs): A novel spraying technique for management of sucking insect pests in cotton ecosystem. *Journal of Biosystems Engineering, 49*(2), 103–111. DOI: 10.1007/s42853-024-00219-y

Subramanian, A. K., Thayalan, D., Edwards, A. I., Almalki, A., & Venugopal, A. (2021). Biomedical waste management in dental practice and its significant environmental impact: A perspective. *Environmental Technology & Innovation, 24*, 101807. DOI: 10.1016/j.eti.2021.101807

Sundaralingam, S., & Ramanathan, N. (2023). A deep learning-based approach to segregate solid waste generated in residential areas. *Eng. Technol. Appl. Sci. Res., 13*(2), 10439–10446. DOI: 10.48084/etasr.5716

Sutikno, T., Fathurrahman, H. I. K., Wahono, T., & Handayani, L. (2024). Robotics Technologies in Urban Smart Waste Management. *High-tech and Innovative Series, 1*, 43–54.

Taseer, A., & Han, X. (2024). Advancements in variable rate spraying for precise spray requirements in precision agriculture using Unmanned aerial spraying Systems: A review. *Computers and Electronics in Agriculture, 219*, 108841. DOI: 10.1016/j.compag.2024.108841

Tonle, F. B., Niassy, S., Ndadji, M. M., Tchendji, M. T., Nzeukou, A., Mudereri, B. T., Senagi, K., & Tonnang, H. E. (2024). A road map for developing novel decision support system (DSS) for disseminating integrated pest management (IPM) technologies. *Computers and Electronics in Agriculture, 217*, 108526. DOI: 10.1016/j.compag.2023.108526

Verma, V. K., Gupta, A. K., Singh, A. K., Raghav, A., Singh, B. P., & Singh, A. K. (2024, October). Real-time Plant Disease Detection and Pesticide Application System Using IoT and CNN. In *2024 International Conference on Computing, Sciences and Communications (ICCSC)* (pp. 1-6). IEEE. DOI: 10.1109/ICCSC62048.2024.10830417

Wakchaure, M., Patle, B. K., & Mahindrakar, A. K. (2023). Application of AI techniques and robotics in agriculture: A review. *Artificial Intelligence in the Life Sciences*, *3*, 100057. DOI: 10.1016/j.ailsci.2023.100057

Wan, J., Li, X., Dai, H. N., Kusiak, A., Martinez-Garcia, M., & Li, D. (2020). Artificial-intelligence-driven customized manufacturing factory: Key technologies, applications, and challenges. *Proceedings of the IEEE*, *109*(4), 377–398. DOI: 10.1109/JPROC.2020.3034808

Wang, C., Qin, J., Qu, C., Ran, X., Liu, C., & Chen, B. (2021). A smart municipal waste management system based on deep-learning and Internet of Things. *Waste Management (New York, N.Y.)*, *135*, 20–29. DOI: 10.1016/j.wasman.2021.08.028 PMID: 34461487

Wolniak, R., & Stecuła, K. (2024). Artificial Intelligence in Smart Cities—Applications, Barriers, and Future Directions: A Review. *Smart Cities*, *7*(3), 1346–1389. DOI: 10.3390/smartcities7030057

Wu, T.-W., Zhang, H., Peng, W., Lü, F., & He, P.-J. (2023). Applications of convolutional neural networks for intelligent waste identification and recycling: A review. *Resources, Conservation and Recycling*, *190*, 106813. DOI: 10.1016/j.resconrec.2022.106813

Yadav, V., & Karmakar, S. (2020). Sustainable collection and transportation of municipal solid waste in urban centers. *Sustainable Cities and Society*, *53*, 101937. DOI: 10.1016/j.scs.2019.101937

Yan, H., Yin, Q., Peng, J., & Bai, B. (2019). Multi-functional tugboat for monitoring and cleaning bottom fouling. *IOP Conference Series. Earth and Environmental Science*, *237*(2), 1–6. DOI: 10.1088/1755-1315/237/2/022045

Yu, K.H., Zhang, Y., Li, D., Montenegro-Marin, C.E., Kumar, P.M., (2021). Environmental planning based on reduce, reuse, recycle and recover using artificial intelligence. Environ. Impact Assess. Rev. 86, 106492 https://doi.org/. eiar.2020.106492. DOI: 10.1016/j

Zhang, Q., Yang, Q., Zhang, X., Bao, Q., Su, J., & Liu, X. (2021). Waste image classification based on transfer learning and convolutional neural network. *Waste Management (New York, N.Y.)*, *135*, 150–157. DOI: 10.1016/j.wasman.2021.08.038 PMID: 34509053

Zhang, Y. W., Liu, M. N., Chen, D., Xu, X. M., Lu, J., Lai, H. R., Wen, C., & Yin, Y. X. (2025). Development and Testing of Row-Controlled Weeding Intelligent Robot for Corn. *Journal of Field Robotics*, *42*(3), 850–866. DOI: 10.1002/rob.22457

Zhao, Z., Hicks, Y., Sun, X., McGuinness, B. J., & Lim, H. S. (2025). A novel self-supervised method for in-field occluded apple ripeness determination. *Computers and Electronics in Agriculture*, *234*, 110246. DOI: 10.1016/j.compag.2025.110246

Zhu, L., Tian, Z., & Du, J. (2023). Spatial–temporal redundancy evaluation of the municipal solid waste incineration treatment capacity: The case study of China. *Environmental Science and Pollution Research International*, *31*(7), 9948–9963. Advance online publication. DOI: 10.1007/s11356-023-26989-0 PMID: 37072590

Zoumpoulis, P., Konstantinidis, F. K., Tsimiklis, G., & Amditis, A. (2024). Smart bins for enhanced resource recovery and sustainable urban waste practices in smart cities: A systematic literature review. *Cities (London, England)*, *152*, 105150. DOI: 10.1016/j.cities.2024.105150

Zyoud, S., & Zyoud, A. (2024). Internet of things supporting sustainable solid waste management: Global insights, hotspots, and research trends. *International Journal of Environmental Science and Technology*, 1–30.

Chapter 4
Regulating the Future:
Policy Frameworks for AI in Environmental Monitoring and Governance

Shashwata Sahu
https://orcid.org/0009-0003-6843-9554
School of Law, KIIT University, India

Navonita Mallick
https://orcid.org/0000-0003-3621-3764
KIIT University, India

Imran Hossain
https://orcid.org/0000-0002-7602-053X
Varendra University, Bangladesh

ABSTRACT

Environmental issues such as climate change, deforestation, and pollution necessitate new solutions beyond conventional governance. Artificial Intelligence improves environmental monitoring by analysing real-time data, predictive modelling, and autonomous decision-making, but it also complicates data privacy, algorithm accountability, ethics, and patchwork regulations. The present study compares global AI regulations and case studies to determine regulatory gaps and ethical issues. It suggests adaptive governance systems focusing on transparency, inclusiveness, and accountability. International cooperation will be fortified to ensure that AI-based solutions aid effectively in equitable and sustainable environmental governance.

DOI: 10.4018/979-8-3373-1892-9.ch004

1.0 INTRODUCTION

The globe is facing a strange environmental emergency consisting of increasing world temperatures, international deforestation at large scales, biodiversity loss, and continually accelerating pollution levels (Hossain et al., 2024a). Climatic change has accelerated the intensities and the level of natural disasters, with humanity's activities continuing to exploit nature at an ultra-accelerating rate (Malhi et al., 2020; Hossain et al., 2024b). The Intergovernmental Panel on Climate Change (IPCC) confirms that human-induced climate change has already led to disruptions at a global scale to societies and ecosystems and, globally, has warmed by about 1.1°C above pre-industrial levels. The last decade was the warmest, and 2024 was the highest-warmest year, with carbon dioxide (CO_2) at its 800,000-year high (Calvin et al., 2023). These dangerous trends imply the need for prompt, evidence-based monitoring, forecasting, and prevention of environmental danger with more powerful administration mechanisms.

To address such demands, Artificial Intelligence (AI) is taking centre stage as an innovative driving force in managing the environment with top-of-the-line tools to trace in real-time, make prediction models, and enforce policy in real-time (Anitha et al., 2025). AI-enabled technology, ranging from satellite imaging data analysis and remote sensing to machine learning patterns, helps manage faster, more accurate, and fact-based responses to environmental threats (Sjödin et al., 2023). Artificial intelligence satellite networks are revolutionizing forest conservation by detecting illegal deforestation in real time so that authorities can act before important damage is caused. AI-based climate models now accurately forecast extreme weather events, giving governments proactive measures to avoid and counter disasters. Sensors for air and water pollution driven by AI facilitate real-time monitoring of contamination, which is helpful to industries and regulatory agencies in enforcing environmental norms more effectively (Raihan, 2023). AI is also playing its part towards energy efficiency, industrial waste reduction, and the best possible utilization of resources, all for the sake of world sustainability (Hossain et al., 2024c). The United Nations Environment Programme (UNEP) also employs AI to identify methane leaks from oil and gas production as a first step towards containing one of the most powerful greenhouse gases responsible for climate change (Dwivedi et al., 2022). Through automatic collection and analysis of data, AI guarantees transparency and accountability in environmental decision-making and facilitates decision-making for policy-making grounded in hard facts. Despite its worth of revolution, the integration of AI into environmental management is beset with several ethical, legal, and regulatory issues.

Most critically, perhaps, is the absence of end-to-end regulatory regimes that address specifically AI applications within this domain. Priors such as outdated, piecemeal, or incomplete legislation are not able to address fundamental questions like ethical handling of AI, algorithm accountability, and data safeguarding. Across nations, AI regulation is haphazard, which translates to uneven enforcement and implementation. The European Union (EU) has enacted stringent regulations on AI, data protection, and AI. India and China are keen to leverage AI for financial development and savvy environmental monitoring. The absence of harmonised rules of governance and ethics for AI holds the use of responsible AI far behind. AI algorithms are "black boxes," and it is not straightforward to know how AI-driven decisions are reached (Rishab Bailey & Jeanne Huang, 2024). Without explainability, being fair, accountable, and meeting ethics on AI-based environmental monitoring will be a problem. Unregulated AI can potentially reify biases unintentionally and disproportionately impact vulnerable ecosystems and marginalised groups. Equal access to AI-based environmental solutions is the key to sustainable inclusivity. Environmental monitoring via AI entails massively collecting vast amounts of big data on an unprecedented scale (Singhal et al., 2024). In the absence of firm legal protection, privacy infringement and potential misuse of environmental data are dangerous. Enforcing the strengthening of the system for governing data is crucial in protecting individual and collective rights.

To realize the full potential of AI with minimal risks, governments, global institutions, and industry stakeholders must collaborate to establish robust AI governance frameworks that promote responsible AI development, advance AI transparency and explainability (XAI), enhance data protection and cybersecurity laws, foster cross-sectoral collaboration between governments, environmental NGOs, academia, and industry, and create ethical AI guidelines for environmental management (Díaz-Rodríguez et al., 2023). This study aims to analyse the state of AI regulation in environmental policy, determine some of the most critical issues with AI-based environmental monitoring, provide actionable policy recommendations on how to increase the utilisation of AI for climate resilience, sustainability, and global environmental justice, and bridge the knowledge divide between AI innovation and legislation to ensure that AI can be appropriately ethically integrated into environmental decision-making. With an international AI regulatory framework, policymakers can ensure AI is a sustainable environmental management tool, climate resilience, and equitable world development.

This research aims to analyze the existing regulatory environment and create an integrated policy framework for incorporating AI in environmental monitoring and management. It will compare the advantages and disadvantages of AI-powered environmental solutions to their potential for increasing data analysis, predictive modelling, decision-making, data privacy, the bias of algorithms, and issues of regulatory incoherence. In addition, the study explores existing paradigms of law and policy dealing with the application of AI in environmental issues and whether they withstand tests and practicality constraints to application. Based on these findings, the research provides policy implications that offer the best balance between technological advancement and regulation in order to ensure that AI supports sustainable, transparent, and equitable environmental governance.

2.0 RESEARCH METHODOLOGY

This research employs a doctrinal method with a focus on the critical analysis of existing laws, policies, and international agreements relating to AI and environmental control. Secondary sources, including reports from institutions like the United Nations Environment Programme (UNEP), World Economic Forum, and OECD, provide a global outlook. Case studies of AI-enabled environmental initiatives, including Google's AI-powered Global Forest Watch and IBM's Green Horizons program, are examined to learn about live issues and policy gaps. Comparative analysis of regulatory regimes across jurisdictions uncovers best practices and areas for improvement.

3.0 ROLE OF AI IN ENVIRONMENTAL MONITORING AND GOVERNANCE

The breakneck evolution of AI is transforming environmental monitoring and management to deal with number one challenges like climate change, deforestation, pollution, and loss of biodiversity. Traditional monitoring schemes do not provide real-time large-scale data. However, AI, with backing from predictive modelling, remote sensing, and automation, improves the capacity to track environmental change, predict threats, and inform evidence-based policy (Onyebuchi Nneamaka Chisom et al., 2024). While AI brings numerous benefits, its widespread uptake is limited by accessibility, expense, and ethics. To equitably and ethically make it available requires overcoming these hindrances through effective policies and laws.

3.1 AI Technologies in Environmental Monitoring

Artificial intelligence technologies have made groundbreaking steps in environmental monitoring by pinpointing ecological variations exactly and opportune. AI technologies are based on the collection of big data, machine learning methodologies, and process automation of interpreting environmental trends and predicting potential dangers. Remote sensing, climate forecast models, air pollution control systems, and nature diversity tracking systems are the most effective AI-fueled tracking technologies (Kuppusamy et al., 2024). Remote sensing is among the most powerful applications of AI in environmental monitoring and is performed by gathering information from the Earth's surface through satellites, drones, and sensor systems. Everything is processed with AI algorithms to detect land use, forest cover, water quality, and urbanisation. With computerised pattern recognition and imagery analysis, artificial intelligence assists in identifying environmental alteration at a historically unprecedented pace and scale (Kuppusamy et al., 2024). Thus, for instance, satellite imaging augmented by machine intelligence can locate illicit deforestation, track desertification processes, and observe climate-change-imposed alterations of glacier shape. Climate change risks planetary ecosystems and human communities with a menacing aura.

Artificial intelligence-based predictive models forecast climate variations, meteorological aberrations, and climatic patterns spanning decades. Machine learning models are learnt on past climate data and atmospheric and oceanic patterns to accurately predict hurricanes, floods, droughts, and heatwaves. Governments and aid agencies take anticipatory measures to minimise the effect of climate risks based on these forecasts. Air pollution rates are also forecast by AI models, which allow the cities to implement steps in emissions abatement and improved air quality (Camps-Valls et al., 2025). AI has also been proven to be a valuable resource in solving environmental pollution by optimising pollution control measures. Smart sensors implanted in cities can monitor air and water pollution levels in real time and alert authorities when pollutants are harmful. AI waste management systems help segregate and recycle materials with maximum effectiveness, reducing the environmental impact of landfills (Hossain et al., 2024c). Likewise, AI is of immense help in tracking deforestation processes. Deforestation through illegal logging, land clearing for agriculture, and urbanisation leads to the loss of diversity and facilitates climate change (Popescu et al., 2024). AI algorithms translate satellite imagery into digital form and recognise trends of forest cover loss to facilitate the government and environmentalists to take action against them. Biodiversity conservation is also simplified through the implementation of AI-based tools, whereby AI-powered drones and camera traps monitor animal activities and detect poaching activities. Conservationists can protect endangered species and maintain ecological harmony by adopting such technologies.

3.1.1 Case Studies of AI in Environmental Monitoring

Google's "Global Forest Watch" is a software program that uses satellite imagery and machine learning algorithms to track global deforestation activity. The system offers real-time deforestation activity detection, and governments, NGOs, and researchers can track illegal logging and prevent it (Saha et al., 2022). The AI-driven system scans large data sets from different sources, identifying those locations with high-risk deforestation and recommending action plans. This project has considerably improved the precision and effectiveness of forest monitoring worldwide.

IBM's Green Horizons project applies big data processing and AI to predict air pollution in urban areas. Based on weather, traffic flows, and factory emissions, the system makes accurate air quality predictions. Projections by Green Horizons educate policymakers and urban planners on what measures like traffic control and limiting factory exhaust should be implemented to enhance air quality (IBM, 2015). This knowledge-driven methodology has been already adopted in urban cities like Beijing, where policy-making based on knowledge has immensely managed air pollution.

3.1.2 Benefits of AI in Environmental Governance:

Environmental monitoring systems equipped with AI have introduced many benefits to policy-making and governance. The real-time processing of massive amounts of data has transformed decision-making in ecology.

Figure 1. Key Benefits of AI in Environmental Governance

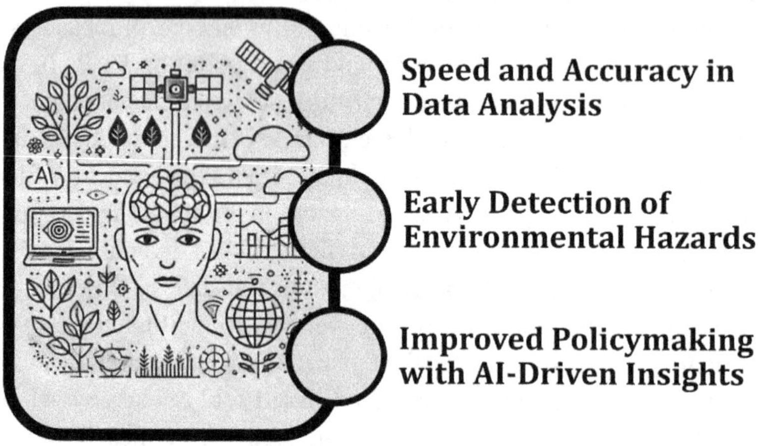

Traditional methods of gathering environmental information using field observation and manual surveys are time-consuming and prone to human mistakes. AI gathers and processes information without human intervention, resulting in real-time and extremely accurate conclusions. The processing speed enables the authorities to identify environmental risks in good time and conduct early interventions (Alotaibi & Nassif, 2024). For example, AI-based air pollutant detection systems can identify contaminated areas in seconds, and the authorities can rectify them simultaneously. The predictive ability of AI is vital in detecting potential environmental hazards before they escalate into a disaster. For instance, AI-based climatic forecasts can forecast hurricanes and fires several weeks in advance, and governments can relocate at-risk people and declare emergency measures accordingly. Likewise, AI-based water quality management systems can identify contaminants in potable water so public health crises are avoided (Giannakidou et al., 2024). AI provides policymakers with evidence-based facts to craft better green policies. By establishing patterns and trends, AI makes it possible for policymakers to put figures on the performance of current green policies and give evidence-based recommendations for reforms. AI simulation has also made it possible to predict the effects of policy actions and the policy actions themselves years and decades ahead so that the government can make greener decisions (Bolón-Canedo et al., 2024). Artificial intelligence has also been applied successfully in planning cities, and AI models help green towns be planned using the most scarce resources.

3.1.3 Challenges in AI Adoption

Although it offers many benefits, environmental monitoring and regulation through the implementation of AI have some challenges. Among these challenges are the high installation costs, limited access, and ethics. Technologies that support AI are expensive, making them unavailable to most developing countries. The deployment of AI-based satellite monitoring systems, the creation of real-time networks for air pollution monitoring, and the maintenance of AI-based data centres are costly and beyond the means of most cash-strapped governments (Peretz-Andersson et al., 2024). Second, AI capacity is located in high-technology countries and is at risk of bypassing developing regions regarding environmental decision-making. There has to be some investment in the digital infrastructure in these regions. There is more data gathering in the environment based on AI, with more concern for surveillance, privacy, and data protection. Satellite imagery photos, near-real-time emission records, and climatic data can infringe on a right to privacy unless utilised appropriately to justify it (Sipola et al., 2023). The decision-making with AI is black-boxed, and as such, holding AI systems responsible if they make mistakes or become biased is problematic. Algorithmic bias is also at risk because AI models that are trained on incomplete or biased data may make the wrong environmental predictions and, consequently, the wrong policy decisions. These problems need to be addressed by making AI-based ecological regulation transparent and accountable.

3.2 Legal and Regulatory Challenges in AI-Driven Environmental Monitoring

AI approach towards environmental monitoring has significantly improved the tracking of climate change, controlling pollution, monitoring forest loss, conserving nature with real-time analysis, predictive models and processing data on the go. While AI aids in data-driven decisions for environmental management, it presents legal and regulatory challenges that include data privacy, accountability for algorithms, regulatory gaps, and ethics (Olawade, Wada, Ige, et al., 2024). Poorly defined policies can lead to AI systems violating privacy rights, reinforcing discrimination, and disproportionately affecting marginalised groups. The best way to achieve equitable and responsible uses of these technologies for environmental governance is to develop strong legal frameworks to address these challenges (Hossain et al., 2023).

Figure 2. AI's Role in Environmental Monitoring (Benefits)

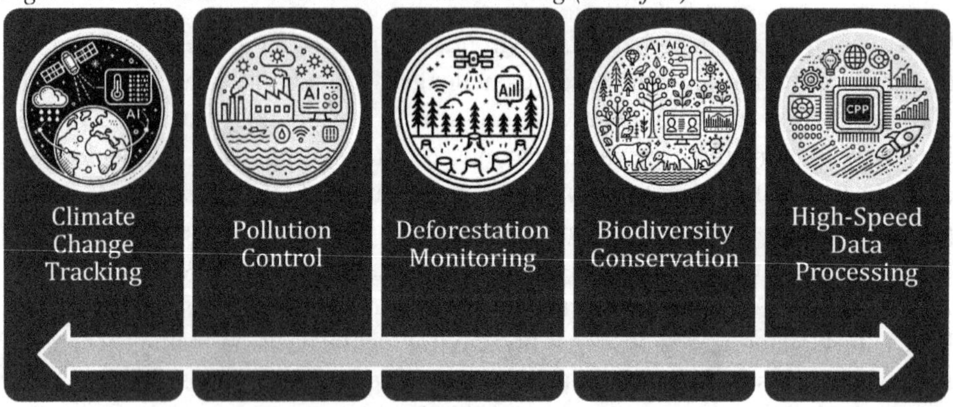

3.2.1 Data Privacy and Security Concerns:

The use of AI to monitor the environment relies largely on the bulk collection of massive amounts of data from a range of sources, from satellites, aerial vehicles, sensors and public databases. Some data is just as important to AI as it is to the accurate analysis of the changes facing the environment and has an equal ability to infringe on data privacy and security. AI will collect sensitive, local data such as ones with geolocation, people's land use, and environmental information (Janga et al., 2023). Without appropriate data protection legislation in place, Governments and third parties can use such data to generate privacy violations and even exploit them. The most significant legal challenge from AI environmental monitoring is that its use must comply, globally, with data protection laws such as the European Union's General Data Protection Regulation (GDPR), the Digital Personal Data Protection Act (DPDPA) in India, etc.

The GDPR, for example, sets strict controls on the gathering of data, so organisations need to secure consent before processing personal data. AI initiatives in the environment need to abide by data collection in accordance with legal needs, particularly when it comes to maintaining private land space or city areas. The DPDPA, enacted in India at present, also safeguards personal data and disallows unauthorised transfer of sensitive information. Nonetheless, environmental AI surveillance mostly draws on large public datasets, thus making tracking the amount of personal data collection challenging (Hoofnagle et al., 2019). AI environmental surveillance relies significantly on bulk data gathering from diverse sources, from satellites, drones, sensors, and public databases. The information is as vital to AI as to properly realizing environmental trends but is equally likely to infringe on data privacy and safety. The volume of data collected by AI systems will encompass sensitive information like geolocation, individual land use conduct, and environmental data at the neighbourhood level.

3.2.2 Algorithmic Accountability and Transparency

Opacity and algorithmic accountability are among the most significant challenges of environmental governance based on AI. AI relies on advanced algorithms based on huge datasets for prediction and recommendation. The algorithms, though, become "black boxes" such that their reasoning is poorly explained. This transparency gap has serious ethical and legal consequences, especially where the outputs of the AI determine environmental policy with effects on entire communities. The biggest risk to AI-environment surveillance is likely algorithmic bias (Yuan & Chen, 2025). Current data sets train AI algorithms and if those data sets have a bent, the AI algorithm can produce misleading or distorted outcomes. For the forest decline-detecting AI system, outdated or poor-quality satellite imagery might be used and resort to the false conclusion that a forest has indeed declined. In some instances, computer programs biased against pollution have misidentified sources of pollution and, at great expense, singled out certain groups or industries to the extent of failing to attend to causes of degradation to the environment (Ghazal et al., 2024). A well-known example of algorithmic bias is the misinterpretation of air monitoring equipment in low-income neighbourhoods. Artificial intelligence algorithms trained on data from wealthier city wards have repeatedly failed to reflect the true levels of pollution among lower-income groups. This failure poses a risk of discriminatory environmental policy, like unfairly singling out particular industries without punishing wealthier districts' pollution. The explainability of

AI-driven decision-making makes it difficult to challenge or correct such biases, thus resulting in discriminatory policy implementations. Governments should require explainable AI (XAI) models in environmental policy to make algorithms accountable (Ozmen Garibay et al., 2023). Explainable AI brings transparency in the sense that policymakers and researchers are able to examine how AI systems arrive at their decisions. Policymakers should also require mandatory auditing for AI-driven environmental sensors to make them equitable, precise, and unbiased in prediction and guidance.

3.2.3 Regulatory Gaps in AI-Driven Environmental Governance:

Even as there is a greater expanded AI presence in environmental monitoring, there is far less global regulatory architecture for the use of AI. Much of the existing environmental law and regulation of AI remains in a silo with few uniform guidelines for using AI in environmental decision-making. This kind of governance failure introduces uncertainty of legality and the heightened risk of AI misuse. It is one of the major barriers that oversight institutions in governing AI greatly vary per jurisdiction. Though countries like the European Union have established ethical terms of use of AI where there is an emphasis on transparency and accountability, there are countries lagging behind without making official legislation regarding AI (Yang et al., 2024). A lack of conformity hinders the establishment of customary standards for the use of AI in environmental surveillance. For instance, an AI system used for global deforestation tracking can comply with the laws of one nation while violating data protection laws in another. The absence of a harmonised international regulatory framework for AI results in uneven environmental policy enforcement. Existing ecological regulations also fail to tackle AI-related issues such as algorithmic failure, data privacy infringement, and liability in case of AI prediction error (Gbadegesin, 2023). Most regulatory systems have so far not dropped the traditional regulatory systems that cannot keep up with the automated and data-centricity of AI-based environmental monitoring. Governments need to come up with AI-specific environmental legislation that addresses these challenges and establish well-delineated liability systems for AI developers and environmental agencies using AI systems.

3.2.4 Ethical Implications of AI in Environmental Decision-Making:

Although AI has useful recommendations on environmental governance, its application is extremely contentious due to its moral concerns, especially in Indigenous and marginalised groups. Policies for AI consider economic efficiency and environmental conservation more than social justice, resulting in unwanted impacts on marginalised communities. The most egregious moral concern might be the manner in which AI-based land and resource management policies affect Indigenous communities (WHO, 2021). AI technology used for tracking deforestation, water use, and land use planning tends to miss out on indigenous culture and practices. For instance, an AI-driven conservation policy could mark certain clusters of forests as "protected areas" that restrict Indigenous communities' access to land they have been relying on for centuries to keep their livelihood afloat. Regardless of the purpose of such policies, they end up resettling communities and infringing on indigenous peoples' rights to land (Khaleel et al., 2024). Additionally, AI-based economic policies cater to corporate and industrial interests ahead of environmental justice. For example, AI-based carbon offsetting schemes allow companies to pay for offsetting their carbon footprint through investments in conservation. The conservation sometimes comprises the purchase of land that resettles communities against their will. Unless human rights guarantees and rigorous ethical limits are placed upon them, AI-powered environmental governance will inevitably foster the interests of industry over those of affected groups' rights. To these ethical challenges, policymakers must integrate human rights considerations into government policy frameworks for AI (Khan et al., 2025; Hossain et al., 2025). AI-driven environmental monitoring must involve participatory decision-making processes that open space for impacted communities to have their voices heard in policy-making. Governments must establish moral AI review boards to assess the social effects of AI environmental plans before they take effect.

4.0 COMPARATIVE ANALYSIS OF AI POLICY FRAMEWORKS ACROSS JURISDICTIONS

The increased application of AI for environmental regulation has created regulatory regimes in jurisdictions to meet national technological capacity, legal systems, and environmental interests. Although AI improves ecological surveillance and decision-making, it must be integrated with open legal and ethical principles. The OECD and UNEP advised that governance practices of AI based on openness, accountability, and sustainability prevail across the globe in very divergent enforcement contexts (Pagallo, 2025). It discusses comparative policy on AI in top jurisdictions and provides best practices for responsible stewardship of sustainable AI for environmental monitoring.

4.1 International AI Governance Frameworks

4.1.1 OECD AI Principles

Figure 3. OECD AI Principles (2019)

The OECD AI Principles, agreed upon in 2019, are a world reference point for AI governance with a focus on transparency, accountability, and inclusivity to ensure that the deployment of AI aligns with human rights and democratic values. The principles are best applied to AI-environmental governance and are applicable to five areas. Explainability and transparency call for AI systems to be architected with structures that allow human intervention and understanding of decision-making. Accountability calls for organisations using AI to be completely responsible for their outputs, while AI-based environmental policies not cause harm in unexpected manners (Corrêa et al., 2023). Fairness and inclusiveness require AI adoption to address many societal impacts, especially environmental justice, where vulnerable groups will most likely suffer disproportionately. Safety and robustness require AI models to stay consistent and impartial, functioning under different conditions without generating inaccuracies. Finally, sustainable development underscores AI's critical role in spurring climate action, conserving natural resources, and regulating pollution (Capraro et al., 2024). Though these guiding principles form a general ethical and regulatory basis, their effectiveness varies with the incorporation of each principle into national AI management systems. They have different implementations, enforcement, and contribution to environmental policy.

4.1.2 United Nations Environment Programme (UNEP) AI Guidelines

Figure 4. 6 Areas of Concentration by UNEP

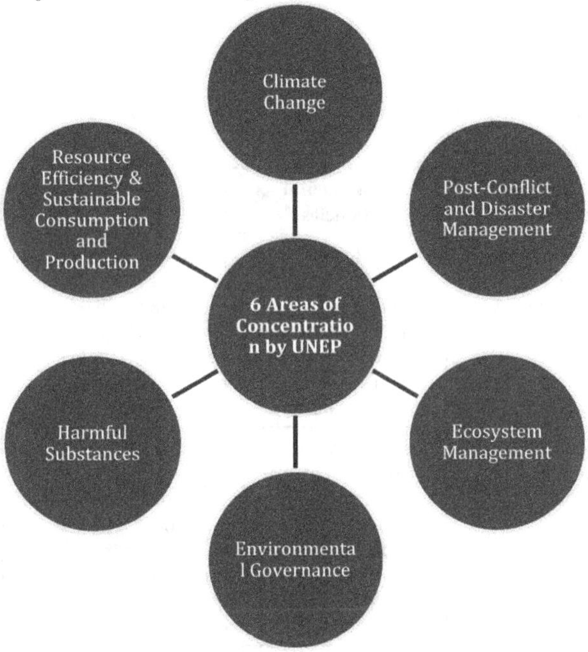

The United Nations Environment Programme (UNEP) has played an instrumental role in shaping AI-based environmental policy, emphasising actualising AI's potential for achieving the Sustainable Development Goals (SDGs) and addressing the priority challenges. The suggestions made by UNEP are towards the ethical and sustainable usage of AI technology to monitor the environment so that there is no manipulation of ecosystems and that AI technologies are being utilised ethically and sustainably (Environment, 2024b). Furthermore, UNEP promotes global AI regulation systems that provide balanced data-sharing practices among countries to prevent data monopolisation and provide balanced access to AI-enabled environmental data. Among the most crucial aspects of UNEP's organisation is assistance in a multi-stakeholder partnership that unites governments, business entities, and civil society organisations to regulate the application of AI-enabled tools in climate action and biodiversity conservation (Environment, 2024a). However, aside from these global initiatives, the regime of AI in environmental protection remains nascent, and states are pushing in different directions based on their level of technological advancement, economic interests, and institutional regulatory arrangements domestically to establish jurisdictional differences with which to regulate AI.

4.2 Comparative Review of AI Policies in Different Countries:

Table 1. Comparative Analysis of AI Policy Frameworks for Environmental Governance Across Key Jurisdictions

Country	AI Governance Framework	Environmental AI Policies	Key Strengths	Challenges
United States	National AI Initiative Act (2020), Executive Order on AI (2019)	AI-driven environmental monitoring projects (NASA, NOAA). AI for wildfire detection and climate risk assessment.	Substantial investment in AI research, robust climate risk modelling, and collaboration with private tech firms.	Lack of federal AI regulations and fragmented policies across states.
European Union	EU AI Act, GDPR, AI Ethics Guidelines (2020)	AI applications in pollution control, deforestation tracking, and carbon emission monitoring.	Strict AI regulations ensure data privacy, strong ethical oversight, and integration with sustainability policies.	Compliance burden for businesses, slow regulatory adoption due to complex EU governance structure.
India	National AI Strategy (2018), Digital Personal Data Protection Act (DPDPA)	AI-driven smart agriculture, water resource management, AI-based air quality forecasting.	Strong focus on AI for sustainable development and government initiatives for AI startups in environmental sectors.	Lack of sector-specific AI regulations and limited AI research funding.
China	AI Development Plan (2017), Smart Environmental Governance Models	AI-powered pollution control, innovative city sustainability projects, AI in carbon neutrality initiatives.	Government-led AI implementation, integration of AI in urban planning, and substantial technological infrastructure.	Concerns over data privacy and lack of transparency in AI decision-making.

All of these jurisdictions have sought various methods of incorporating AI in environmental control with varying degrees of regulatory authority and policy effectiveness.

4.3 Lessons Learned and Best Practices

Comparative examination of the AI policy regimes within the jurisdictions offers lessons and lessons learned to inform future regulation of environmental surveillance by AI.

4.3.1 Key Strengths of Existing Frameworks

End-to-end AI regulation in countries such as the European Union (EU) and China signifies that government-led regulation of AI should strike a balance between ethical reasoning and AI-powered environmental solutions. EU AI Act and Chinese AI policy prioritise intense regulation, and AI technologies need to be harmonised to sustainable ends with careful planning so that the likely hazards, such as abuse and bias, are not adopted. The harmonisation of the EU to GDPR also helps regulate AI by imposing stringent data security and privacy regulations to eliminate misuse of environmental data (Panait et al., 2021). Meanwhile, investment in AI research has been the main driving force for AI application in climate modelling and pollution control, particularly in the United States and China, where massive government expenditures and public-private partnerships have fueled AI-based environmental technologies. NASA's AI-based wildfire detection system is of particular interest as a model for applying AI in disaster preparedness and ecological surveillance. So is India's National AI Strategy that highlights the uses of AI for sustainable development in climate resilience, agriculture, water harvesting, and management of pollution (Bibri et al., 2024). Globally, UNEP AI Guidelines allow multi-stakeholder collaboration toward global cooperation in AI-based sustainability projects. They tend to broadly encourage strategic AI regulation, ethical innovation, and transborder collaboration to maximise the application of AI potential to address worldwide environmental issues.

4.3.2 Key Weaknesses and Challenges

The unbalanced regulatory framework of countries like the United States and India poses an important challenge for AI regulation in environmental monitoring as the difference in federal regulations regarding AI introduces differing policies between states that make it difficult to have equal ecological standards. Second, India's poor investment in research on AI confines its capacity for creating more evolved AI solutions in addressing environmental issues, hindering the pace of technological progress toward climate resilience and pollution reduction (Bill Whyman, 2023). In contrast, the over-regulation by the European Union (EU) to guarantee that AI is utilised ethically has unwittingly levied a cost of compliance against business organisations that are preventing organisations from adopting AI-based environmental monitoring technology. Moreover, the complex system of governance in the EU has been a factor in the slowness of policy enforcement, which has resulted in the lag in adopting AI in environmental governance despite its possible contributions. At the same time, China's advanced AI-facilitated environmental system of government has been discovered to be non-transparent and to have state-controlled data collection systems whose subsequent ethics issues on data privacy and AI surveillance need to be addressed (Konrad Meier & Roger Spichiger, 2024). AI-powered environmental monitoring implications in China have raised issues over data security and human rights with higher, superior calls for higher openness and transparency in the control of AI. All of these challenges cumulatively depict the fact that while AI holds huge promise for the management of the environment, actualisation is subject to policy frameworks for harmonisation tackling regulatory lacuna, burden of compliance, and ethical issues in various jurisdictions.

5.0 POLICY RECOMMENDATIONS FOR AI IN ENVIRONMENTAL GOVERNANCE

As AI transforms the management and monitoring of the environment, open regulatory and policy guidelines have become ever more important. As AI enhances climate models, air and water pollution tracking, and forest cover mapping, AI creates concerns regarding data confidentiality, accountability, ethics, and the absence of international regulatory harmonisation (Rautela & Goyal, 2024). Policy soundness must balance technological advancement with security legal and ethical safeguards for policymakers to protect the use of AI. This chapter presents essential policy recommendations to ensure the maximum utilization of AI in environmental management without exposing policymakers to potential harm.

5.1 Crafting a Global AI Governance Framework for Environmental Monitoring:

One of the urgent challenges of AI-environmental governance is that there is no single global regulatory framework because AI technologies are operating transnationally, and environmental issues such as climate change, loss of biodiversity, and pollution are global problems. Global harmonised policies need to be adopted so that AI for environmental monitoring can be regulated effectively (Emma Klein & Stewart Patrick, 2024). Since deforestation due to Amazon affects climatic phenomena on a global scale and air pollution may extend beyond state lines as well, international cooperation is needed to develop harmonized control mechanisms of

AI. Policymakers should aim for cross-border cooperation that subjects uniform AI processes to ensure that AI applications are aligned with international environmental conservation objectives. Also, institutions such as the United Nations (UN), OECD, and the World Economic Forum (WEF) should create global AI governance policies on vital domains to avoid moral AI crimes and data privacy infringement. These encompass AI ethics in environmental decision-making, such that AI applications ensure sustainability and not economic exploitation; data governance and privacy norms, where collection, storage, and sharing of environmental data rules are clearly defined; and accountability measures, where AI developers and policymakers are held responsible for AI-driven decisions affecting ecological policies (Rafael Araujo & João Mourão, 2023). To further an even more formal and efficacious structure, countries should work towards a binding worldwide agreement on AI and environmental governance in the Paris Climate Change Agreement style, which would enable countries to adopt uniform AI policy while providing regional adaptation space.

5.2 Ensuring Transparency and Accountability in AI Algorithms:

AI-based environmental regulation systems need to be transparent, understandable, and fair. Still, virtually all AI-based decisions are "black boxes" whose inner workings cannot be deciphered, which makes it difficult to know how they arrive at their conclusions. This transparency can lead to defective environmental policy that is difficult to appeal or correct. For this, Explainable AI (XAI) models must be employed to generate explicit and transparent explanations of what they generate. Governments and organisations that operate to save the environment must force the use of XAI models so that AI-based recommendations adhere to environmental sustainability standards (Hassija et al., 2024). For instance, if an AI system reports deforestation risks, it will have to report on what input led to this prediction, i.e., satellite images, climate or forest activity trends, so that they can provide better accountability and responsibility of AI technology for environmental monitoring. Moreover, to avoid untrustworthy or biased AI tools, responsibility measures must be adhered to stringently for AI developers and organisations using AI towards ecological management. All of these would have to require compulsory AI audits to assess the validity, equity, and accuracy of AI algorithms, legal accountability for AI mistakes, assigning blame to the creators of flawed AI-aided environmental policy decisions, compulsory disclosure procedures with guarantee of AI-aided environmental policy justification in understandable, straightforward language (Radanliev, 2025). By employing these responsibility instruments, policymakers ensure that AI technologies benefit the environment within the government and limit the potential dangers of black-boxed decision-making and biased AI predictions.

5.3 Addressing Data Privacy and Ethical Challenges:

AI environmental monitoring largely depends on big data, which has led to profound concerns regarding privacy, security, and responsible use of data. In the absence of robust regulatory measures, AI applications potentially endanger data misuse, surveillance, and discriminatory policy-making with disproportionate effects on already vulnerable groups. Governments need to, therefore, implement stringent data protection laws that govern how AI environmental systems process, store, and use data. These laws must provide that environmental data of individuals and communities is protected, not disclosed, and not readable or exploitable to third parties. In addition, all AI-based data collection must be done in conformity with moral codes, particularly when mapping Indigenous lands, reserved forests, or private estates to avoid violating land rights and cultural heritage (Albahri et al., 2024). Interoperability data-sharing rules must likewise be embraced, as is being mandated under global data privacy laws such as the EU's General Data Protection Regulation (GDPR), India's Digital Personal Data Protection Act (DPDPA), and other global AI laws. In addition, AI ethics review boards should be established to determine the ethical salience of AI-environmental activity and track the fairness and inclusiveness of AI policy. These boards would be essential in addressing the impact of AI-based environmental choice on vulnerable populations so that AI models are not making decisions that create assumptions that culminate in discriminatory ecological policy (Sunidhi Tyagi, 2024). Review boards would also have to establish the possibility of AI bias in environmental choice and coordinate usage of AI to ecological justice. Through the integration of firm ethical leadership into AI regulatory regimes, policymakers can see to it that AI-informed environmental policy won't disproportionately influence vulnerable parts of society and AI will be applied equally, responsibly, and in an open way in environmental choice.

5.4 Encouraging Multi-Stakeholder Collaboration:

AI environmental policy needs coordination among government, business, research institutions, and civil society to include a range of voices in AI policy and greater emphasis on sustainability compared to business. Multiple-stakeholder partnerships are crucial in driving responsible AI innovation through the relation of technology development with ecological defence objectives. Governments need to create public-private partnerships that will enable the creation of AI environmental solutions within regulatory norms. Technology firms developing AI solutions need to work with ecological regulators to introduce AI solutions to conservation efforts meaningfully (Francisco & Linnér, 2023). Besides, global organisations like the United Nations Environment Programme (UNEP), the Organisation for Economic

Co-operation and Development (OECD), and the United Nations Development Programme (UNDP) need to enable cross-border knowledge transfer so that AI-based sustainability initiatives leverage global experience and best practices. Furthermore, there should be public participation and accountability in AI-based environmental policy-making to make sure that AI systems are in the public interest instead of corporate or governmental interests (UNDP, 2024). Thus, governments must carry out public consultations prior to suggesting AI-based environmental policies to allow environmental movements, local communities, and scientific experts to raise concerns and be involved in policy-making. Transparency must be a core principle to AI environmental decision-making in order to build public trust and accountability. In addition, open-access AI models should be promoted, where environment data are made available publicly to research and academia to stimulate innovation and independent analysis of AI-founded policies. By engaging multiple stakeholders, AI governance can be made more public, ethical, and public-interest-oriented, so that AI is harnessed as an environmental justice and sustainability tool rather than a corporate or political tool.

5.5 Capacity Building and AI Literacy for Policymakers:

Low AI literacy among regulators and policymakers is one of the key impediments to efficient regulation of environmental monitoring using AI. AI technology is highly sophisticated, and regulators do not possess knowledge to formulate sufficient AI policies or assess AI-driven ecological programs. Governments should therefore invest in policymaker, regulator, and environment officer AI capacity-building programs. Such training modules need to cover the basics of machine learning and AI, with emphasis on environmental monitoring areas like biodiversity conservation, climate modelling, and air pollution detection (König et al., 2023). Policymakers should also better understand AI governance's ethical and legal requirements, such as AI-based solutions to meet data protection requirements, transparency, and accountability. In addition, regulatory programs must include case studies of AI-enabled environmental projects to inform regulators of actual uses, challenges, and regulatory implications (Papagiannidis et al., 2025). That will raise the AI literacy of decision-makers and thus make governments better equipped to regulate AI, regulate suitably, and pursue ethical AI applications in the environment, finally leading to a more precise, smoother, and cleaner AI-based environmental policy process.

6.0 CONCLUSION

The application of AI in environmental management has facilitated significant developments in mitigating climate change, pollution control, deforestation monitoring, and biodiversity conservation. AI tools like remote sensing, predictive modelling, and computerised data analysis have facilitated the enhanced ability of policymakers and environmental institutions to identify, monitor, and react to ecological problems in real-time. Nonetheless, in addition to its potential to transform, the use of AI in environmental monitoring has also brought different legal, regulatory, and ethical issues (R. C. Santos & Cagica Carvalho, 2025). These range from data privacy to algorithmic transparency, accountability, and differences in international regulation across nations, all of which call for robust AI governance frameworks. Maybe one of the most potent findings of this research is that even though nations have made advances in the regulation of AI, there is still no one world framework for environmental management using AI. OECD Principles of Artificial Intelligence and UNEP AI Guidelines offer some initial regulations, yet they are fragments and pieces of country policy. Comparative regulation of AI across jurisdictions shows that. In contrast, some places, such as the European Union, strictly enforce AI and data protection laws, and some countries, such as India and China, favour AI to drive economic growth and intelligent environmental governance (Zaidan & Ibrahim, 2024). The lack of standardised AI ethics and data governance principles poses a significant barrier to AI's responsible and equitable use in environmental decision-making.

Artificial intelligence will likely play an even more significant role in green sustainability as machine learning, automation, and big data analysis improve. The future application of AI to environmental policymaking will be heightened climate change prevention, where AI-simulated climate models will foresee with better accuracy the occurrences of extreme climate phenomena, rising sea levels, and carbon emissions trends and tendencies such that proactive disaster preparedness and efficient emissions cuts can become a reality. Furthermore, AI will further drive ecosystem conservation and biodiversity preservation by tracking species and watching over habitats to enable conservationists to detect species facing threats and take action early to avoid biodiversity loss (Olawade, Wada, David-Olawade, et al., 2024). AI will also increasingly be utilised to fight pollution, with intelligent AI sensors monitoring air and water in real-time, enabling quick response to environmental pollution and satisfying emissions requirements. Additionally, AI will support green resource management and circular economy by optimising sustainable supply chain management, recycling operations, and waste sorting and minimising environmental degradation while facilitating green economic models (Konya & Nematzadeh, 2024; Hossain et al., 2024d). All these AI-driven innovations, however, will be founded upon aggressive regulatory policies, ethical applications of AI, and encouraging cross-sector coordination among environmental groups, industries, and governments to maximise AI use while building a sustainable world.

To leverage AI potential in environmental management without encountering associated risks, policymakers must shift from a reactive to a proactive approach. Governments and global and non-state actors must collaborate to develop international AI governance structures that ensure fair, transparent, and accountable AI applications in environmental monitoring. An integral component of this venture is necessitating AI accountability and transparency features, including using Explainable AI (XAI) models that will eliminate algorithmic bias and enhance the accuracy of decisions. Also, data protection laws must be bolstered to guard against misuse of environmental data while facilitating the efficient application of AI toward sustainability practices (Alhosani & Alhashmi, 2024). Equally important is creating AI literacy and capacity-building programs for regulators and policymakers to provide them with appropriate knowledge to manage and regulate AI-facilitated environmental activities. Furthermore, multi-stakeholder forums involving governments, environmental non-governmental organisations, academia, and the private sector must be promoted to collaborate on AI-based sustainability opportunities in line with global climate goals. Unless urgent policy action is taken, the uncontrolled proliferation of AI would release uncontrolled and potentially malevolent environmental applications into the world, with a disproportionate impact on marginalised groups and vulnerable ecosystems, necessitating the argument for responsible, equitable, and forward-looking AI policymaking. AI can transform environmental governance through data-driven intelligence, predictive modelling, and automation potential to fuel sustainability initiatives (Ali et al., 2024). Unregulated use of AI has devastating effects in the form of algorithmic bias, ethical concerns, data privacy infringement, and loopholes in regulations. Governments need to establish strong legal frameworks, ethical standards, and accountability structures to make sure AI is an instrument of environmental advancement and not exploitation (Chavali et al., 2024). The fate of AI and environmental governance is at stake between environmental justice, sustainability, and equity on the one hand and adopting innovation on the other. It is a choice that can be made by policymakers by reconciling AI technology with well-delineated legal protections. It can be made to derive its full potential, limit its harms, and bring AI to a greener, more sustainable, and ethically governed world.

REFERENCES

Albahri, A. S., Khaleel, Y. L., Habeeb, M. A., Ismael, R. D., Hameed, Q. A., Deveci, M., Homod, R. Z., Albahri, O. S., Alamoodi, A. H., & Alzubaidi, L. (2024). A systematic review of trustworthy artificial intelligence applications in natural disasters. *Computers & Electrical Engineering*, *118*, 109409. DOI: 10.1016/j.compeleceng.2024.109409

Alhosani, K., & Alhashmi, S. M. (2024). Opportunities, challenges, and benefits of AI innovation in government services: A review. *Discover Artificial Intelligence*, *4*(1), 18. DOI: 10.1007/s44163-024-00111-w

Ali, O., Murray, P. A., Momin, M., Dwivedi, Y. K., & Malik, T. (2024). The effects of artificial intelligence applications in educational settings: Challenges and strategies. *Technological Forecasting and Social Change*, *199*, 123076. DOI: 10.1016/j.techfore.2023.123076

Alotaibi, E., & Nassif, N. (2024). Artificial intelligence in environmental monitoring: In-depth analysis. *Discover Artificial Intelligence*, *4*(1), 84. DOI: 10.1007/s44163-024-00198-1

Anitha, K., Ghosal, I., Amala, J., & Hossain, I. (2025). Future Trends and Innovations in Environmental Governance for Sustainable Development. In *Intersecting Environmental Governance With Technological Advancements* (pp. 381–412). IGI Global Scientific Publishing.

Bibri, S. E., Krogstie, J., Kaboli, A., & Alahi, A. (2024). Smarter eco-cities and their leading-edge artificial intelligence of things solutions for environmental sustainability: A comprehensive systematic review. *Environmental Science and Ecotechnology*, *19*, 100330. DOI: 10.1016/j.ese.2023.100330 PMID: 38021367

Bolón-Canedo, V., Morán-Fernández, L., Cancela, B., & Alonso-Betanzos, A. (2024). A review of green artificial intelligence: Towards a more sustainable future. *Neurocomputing*, *599*, 128096. DOI: 10.1016/j.neucom.2024.128096

Calvin, K., Dasgupta, D., Krinner, G., Mukherji, A., Thorne, P. W., Trisos, C., Romero, J., Aldunce, P., Barrett, K., Blanco, G., Cheung, W. W. L., Connors, S., Denton, F., Diongue-Niang, A., Dodman, D., Garschagen, M., Geden, O., Hayward, B., Jones, C., . . . Péan, C. (2023). *IPCC, 2023: Climate Change 2023: Synthesis Report. Contribution of Working Groups I, II and III to the Sixth Assessment Report of the Intergovernmental Panel on Climate Change [Core Writing Team, H. Lee and J. Romero (eds.)]. IPCC, Geneva, Switzerland.* (First). Intergovernmental Panel on Climate Change (IPCC). DOI: 10.59327/IPCC/AR6-9789291691647

Camps-Valls, G., Fernández-Torres, M.-Á., Cohrs, K.-H., Höhl, A., Castelletti, A., Pacal, A., Robin, C., Martinuzzi, F., Papoutsis, I., Prapas, I., Pérez-Aracil, J., Weigel, K., Gonzalez-Calabuig, M., Reichstein, M., Rabel, M., Giuliani, M., Mahecha, M. D., Popescu, O.-I., Pellicer-Valero, O. J., & Williams, T. (2025). Artificial intelligence for modeling and understanding extreme weather and climate events. *Nature Communications*, *16*(1), 1919. DOI: 10.1038/s41467-025-56573-8 PMID: 39994190

Capraro, V., Lentsch, A., Acemoglu, D., Akgun, S., Akhmedova, A., Bilancini, E., Bonnefon, J.-F., Brañas-Garza, P., Butera, L., Douglas, K. M., Everett, J. A. C., Gigerenzer, G., Greenhow, C., Hashimoto, D. A., Holt-Lunstad, J., Jetten, J., Johnson, S., Kunz, W. H., Longoni, C., & Viale, R. (2024). The impact of generative artificial intelligence on socioeconomic inequalities and policy making. *PNAS Nexus*, *3*(6), pgae191. Advance online publication. DOI: 10.1093/pnasnexus/pgae191 PMID: 38864006

Chavali, D., Baburajan, B., Gurusamy, A., Dhiman, V. K., & Katari, S. C. (2024). Regulating Artificial Intelligence: Developments And Challenges. *IJPS Journal*. DOI: 10.5281/zenodo.10898480

Chisom, O. N., Biu, P. W., Umoh, A. A., Obaedo, B. O., Adegbite, A. O., & Abatan, A. (2024). Reviewing the role of AI in environmental monitoring and conservation: A data-driven revolution for our planet. *World Journal of Advanced Research and Reviews*, *21*(1), 161–171. DOI: 10.30574/wjarr.2024.21.1.2720

Corrêa, N. K., Galvão, C., Santos, J. W., Del Pino, C., Pinto, E. P., Barbosa, C., Massmann, D., Mambrini, R., Galvão, L., Terem, E., & De Oliveira, N. (2023). Worldwide AI ethics: A review of 200 guidelines and recommendations for AI governance. *Patterns (New York, N.Y.)*, *4*(10), 100857. DOI: 10.1016/j.patter.2023.100857 PMID: 37876898

Díaz-Rodríguez, N., Del Ser, J., Coeckelbergh, M., López De Prado, M., Herrera-Viedma, E., & Herrera, F. (2023). Connecting the dots in trustworthy Artificial Intelligence: From AI principles, ethics, and key requirements to responsible AI systems and regulation. *Information Fusion*, *99*, 101896. DOI: 10.1016/j.inffus.2023.101896

Dwivedi, Y. K., Hughes, L., Kar, A. K., Baabdullah, A. M., Grover, P., Abbas, R., Andreini, D., Abumoghli, I., Barlette, Y., Bunker, D., Chandra Kruse, L., Constantiou, I., Davison, R. M., De', R., Dubey, R., Fenby-Taylor, H., Gupta, B., He, W., Kodama, M., & Wade, M. (2022). Climate change and COP26: Are digital technologies and information management part of the problem or the solution? An editorial reflection and call to action. *International Journal of Information Management*, *63*, 102456. DOI: 10.1016/j.ijinfomgt.2021.102456

Emma Klein & Stewart Patrick. (2024). *Envisioning a Global Regime Complex to Govern Artificial Intelligence*. Carnegie Endowment for International Peace. https://carnegieendowment.org/research/2024/03/envisioning-a-global-regime-complex-to-govern-artificial-intelligence?lang=en

Environment, U. N. (2024a, May 13). *AI Solutions for the Environment | UNEP - UN Environment Programme*. https://www.unep.org/topics/digital-transformations/digital-accelerator-lab/ai-solutions-environment

Environment, U. N. (2024b, July 8). *Artificial Intelligence (AI) end-to-end: The Environmental Impact of the Full AI Lifecycle Needs to be Comprehensively Assessed | UNEP - UN Environment Programme*. https://www.unep.org/resources/report/artificial-intelligence-ai-end-end-environmental-impact-full-ai-lifecycle-needs-be

Francisco, M., & Linnér, B.-O. (2023). AI and the governance of sustainable development. An idea analysis of the European Union, the United Nations, and the World Economic Forum. *Environmental Science & Policy*, *150*, 103590. DOI: 10.1016/j.envsci.2023.103590

Gbadegesin, O. A. (2023). Leveraging Artificial Intelligence (AI) in Strengthening the Legal Framework for Regulation of Wildlife and Forest Crimes in Nigeria. *Environmental Policy and Law*, *53*(4), 259–274. DOI: 10.3233/EPL-230011

Ghazal, S., Munir, A., & Qureshi, W. S. (2024). Computer vision in smart agriculture and precision farming: Techniques and applications. *Artificial Intelligence in Agriculture*, *13*, 64–83. DOI: 10.1016/j.aiia.2024.06.004

Giannakidou, S., Radoglou-Grammatikis, P., Lagkas, T., Argyriou, V., Goudos, S., Markakis, E. K., & Sarigiannidis, P. (2024). Leveraging the power of internet of things and artificial intelligence in forest fire prevention, detection, and restoration: A comprehensive survey. *Internet of Things : Engineering Cyber Physical Human Systems*, *26*, 101171. DOI: 10.1016/j.iot.2024.101171

Hassija, V., Chamola, V., Mahapatra, A., Singal, A., Goel, D., Huang, K., Scardapane, S., Spinelli, I., Mahmud, M., & Hussain, A. (2024). Interpreting Black-Box Models: A Review on Explainable Artificial Intelligence. *Cognitive Computation*, *16*(1), 45–74. DOI: 10.1007/s12559-023-10179-8

Hoofnagle, C. J., Van Der Sloot, B., & Borgesius, F. Z. (2019). The European Union general data protection regulation: What it is and what it means. *Information & Communications Technology Law*, *28*(1), 65–98. DOI: 10.1080/13600834.2019.1573501

Hossain, I., Haque, A. M., Kadir, A., & Rana, M. S. (2025). Achieving sustainable development goals through knowledge management in Industry 4.0: A critical review from the perspective of Bangladesh. *Artificial Intelligence and Machine Learning Applications for Sustainable Development*, 82-102.

Hossain, I., Haque, A. M., Kılıç, Z., Ullah, S. A., Azrour, M., & Mabrouki, J. (2024d). Exploring Household Waste Management Practices and IoT Adoption in Barisal City Corporation. In *Smart Internet of Things for Environment and Healthcare* (pp. 27–45). Springer Nature Switzerland. DOI: 10.1007/978-3-031-70102-3_2

Hossain, I., Haque, A. M., & Ullah, S. A. (2023). Role of Government Institutions in Promoting Sustainable Development in Bangladesh: An Environmental Governance Perspective. *Journal of Current Social and Political Issues*, *1*(2), 42–53. DOI: 10.15575/jcspi.v1i2.485

Hossain, I., Haque, A. M., & Ullah, S. A. (2024c). Assessing Sustainable Waste Management Practices in Rajshahi City Corporation: An Analysis for Local Government Enhancement using IoT and AI Technology. *Environmental Science and Pollution Research International*. Advance online publication. DOI: 10.1007/s11356-024-33171-7 PMID: 38581631

Hossain, I., Mahmudul Haque, A. K. M., & Akram Ullah, S. M. (2024b). Socio-economic Dimensions of Climate Change in Urban Bangladesh: A Focus on the Initiatives of Local Governing Agencies. In *Climate Crisis, Social Responses and Sustainability: Socio-ecological Study on Global Perspectives* (pp. 293-316). Cham: Springer Nature Switzerland. DOI: 10.1007/978-3-031-58261-5_13

Hossain, I., Rana, M. S., Haque, A. K. M., & Al Masud, A. (2024a). Urban household adaptation to extreme heatwaves: Health impacts, socio-economic disparities and sustainable strategies in Rajshahi. *Discover Sustainability*, *5*(1), 1–18. DOI: 10.1007/s43621-024-00697-2

IBM. (2015). *IBM Expands Green Horizons Initiative Globally To Address Pressing Environmental and Pollution Challenges*. IBM UK Newsroom. https://uk.newsroom.ibm.com/2015-Dec-09-IBM-Expands-Green-Horizons-Initiative-Globally-To-Address-Pressing-Environmental-and-Pollution-Challenges

Janga, B., Asamani, G., Sun, Z., & Cristea, N. (2023). A Review of Practical AI for Remote Sensing in Earth Sciences. *Remote Sensing (Basel)*, *15*(16), 4112. DOI: 10.3390/rs15164112

Khaleel, M., Murtaza, N., Mueen, Q. H., Ahmad, S. A., & Qadri, S. F. (2024). Use of AI in conservation and for understanding climate change. In *A Biologist s Guide to Artificial Intelligence* (pp. 201–240). Elsevier., DOI: 10.1016/B978-0-443-24001-0.00013-0

Khan, M. I., Yasmeen, T., Khan, M., Hadi, N. U., Asif, M., Farooq, M., & Al-Ghamdi, S. G. (2025). Integrating industry 4.0 for enhanced sustainability: Pathways and prospects. *Sustainable Production and Consumption*, *54*, 149–189. DOI: 10.1016/j.spc.2024.12.012

König, P. D., Wurster, S., & Siewert, M. B. (2023). Sustainability challenges of artificial intelligence and Citizens' regulatory preferences. *Government Information Quarterly*, *40*(4), 101863. DOI: 10.1016/j.giq.2023.101863

Konrad Meier & Roger Spichiger. (2024). *The EU AI Act: What it means for your business.* https://www.ey.com/en_ch/insights/forensic-integrity-services/the-eu-ai-act-what-it-means-for-your-business

Konya, A., & Nematzadeh, P. (2024). Recent applications of AI to environmental disciplines: A review. *The Science of the Total Environment*, *906*, 167705. DOI: 10.1016/j.scitotenv.2023.167705 PMID: 37820816

Kuppusamy, S., Meivelu, M., Praburaman, L., Mujahid Alam, M., Al-Sehemi, A. G., & K, A. (2024). Integrating AI in food contaminant analysis: Enhancing quality and environmental protection. *Journal of Hazardous Materials Advances*, *16*, 100509. DOI: 10.1016/j.hazadv.2024.100509

Malhi, Y., Franklin, J., Seddon, N., Solan, M., Turner, M. G., Field, C. B., & Knowlton, N. (2020). Climate change and ecosystems: Threats, opportunities and solutions. *Philosophical Transactions of the Royal Society B: Biological Sciences, 375*(1794), 20190104. DOI: 10.1098/rstb.2019.0104

Olawade, D. B., Wada, O. Z., David-Olawade, A. C., Fapohunda, O., Ige, A. O., & Ling, J. (2024). Artificial intelligence potential for net zero sustainability: Current evidence and prospects. *Next Sustainability*, *4*, 100041. DOI: 10.1016/j.nxsust.2024.100041

Olawade, D. B., Wada, O. Z., Ige, A. O., Egbewole, B. I., Olojo, A., & Oladapo, B. I. (2024). Artificial intelligence in environmental monitoring: Advancements, challenges, and future directions. *Hygiene and Environmental Health Advances*, *12*, 100114. DOI: 10.1016/j.heha.2024.100114

Ozmen Garibay, O., Winslow, B., Andolina, S., Antona, M., Bodenschatz, A., Coursaris, C., Falco, G., Fiore, S. M., Garibay, I., Grieman, K., Havens, J. C., Jirotka, M., Kacorri, H., Karwowski, W., Kider, J., Konstan, J., Koon, S., Lopez-Gonzalez, M., Maifeld-Carucci, I., & Xu, W. (2023). Six Human-Centered Artificial Intelligence Grand Challenges. *International Journal of Human-Computer Interaction, 39*(3), 391–437. DOI: 10.1080/10447318.2022.2153320

Pagallo, U. (2025). On Twelve Shades of Green: Assessing the Levels of Environmental Protection in the Artificial Intelligence Act. *Minds and Machines, 35*(1), 10. DOI: 10.1007/s11023-025-09713-4

Panait, C., Ljubenkov, D., & Alic, D. (2021). *Striking the balance between innovation and regulation in AI – is Europe leading the way or lagging behind? 1*, 27–45.

Papagiannidis, E., Mikalef, P., & Conboy, K. (2025). Responsible artificial intelligence governance: A review and research framework. *The Journal of Strategic Information Systems, 34*(2), 101885. DOI: 10.1016/j.jsis.2024.101885

Peretz-Andersson, E., Tabares, S., Mikalef, P., & Parida, V. (2024). Artificial intelligence implementation in manufacturing SMEs: A resource orchestration approach. *International Journal of Information Management, 77*, 102781. DOI: 10.1016/j.ijinfomgt.2024.102781

Popescu, S. M., Mansoor, S., Wani, O. A., Kumar, S. S., Sharma, V., Sharma, A., Arya, V. M., Kirkham, M. B., Hou, D., Bolan, N., & Chung, Y. S. (2024). Artificial intelligence and IoT driven technologies for environmental pollution monitoring and management. *Frontiers in Environmental Science, 12*, 1336088. DOI: 10.3389/fenvs.2024.1336088

Radanliev, P. (2025). AI Ethics: Integrating Transparency, Fairness, and Privacy in AI Development. *Applied Artificial Intelligence, 39*(1), 2463722. DOI: 10.1080/08839514.2025.2463722

Rafael Araujo & João Mourão. (2023). *The Amazon Domino Effect: How Deforestation Can Trigger Widespread Degradation*. CPI. https://www.climatepolicyinitiative.org/publication/the-amazon-domino-effect-how-deforestation-can-trigger-widespread-degradation/

Raihan, A. (2023). Artificial intelligence and machine learning applications in forest management and biodiversity conservation. *Natural Resources Conservation and Research, 6*(2), 3825. DOI: 10.24294/nrcr.v6i2.3825

Rautela, K. S., & Goyal, M. K. (2024). Transforming air pollution management in India with AI and machine learning technologies. *Scientific Reports*, *14*(1), 20412. DOI: 10.1038/s41598-024-71269-7 PMID: 39223178

Rishab Bailey & Jeanne Huang. (2024, June 18). *Briefing Note: AI Regulation in the European Union and China*. Digital Trade Alliance. https://dtalliance.org/2024/06/18/briefing-note-ai-regulation-in-the-european-union-and-china/

Saha, S., Bhattacharjee, S., Shit, P. K., Sengupta, N., & Bera, B. (2022). Deforestation probability assessment using integrated machine learning algorithms of Eastern Himalayan foothills (India). *Resources, Conservation & Recycling Advances*, *14*, 200077. DOI: 10.1016/j.rcradv.2022.200077

Santos, R. C., M., & Cagica Carvalho, L. (. (2025). AI-driven participatory environmental management: Innovations, applications, and future prospects. *Journal of Environmental Management*, *373*, 123864. DOI: 10.1016/j.jenvman.2024.123864 PMID: 39752951

Singhal, A., Neveditsin, N., Tanveer, H., & Mago, V. (2024). Toward Fairness, Accountability, Transparency, and Ethics in AI for Social Media and Health Care: Scoping Review. *JMIR Medical Informatics*, *12*, e50048. DOI: 10.2196/50048 PMID: 38568737

Sipola, J., Saunila, M., & Ukko, J. (2023). Adopting artificial intelligence in sustainable business. *Journal of Cleaner Production*, *426*, 139197. DOI: 10.1016/j.jclepro.2023.139197

Sjödin, D., Parida, V., & Kohtamäki, M. (2023). Artificial intelligence enabling circular business model innovation in digital servitization: Conceptualizing dynamic capabilities, AI capacities, business models and effects. *Technological Forecasting and Social Change*, *197*, 122903. DOI: 10.1016/j.techfore.2023.122903

Tyagi, S. (2024). A CRITICAL ASSESSMENT OF THE DIGITAL PERSONAL DATA PROTECTION ACT IN LIGHT OF THE GDPR FRAMEWORK - Sunidhi Tyagi. *Ijalr*. https://ijalr.in/volume-4-issue-3/a-critical-assessment-of-the-digital-personal-data-protection-act-in-light-of-the-gdpr-framework-sunidhi-tyagi/

UNDP. (2024). *The Role of International Cooperation in the Responsible Use of AI Usage in Serbia and Globally | United Nations Development Programme*. https://www.undp.org/serbia/news/role-international-cooperation-responsible-use-ai-usage-serbia-and-globally

WHO. (2021). *Ethics and Governance of Artificial Intelligence for Health: WHO Guidance* (1st ed.). World Health Organization.

Whyman, B. (2023). *AI Regulation is Coming- What is the Likely Outcome?*https://www.csis.org/blogs/strategic-technologies-blog/ai-regulation-coming-what-likely-outcome

Yang, J., Blount, Y., & Amrollahi, A. (2024). Artificial intelligence adoption in a professional service industry: A multiple case study. *Technological Forecasting and Social Change*, *201*, 123251. DOI: 10.1016/j.techfore.2024.123251

Yuan, Q., & Chen, T. (2025). Holding AI-Based Systems Accountable in the Public Sector: A Systematic Review. *Public Performance & Management Review*, •••, 1–34. DOI: 10.1080/15309576.2025.2469784

Zaidan, E., & Ibrahim, I. A. (2024). AI Governance in a Complex and Rapidly Changing Regulatory Landscape: A Global Perspective. *Humanities & Social Sciences Communications*, *11*(1), 1121. DOI: 10.1057/s41599-024-03560-x

Chapter 5
Augmented Reality as a Catalyst for Environmental Awareness and Sustainability Education

Anisha Biswas
https://orcid.org/0009-0007-3663-4020
Brainware University, India

Indrajit Ghosal
https://orcid.org/0000-0003-0744-2672
Brainware University, India

ABSTRACT

AR is changing how we learn by mixing digital and real-life experiences. This study looks at how AR can help us understand the environment and promote sustainability. The researchers reviewed 86 articles from sources like Scopus, Web of Science, and IEEE Xplore, focusing on studies from 2014 to 2024. They explored how AR is used in varioul topics like biodiversity, climate change, waste management, and disaster preparedness. AR helps students stay engaged, remember information better, and understand complex environmental issues. Using 3D models, real-time simulations, and personalized experiences helps students think more deeply and critically. The research shows that AR can encourage lasting changes in behavior and promote sustainable habits when used in education. Policymakers and teachers are encouraged to add AR to school programs, invest in resources and training for teachers, and create rules to ensure data is used ethically. The study concludes that

DOI: 10.4018/979-8-3373-1892-9.ch005

AR has great potential to help create environmentally aware citizens ready to face sustainability challenges in the 21st century.

1. INTRODUCTION

The application of digital technologies in learning has come a long way in the last few decades, moving from teacher-led classroom sessions to interactive and engaging forms of learning. Some of the early digital education tools, including computer-aided learning (CAL) and multimedia-based instruction, paved the way for contemporary e-learning environments. Following the development of virtual learning environments (VLEs) and mobile learning apps, educators have also searched for novel ways of engaging learners with interactive content. A new technology called augmented reality (AR) projects digital data—like text, pictures, movies, and three-dimensional objects—into the physical world. Employees of Boeing originally coined the phrase "augmented reality" in 1990 (Akçayır et al., 2017). "A situation in which a real-world context is dynamically overlaid with coherent location or context-sensitive virtual information" is a wide definition of augmented reality, according to Klopfer and Squire (2008). These days, augmented technologies are widely used in many different industries and are rapidly being recognized for their exceptional performance in giving users an immersive and interactive environment (Alam et al., 2017). Students can connect the virtual environment (VE) and the real world (RE) with the aid of AR applications (Fazel & Izadi, 2018). AR is one of the most important new educational technologies over the next five years because of these advantages (Martin et al., 2011). Augmented Reality (AR) is the newest innovation in education technology, providing a link between physical and virtual learning environments. Differing from other traditional digital tools that are based on passive consumption of content, AR provides real-time interaction with virtual objects within real environments. This makes AR most effective for environmental education where students can see and interact with environmental phenomena that are otherwise challenging to experience in real life, including climate change effects, pollution effects, and loss of biodiversity.AR technologies help facilitate real-world research by students, and virtual things such as texts, videos, and images become some extra sources to help students investigate their real environment (Dede, 2009). According to Bronack (2011), educational academics do not care about AR technologies per se, despite the fact that they need expensive devices and complex tools.In recent years, there has been a marked increase in the necessity of promoting environmental awareness and sustainability education. A growing number of communities around the world are dealing with serious problems related to resource depletion, biodiversity loss, and climate change. There is currently a search for a new strategy to engage and educate people about sustainable practices. Augmented reality (AR), which bridges the gap between the virtual and real worlds, produces an enhanced and augmented experience (Bronack, 2011). Citizen science is public involvement in scientific experiments and environmental monitoring, and AR adds to it by increasing the interactivity of data collection and interpretation. With AR-based citizen science platforms, users can submit useful real-time environmental information without specialized knowledge.As long ago as 1980, IUCN advocated the incorporation of education programs in all conservation and resource-use programs “in order to improve local understanding and support for conservation and to enhance the projects' prospects of lasting success” (IUCN 1980, p. 44). Besides the many facets of environmental education, it is always dynamic. More contemporary approaches focus on being more active in changing behaviour, allowing for learning rather than dictating and focusing on the social dimension of environmental concerns. Charting the development from an emphasis on natural science and conservation to a more social and political perspective, Mappin and Johnson (2005) trace the perceived role, location, and practice of environmental education. The agreements and policies made at the various summits also showed that it was realized that the future of the human species and all other species that make up the earth's biodiversity cannot be guaranteed unless environmental considerations are incorporated into the planning, implementation, and assessment of global population development (UNEP 1992; UNESCO 2014).Augmented Reality (AR) and other transformative technologies are the vanguard of a new era in education (Muñoz-Saavedra et al., 2020).Shankar et al. (2023) drew attention to these technologies' potential to create powerful learning environments; they might also be a transformation—educational technologies.In 2023, de Moraes Rossetto et al. evaluated the usefulness of AR in teaching the complicated subject of the solar system.The research discovered that an augmented reality (AR) has the potential to bolster the educational process greatly by providing a contextualized and captivating learning experience.

Focusing on learners who are blind or have low vision, Brown et al. (2023) determine how valuable several nonvisual modalities are in the AR/VR learning spaces when it comes to imparting knowledge in the STEM fields. Every student gets an advantage from these options and their inherent flexibility. And these options help to adroitly accommodate a wider range of learning styles and speeds. Yet when we have an ostensibly egalitarian educational setting, a libretto de la musique that's governed by fair learning environments, it's hard to see how we have vast and abiding historic public policy injustices in this almost universal access to many of the pathways, options, and tools we're discussing. Furthermore, AR and AI promote the cultivation of the essential skills of the 21st century—creativity, problem-solving, and collaboration. In a global economy that is changing rapidly, these skills are more critical than ever. The narrative that these technologies somehow inhibit skill development has simply got to go. The study aims to explore the younger generation's potential to develop environmental awareness and be educated about sustainability through the use of augmented reality (AR), a leading-edge technology with the potential for a huge impact in a relatively new medium.

2. TECHNOLOGICAL OVERVIEW

Azuma et al. (2001) define augmented reality as a system that overlays virtual things in the real world using computer-generated sensory input, such as pictures, videos, music, and GPS data.. It can give information not possible to access in the actual world with the fusion of a virtual object and a real-world setting. Since technology is changing fast, the concept of augmented reality may be developed further as more hardware and software tools can be used to produce augmented reality. For instance, the possibility of augmented reality can be enhanced through the development of mobile computers (Martin et al., 2011). Students can also remotely control and interact with both virtual and actual equipment using superimposed virtual components on remote devices. These combined AR technologies facilitate ubiquitous learning improved by remote laboratories, computer simulations, physical models, and virtual or three-dimensional objects (Dunleavy et al., 2009). Pervasive AR systems utilize mobile computers with location-registered technologies, such as the Global Positioning System (GPS), in place of head-mounted displays. Being less obtrusive than others, pervasive or mobile AR systems focus more on actual situations. Another use case is AR in conjunction with remote labs and hybrid worlds (Andújar, Mejías, & Márquez, 2011). Cellphones are one of the most popular application intermediaries for applying AR/VR in environmental science because of their widespread use. Several studies have demonstrated the possibilities and advantages of adopting augmented reality (AR) on smartphones to access and display catastrophe data in a geographical environment (Itamiya et al. 2019).

A growing component of effective teaching is the use of technological tools into the curriculum (Pierson, 2001). According to Robertson et al. (2008), animation together with interesting data and an interesting presenter can enable an audience to better understand information analysis findings. These visualization tools can be applied to solve the problem of misperception and enable pupils to better understand. Through the use of 3D synthetic objects that students can interact with, AR may improve learning experiences. With AR, students may enhance their visual impression of the target system or environment by using 3D synthetic objects t (Arvanitis et al., 2007). An example of the use of 3D AR in astronomy education was presented by Kerawalla et al. (2006). AR and conventional teaching sessions were the two session types considered in the study. Teachers and primary school pupils observed and rotated a virtual 3D spinning world to learn about the earth and sun, as well as day and night, using a whiteboard, projector, webcam, AR tile, and virtual 3D modelling program.A mobile augmented reality application for on-site improved environmental monitoring was presented by Veas et al. (2012), who also provided two case studies with encouraging participant outcomes. An open-source program for 3D graphics production (i.e., Blender) was used by Macchione et al. (2019) to present a virtual environment where an urban environment (e.g., buildings, streams, roads, levees, textures) could be replicated. In this environment, they could model the hydraulic dynamics that would occur under various flood scenarios.

Real-world environmental education is helped by augmented reality (AR), which directly interfaces with real-time interactivity involving environmental simulations like ecosystems and natural processes. This establishes immersive learning environments!Using interactive 3D models and geospatial mapping, students can now visualize intricate environmental systems, delving into geographic data in ways not previously possible. Some exciting examples of applied AR in environmental education are occurring at the University of Illinois, where researchers are using Google. According to AR researchers at the University of Illinois, this working into augmented reality has profound implications for far more than just immersive meetings. "It allows us to go places that are otherwise inaccessible to us, without the need for a physical construction. Building on previous work done in vocational interest areas in science, environmental education researchers are finding new uses for augmented reality in visualizing ideas and depicting scenarios that can be harmful to the environment.

3. AR'S INFLUENCE ON EDUCATION AND ENVIRONMENTAL CONSCIOUSNESS

AR has become a standard educational tool by simplifying complicated ideas and making them more interactive. Studies indicate that AR increases engagement, enhances retention of knowledge, and develops critical thinking in students (Ibáñez & Delgado-Kloos, 2018). In environmental education, AR simulations enable users to see pollution levels, deforestation effects, and climate change effects, thus developing a stronger emotional attachment to sustainability challenges (Rahman et al., 2022).Citizen science projects based on AR have enabled communities to gather and interpret environmental data to support global sustainability initiatives. By combining real-time data with interactive visualizations, AR has contributed significantly to informing and motivating people to act for environmental protection (Nisi et al., 2019).

4. COMPARATIVE ANALYSIS WITH OTHER TECHNOLOGIES

Augmented Reality (AR) and Virtual Reality (VR)

AR and VR are often confused for one another due to their nearly shared technology and target end-users who want even greater excitement in their interactive experiences. But these two forms of educational extreme makeovers are quite different. And they're not interchangeable.

Augmented Reality (AR): This actually existed before VR and has been around for a long time. AR allows you to interact with the real world and virtual worlds simultaneously.

Virtual Reality (VR): Crazy as it might sound, VR works best when users can't get to the real world (for whatever reason) and need a super immersive and controlled experience instead.

Augmented Reality (AR) and Mixed Reality (MR)

Mixed reality (MR) is a combination of augmented reality and virtual reality. It involves digital components that not only overlay but also interact in real-time with the physical world (Billinghurst et al., 2015). Consider a sustainability seminar where instructors adjunct with AR to help students visualize a lesson. They might use the AR computer to show students how different kinds of pollution affect a city; in this case, overlays of pollution changes affecting such a city's ecosystem might be projected onto the real-world model of the city's plan. Instructors could take advantage of this moment, leading to a teachable instance, by asking students to consider what might happen if different kinds of emissions were lowered and then following up such a question with their observations (Preece et al., 2021). This sort of scenario may be easier to create with AR, as MR typically requires a more powerful set of real-time processors. To a large extent, this limits access to MR unless one is using a very powerful computer.

Augmented Reality (AR) and Artificial Intelligence (AI)

Augmented reality (AR) is enhanced by artificial intelligence (AI) when it comes to providing data-driven discussions that inform how to personalise the learning experience. Suppose a student learning about sustainable practices is using an AI-powered AR application. In that case, the application can assess their behaviour and modify the material in real-time to accommodate any number of variables. AI can understand learning styles; it can also understand when a sustainable practicum is too easy or too difficult for a student and modulate accordingly. And those are just some of the ways that AI understands human behaviour in order to enhance AR.

Augmented Reality (AR) and Other Educational Technologies

Many studies show that AR is good for education. When compared to other educational tools, AR has unique benefits and drawbacks, especially in teaching about the environment. Unlike regular digital tools like videos or images, AR allows real-time interaction by combining digital information with the real world (Abad-Segura et al., 2020). This helps students learn from their surroundings. Unlike VR, which puts learners in fake environments, AR keeps them connected to the real world, which is great for outdoor learning and exploring the environment (Altinpulluk, 2019). AR helps students see complex issues like deforestation or climate change right where they are. While Artificial Intelligence can adapt learning to how a user behaves, it doesn't provide the same hands-on experience that AR does with real spaces and ecosystems (Majeed & Ali, 2020). However, using AR can be hard because it can be expensive to develop, complicated to use, and not always accessible for everyone. Despite these challenges, AR is very effective in teaching environmental concepts because it helps students interact with real ecological data, leading to a better understanding and encouraging change in behaviour (Wu et al., 2013).Augmented Reality has special benefits that make it different from other educational tools, especially in teaching about the environment. For example, Virtual Reality (VR) creates complete simulations for training, but it can feel disconnected from real life, making it less useful for outdoor learning where seeing real environmental changes is important (Geroimenko, 2020). Mixed Reality (MR) combines AR and VR, allowing users to interact with both real and virtual objects, but it often needs expensive equipment, which can limit its use in schools with fewer resources (Cabero & Barroso, 2016). On the other hand, traditional tools like 2D animations, slides, and videos are cheaper and easier to find, but they don't provide the same depth and emotional connection that AR does (Garzón, 2021) . In environmental education, AR helps show important concepts like rising sea levels or animal migrations right in front of students, making it very effective. AR also encourages hands-on learning by letting students work with data, take part in simulations, and see how different factors in nature affect each other (Buchner & Kerres, 2023).This active way of learning helps students understand concepts better and feel more committed to sustainability, something that regular teaching methods struggle to achieve.

5. METHODOLOGY

This review looks at how Augmented Reality (AR) can help people understand the environment and learn about sustainability. The aim was to bring together important studies, case examples, and ideas that connect AR technology and environmental education.

Literature Search Strategy

To get a complete and fair view, we found important academic sources by searching through databases like Scopus, Web of Science, Google Scholar, and IEEE Xplore. We used keywords like "augmented reality", "environmental education", "sustainability", "AR in education", "citizen science AR", and "interactive learning environment".Mainly included studies from 2014 to 2024 but also added important works from earlier years to give some background.

Inclusion & Exclusion Criteria

From 312 records in databases like Scopus, Web of Science, IEEE Xplore, and Google Scholar. After removing duplicates, we had 275 records to check by title and abstract. Excluded 90 records because they were not about augmented reality (AR) in education, and 43 were excluded because they were commercial or not peer-reviewed. This left with 142 full-text articles to review in detail. Excluded 56 of these because their methods were unclear or not relevant to our research. In the end, we included 86 studies in our final review.

Data Extraction and Analysis

We looked at chosen studies to understand their goals, methods, main results, and how they relate to AR (augmented reality) in environmental education. We grouped the findings into common themes like teaching benefits, technology advantages, effects on behavior, and challenges in using the technology. We also compared AR to other educational technologies like VR, MR, and AI tools when it made sense.

Figure 1. Flowchart illustrating the study selection process

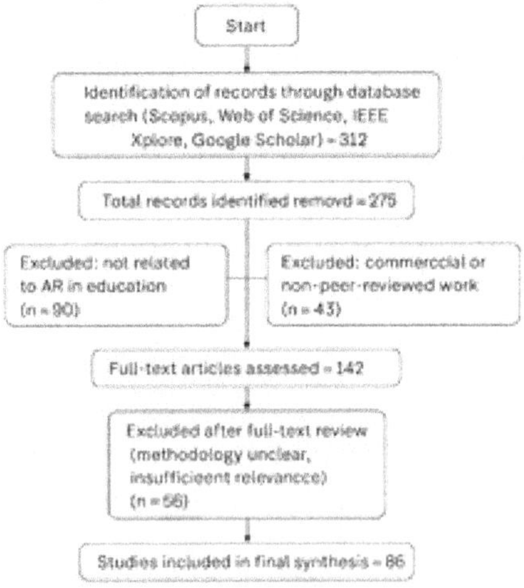

6. APPLICATIONS OF AR IN ENVIRONMENTAL EDUCATION

Biodiversity Conservation

Many people are still concerned about habitat loss and biodiversity declines, which has led to calls for interdisciplinary solutions to halt the decline of species (McElwee et al., 2020). Ironically, instead of advocating less dependency, environmental organisations and academics have recommended that accepting developing technology can prove to be a valuable strategy in order to teach people about the environment (Berger-Tal & Lahoz-Monfort, 2018). Wu et al. (2018) focused on gamification for effective Araujo et al. (2021) highlighted that AR has the potential to overcome accessibility barriers in biodiversity education. For instance, the virtual access through AR allows populations in urban cities or those who are mobility-challenged the opportunity to access and learn about biodiversity, ensuring inclusivity in conservation education.

Climate Change Education

All levels of learning value teaching and learning, but the academic institution holds the most significant role. Therefore, it is vital that climate change be brought to the forefront of academia. Tew et al. (2021) argue that when this subject is "mainstreamed" in the curriculum, it ensures that every professor and student is "abreast of" climate change. They contend that climate change education happens through two main pathways: perhaps what is more commonly thought of, which is direct teaching in courses about climate change, and also what they consider to be a more insidious approach, which is that climate change is integrated into all sorts of different classes across many disciplines. Case studies, focus groups, role plays, simulations, and skill workshops all serve as direct pathways to solving sustainability problems in interdisciplinary groups (Lopez-Medina et al., 2019). Mill et al. (2020) emphasized the potential of augmented reality (AR) as a tool to assess climate change education. ...

Figure 2. Pie chart illustrating the various aspects of augmented reality as a catalyst for environmental awareness and sustainability education

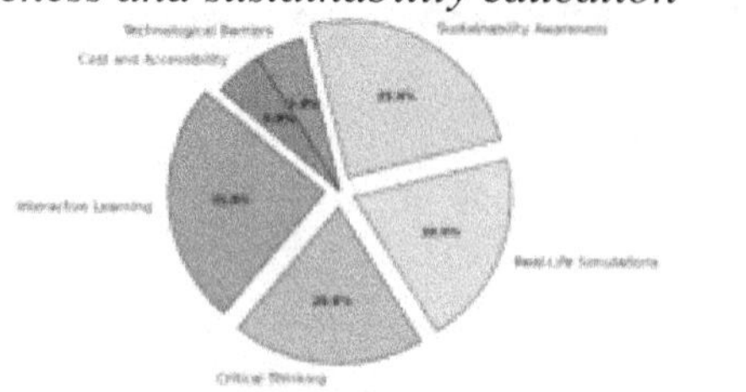

Climate change cannot be ignored, and today even the most skeptical individuals recognize that human actions have altered the natural climate of our planet. A number of significant consequences are now being experienced, and even the IPCC (Intergovernmental Panel on Climate Change), which is often accused of understatement, has warned that the overall effects are more serious than was forecast a few years ago.The most profound impacts from climate change are expected to occur in that part of the world where the poorest of the poor live—the tropics. July 2019 marked a grim milestone when the world's average surface temperature for that month exceeded 1 °C above preindustrial levels; the year 2019 was remarkable for the number and intensity of weather extremes it produced.Coastal communities have been targeted by augmented reality (AR) as a means to portray the potential hazards of rising sea levels. To assist in this effort, Climate Central unveiled the Surging Seas AR app in 2019, utilizing it as a tool to help not just residents but also politicians visualize the kinds of moderate, severe, and even extreme flooding risks we can expect in the next few decades if we continue on our present path. This app stands out because it offers real-time projections for places like Miami and New York City, which helps make the conversation about climate change a lot more urgent because those in power can directly see the kinds of risks we are facing.

Waste Management and Recycling

Recycling and garbage management can be improved by augmented reality technology that helps sort, educate, and give real-time advice to the public. This is done through immersive and entertaining experiences that have the power to intensify the attention and interest of a distracted public, and that may even retain the audience's involvement after the initial novelty wears off. AR can also be a tool for raising public consciousness, in ways that are direct and indirect, and that are likely to get people to take up and keep good waste management and recycling practices. Smith et al. (2020) claim that R-based applications, such as virtual garbage sorting simulators, are an effective means of teaching the public to perform recycling correctly. Miller et al. (2020) talked of using augmented reality (AR) to visualize trash data, such as the recyclables we put in our trash cans that can't be traced to landfills.

Disaster Management Training

Figure 3. Augmented Reality (AR) in promoting environmental awareness and sustainability education

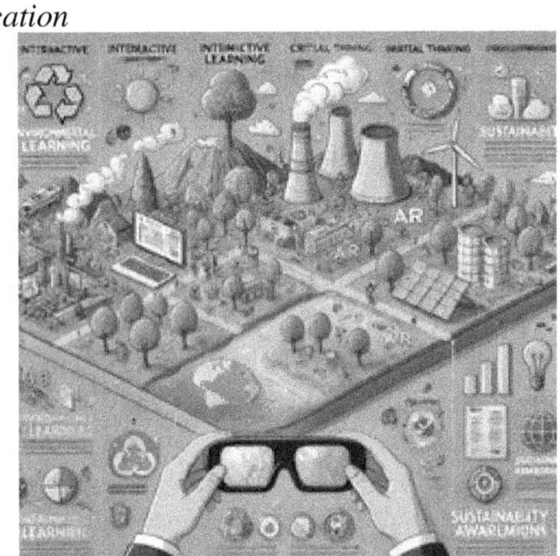

Disaster management training and individual growth may be advanced through augmented reality. Decision-making may be improved, dangers identified, and tactics developed for efficient disaster response through what are termed as simulated real-life situations that could arise during actual disasters in safe, controlled environments. According to Sermet and Demir (2020), these applications are powered by a variety of data sources, including hydrological, geographical, and meteorological data obtained from organizations such as the National Oceanic and Atmospheric Administration (NOAA), the United States Geological Survey (USGS), The Weather Channel (TWC), the Iowa Flood Centre (IFC), and ArcGIS City Engine.Yet, it has been demonstrated that augmented reality (AR) is a powerful tool for not just education, but also emergency response and disaster preparation training. First responders and the general public, too, may practice their reaction tactics in a safe, simulated environment using AR-based applications that not only depict floods, for example, but do so in a realistic manner (Mills et al., 2020).At-risk populations, too, can access these programs by virtue of them being arguably more affordable, scalable, and adaptable to different contexts than traditional catastrophe simulations (Preece et al., 2021).The training that could most benefit from the use of AR within the disaster preparedness realm is something called situational awareness training. This is where juniors in the fire department may see a virtual fire in a safe setting and react accordingly. This is, of course, taking it up a notch from the traditional use of virtual reality in training.

Urban Sustainability

Urban sustainability promotes, above all, economic development, environmental protection, and social well-being. The smart city, in turn, is one where sustainable practices rule the day and technology is used as a means to an end; this end is not only efficiency but also a sharply reduced ecological footprint (Ahvenniemi et al., 2017). AR has emerged as a serious tool for raising public awareness of environmental issues, since it can offer truly immersive experiences (Geroimenko, 2020). Investigations have shown that the up-and-coming medium of augmented reality (AR) can be put to good use in the rather traditional field of urban planning, where something compatible with the standards of the visualisation age we live in—something that could replace the "enlightened cubic model of the 1960s"—is needed (Preece et al., 2021). Nonetheless, a number of unresolved issues linger, including whose problem it is when not everyone has the same access to the relevant technology.

Using AR to Simulate Wildfires

AR Technology has been incorporated into the training of California's firefighters who are on the frontlines of wildfires. The Wildfire AR Training Simulator (WATS) was developed to carry out this function, with realistic fire spread modeling serving as its raison d'être. For such modeling to be useful, meteorological data providing real-time updates were needed, and these were incorporated into WATS. Firefighters are now able to use augmented reality headsets to do the calibrations necessary to interfacing the model with actual conditions and locus (Huang et al., 2022).

Water Quality and Pollution Tracking

Augmented reality is changing how people carry out community-based water quality monitoring. Using Earthwatch's FreshWater Watch AR app, volunteers can obtain real-time assessments of the pollution levels in their community water bodies. They scan the water body and instantly get the AR readout of the important health indicators: clarity (turbidity), contaminants, and pH levels. This app is an essential tool of the international collective research effort, to which those who live near the monitored water bodies have been contributing.In the same way, Plastic Tide AR keeps an eye on plastic refuse around beaches, employing augmented reality (AR) tools and drone-shot footage. By letting people interact with a set of data points rendered in AR, the project permits scientists to plot the ocean's hotspots for plastic refuse. This kind of AR-based citizen science provides researchers and environmental organizations with vital feedback while simultaneously reaching out to and engaging individuals more effectively than many other strategies.

Biodiversity Monitoring

One of the most important applications of augmented reality (AR) for citizen science is the conservation of biodiversity. Two applications that utilize AR to facilitate the identification of plant and animal species in a given area are iNaturalist and Seek by iNaturalist. These apps provide identification and ecological data in real time via AI-driven AR overlays when a person pictures a species (Garzón & Acevedo, 2019). This makes ecological research accessible to everyone and allows the conservation of species to become a participatory activity in which local populations can engage.

Another example, the Wild Me AR Platform, lets users track at-risk animals using augmented overlays of their migration paths. By enlisting citizen scientists to help collect and maintain these data, such technologies allow understaffed conservation researchers to understand these volatile ecosystems much better than they would be able to otherwise.

7. PEDAGOGICAL BENEFITS

The most commonly cited AR affordance is the capability to display for a group of students many partial but complementing viewpoints on an issue located inside a physical location (Dunleavy et al., 2009)._Furthermore, AR enables teachers to use the actual world as an extra layer of content that students can view, interact with, and evaluate (Perry et al., 2008). AR allows the students to make personal judgment, and it also allows the teacher to delegate the responsibility to the class. These contributions enhance the student engagement (Munoz-Cristobal et al., 2015). According to Fernandez-Robles (2018), augmented reality can be presented as a truly helpful technology for higher education because it makes possible the use of active methodologies that provide the ability to visualize the object from different angles, facilitates the acquisition of difficult information, makes possible the presentation of simulated scenarios, and strengthens printed material. Environmental education can now employ Augmented Reality (AR), which would immerse learners within real environmental concerns, enhance knowledge and memory for complex issues due to simulations of ecological processes, visualizations of ecosystems, and an interactive form that involves real engagement with these ecological issues by learners. According to Shakirova et al. (2023), AR technology aided in the environmental literacy development of high school pupils. Ardyansyah and Rahayu's study (2023) explained that a group of high school students had an improved understanding of the chemistry subject of carbon compounds. This was after they used an environmental literacy and augmented reality learning resource made available to them by the researchers. The very nature of the carbon compounds involved in the assembly and disassembly of all biological entities makes this a subject of particular environmental interest. Biology, which is the natural science that studies life and living organisms, especially their structure, function, growth, evolution, and behavior, helps students understand these fundamental principles. But using the AR platform as a learning tool offers something beyond mere memorization. It offers students the opportunity to explore the virtual ecosystems of these living organisms, carry out experiments and with the use of their iPhones, they can make interactive models of all these experiments they have worked on. This is what the researchers mean by making a pedagogy of the interpretive ecology of their learning experiences.

8. THE PSYCHOLOGICAL AND COGNITIVE EFFECTS OF AR IN LEARNING

Augmented Reality (AR) is transforming education by enabling interesting and interactive learning experiences.Combining real-world and virtual components promotes enhanced problem-solving skills, superior memory recall, and greater engagement. However, despite its advantages, AR brings with it cognitive difficulties that require thoughtful instructional design—cognitive load, for one. This section reviews what we know so far about how AR affects cognitive load, memory, and critical thinking in the classroom. Remembering is a key ingredient of successful learning. The more ways you present something, the better the chances that it will stick in someone's memory.Research has shown that AR learning environments are much more engaging than conventional ones. Because of this, the environments in which we now teach our students have to be rich in context and kind of a "wow" effect, like almost having a star in the constellations. If AR has a better return on investment of engagement per mile in the interaction space, then it also has better returns on memory than conventional teaching methods.For example, a study by Garzón and Acevedo (2019) shows that students who used augmented reality (AR) applications to learn about biodiversity were 30% better at remembering information about the specifications and traits of different species than students taught with traditional, 2D mediums like textbooks. Compared to the passive reception of information we see with traditional teaching tools, AR allows for an experience that is much closer to active learning. Working with digital 3D models of organisms in real time is not so different from what a biologist might be doing in their lab. This gain in cognition has also been enabled.

9. SOCIOCULTURAL EFFECTS OF AR ON ENVIRONMENTAL EDUCATION

Augmented reality (AR) has emerged as a powerful technology in the context of environmental education. AR not only makes the learning experience better but also molds the ideas and perceptions of society regarding sustainability.

AR's Role in Influencing Social Perceptions of Sustainability

One of AR's biggest benefits for environmental education is its ability to interactively and clearly illustrate complex ecological processes. Studies have shown that immersive augmented reality experiences can evoke the kind of strong emotional responses that lead to increased awareness and proactive sustainability behaviors (Wu et al., 2020). When it comes to comprehending the symptoms and causes of climate change, AR applications that simulate such effects (e.g., sea level rise, forest loss) are serving a similar purpose to that of the best science teachers: enhancing the user's grasp of the urgency of these concerns (Mills et al., 2020).AR enables students to link up abstract ideas with real-world applications by incorporating real-world environmental troubles into interactive simulations.AR also makes problem-solving and collaborative learning easier. Communities may have online discussions about sustainability solutions and harness a collective sense of environmental stewardship using multi-user augmented reality systems (Nisi et al., 2019). For example, augmented reality (AR) applications in urban planning enable local stakeholders to visualize green infrastructure projects and their outcomes well before any actual implementation occurs, thus facilitating a decision-making process that is both more participatory and more inclusive (Nisi et al., 2019).

Cultural Adaptations and Inclusivity in AR-Based Environmental Programs

Cultural inclusion in AR-based environmental education is essential for global sustainability initiatives. Different cultures, with their unique worldviews, approaches, and lifeways, have varied ways of perceiving and dealing with environmental problems. Consequently, when developing AR applications, localized materials that resonate with particular cultures must be used. For example, an AR-based education program is using TEK in its curriculum—developed in partnership with the Zuni (a Native American) tribe—so that Zuni children can learn about the AR-based environmental education program (Huerta et al., 2021). Since TEK has been used for generations as a way to manage natural resources sustainably, it is imperative that what the Zuni tribe knows about local ecology serves as a foundation for that education program.Indigenous stories can close the gap between traditional and modern environmental research, suggest the authors of a new paper. Using augmented reality (AR) applications to tell environmental stories—involving not just modern science but also ancient knowledge—could be a way to make environmental education more inclusive and effective, the researchers say. Accessibility is also a key part of this newfound inclusivity. Currently, the vast majority of AR applications are developed in English and rely on voice input, which limits access to non-English speakers and people who are deaf.Moreover, AR can afford voiceless segments of society the opportunity to express themselves in support of environmental action. Elbowed aside by the growth machine, the urban poor often live with dubious waste management and high levels of air pollution. Augmented reality applications, at their most basic level, can empower citizens by giving them the opportunity to make a public record of their private environmental experiences—complaining to the cops, the City Hall, and not a few reporters. Better yet, AR allows for community-driven environmental action by letting residents map their locales (in) enviable, a topless we're-on-the-street experience that happens in a city of lots. AR also has the power, when used in this way, to reach beyond the individual citizen for whom the experience is life-changing.

10. EMERGING AR TECHNOLOGIES AND THEIR POTENTIAL IN ENVIRONMENTAL EDUCATION

New developments in Augmented Reality (AR) are changing how we teach, making environmental education better, more interactive, and inclusive. One exciting change is combining AR with Artificial Intelligence (AI). AI can adjust lessons based on how students learn, helping those who find certain topics hard. For example, an AI AR app on biodiversity can change the difficulty of lessons and give personalized help to students (Kamarainen et al., 2013).Another big change is the improvement of AR devices. Lighter and cheaper AR headsets and mobile platforms are making it easier for more schools to use these tools. Devices like Apple's Vision Pro and advanced Android AR products now support high-quality 3D models and allow users to interact using voice and gestures. This helps students learn about complicated subjects like climate change and deforestation in a more hands-on way (Sermet & Demir, 2020).Also, linking AR with real-time data from the Internet of Things (IoT) can make learning more relevant. An AR app that connects to local air or water quality sensors could show students live data in their environment, making the lessons more real and focused on action (Al-Ansi et al., 2023).To make AR useful for everyone in environmental education, it's important to improve accessibility. New designs that include features like touch feedback, voice commands, and audio descriptions are making AR easier to use for students with disabilities. This inclusive approach is crucial for ensuring that AR can help everyone learn about sustainability(Acosta Castellanos & Queiruga-Dios, 2022).

11. CHALLENGES

Technical difficulties

Imminent immersive technologies such as augmented reality provide a great lot of potential to reduce the digital divide for persons with disabilities by delivering fresh possibilities to foster social engagement in virtual environments. Research has focused on the ability of AR to offer innovative solutions that have the potential to alter and target specific cognitive, sensory, interpersonal, and motor processes that are implicated in abnormal developmental pathways for neurodiverse users and individuals with cognitive disabilities (Farroni et al., 2022). Although some preliminary study has explored the barriers for people with physical disabilities or neurodiverse people, much less has been done to investigate the barriers that exist for users who have visual or aural impairments. For instance, it has been observed that existing virtual reality systems are not accessible for blind or low vision users and that adaptation in interactive experiences is essential (Gkini et al., 2021, Teófilo et al., 2018).

According to Dunleavy et al. (2009), students often felt overloaded and confused while participating in a multi-user augmented reality simulation because they had to cope with both hard tasks and foreign technology. The authors of this research, Rosenbaum et al. (2007), note that learners' understanding of dynamic models and complex causation improved due to the authenticity associated with an AR learning environment. Besides, an AR environment may make students more interested and motivated in which they would learn better by using research methods and gain information related to the topic more precisely. Learning difficulties that have been identified in previous research can be mitigated by AR systems. For example, students generally find it difficult to visualize intangible processes such as the earth's rotation (Kerawalla et al., 2006). More scaffolding and support would be needed for younger students, as well as those who do not have extensive experience in performing open-ended investigations, to help them develop a suitable plan of action, identify possible solutions to their problem, and interpret clues from the technological tools and embedded within the real world (Klopfer & Squire, 2008). The setup, distribution, and maintenance of AR apps may require a significant investment. This is challenging, especially for underfunded schools and other similar organizations that focus on environmental studies (Bacca et al., 2014). Moreover, it could be prohibitively expensive to amend AR materials to convey current research and to express regional environmental issues of concern. The complaints about the quality and the quantity of the AR learning experience were mostly about the learning opportunities being too short or too long, and for both kinds of complaints, AR learning was said mostly to be too difficult (Billinghurst et al., 2015). AR human factors research is on the right track if it is concentrating on the problems of the hardware, the latency, and the poor interface. Useful research is done when the real problems are identified and examined with decent human factors standards and methods.

Economic issues

There are substantial financial obstacles to incorporating augmented reality (AR) into sustainability and environmental awareness teaching. Accessibility may be curtailed by the high development and maintenance costs of AR apps, as well as the sophisticated technology needed to use them—smartphones, AR headsets, etc.—that simply aren't in the budgets of many resource-strapped locales. Compounding matters is the need for investment in the technology infrastructure, trained personnel, and research that any AR initiative would demand just to get off the ground. And what these results suggest is that AR is unlikely to become a widespread game-changer for environmental education any time soon. Prior studies support this conclusion, going so far as to say that AR development and maintenance costs may outweigh any potential savings in energy (and time) that could result from using this new technology.

Students sometimes utilize technology to learn about a range of subjects for their studies. Under the conditions of civic engagement and critical thinking, it is productive for skill development and is good for mental health when trying to figure out the future (Daniela et al., 2018). If one is to be sustainable, it has been shown (Issa et al., 2017) that individuals and organizations must include sustainability policies into their operations and must "act with and correlate to sustainability in mind" (p. 2) when making decisions and controlling their day-to-day activities. This is something that can be done and executed policy-wise using a range of tools and technologies, one step at a time.

Table 1. Insights and Challenges of Using Augmented Reality in Environmental Awareness and Sustainability Education

Aspect	Key Insights	References
Educational Benefits	Enhances engagement, interactivity, and retention of sustainability concepts.	Akçayır & Akçayır (2017); Wu, Lee, Chang, & Liang (2020)
Immersive Learning	Simulates real-life environmental scenarios, making abstract concepts tangible.	Ibáñez & Delgado-Kloos (2018); Yoon, Anderson, Lin, & Elinich (2017)
Active Learning Promotion	Encourages critical thinking and problem-solving for complex environmental challenges.	Yoon et al. (2017)
Cost and Accessibility	High costs of development and hardware limit widespread adoption in underprivileged areas.	Cheng & Tsai (2020); Arici, Yildirim, Caliklar, & Yilmaz (2019)
Technological Barriers	Requires technical expertise and reliable infrastructure for effective implementation.	Cheng & Tsai (2020)
Future Potential	Mobile AR applications offer scalable solutions for broader curriculum integration.	Arici et al. (2019)

12. ETHICAL AND PRIVACY CONCERNS IN AUGMENTED REALITY (AR)

Data security and privacy are two of the main issues surrounding the usage of augmented reality in schools. Numerous augmented reality apps provide a substantial amount of user data, such as facial recognition, geolocation, and real-time behaviour analytics (Billinghurst & Duenser, 2012). Location-based learning applications for environmental education, including those that measure pollution levels or pinpoint biodiversity hotspots, have the unintended consequence of exposing users to privacy violations.

Risks for Location and Personal Data

Most AR-based environmental education applications require GPS and can give users an interactive outdoor experience. The problem is that when apps are used for location-based tracking, there's a risk of unauthorized data harvesting, monitoring, and misuse of personal data (Cheng & Tsai, 2020). Apps that are supposed to educate students about local ecosystems may, in the process, be sending sensitive location information to third parties (because that's how many apps operate these days), and in the process, they could be compromising user privacy and security (Wu et al., 2020).In addition, AR devices can incorporate biometric data collection—such as eye tracking and facial recognition—to create more immersive user experiences. But this means AR could become a vector for surveillance and identity theft. Unless properly encrypted and anonymized, the vast amount of data AR apps generate risks being misused by companies for profit, whether through targeted advertising or direct sale of the information to third parties.

User Control and Consent

Privacy concerns become more acute when explicit consent is not obtained from users. Most augmented reality applications used in environmental education target young students, making it essential for developers and educators to offer clear privacy policies and parental controls (Chai et al., 2019). Otherwise, how can children—and by extension, the applications' other potential users—know when they've forfeited the right to control any number of their private lives' dimensions? The Children's Online Privacy Protection Act (COPPA) and the General Data Protection Regulation (GDPR) are two sets of laws that can help reduce this risk. Both mandate strict rules about what developers can and cannot do with the intimate data they gather from users, and both affect a lot of what happens in the digital world.

13. FINDINGS

The review shows that using AR (augmented reality) in environmental education helps students learn better and feel more connected to sustainability. AR makes tough topics like climate change and pollution easier to understand than regular teaching. Many studies found that AR helps students remember things and think better because it is interactive and involves different senses. It also supports various learners, including those with disabilities, by providing adjustable content. Adding AI (artificial intelligence) can improve these experiences by personalizing learning for each student. However, the findings also highlight some important issues, like not having enough long-term data on behavior changes, unequal access to technology, and fewer opportunities for students from underrepresented groups.

14. LIMITATIONS AND POSSIBLE BIASES IN THE STUDIES REVIEWED

Studies show that Augmented Reality (AR) can help people understand the environment and learn about sustainability. However, there are some problems. One main issue is that many studies are small and only look at short programs in controlled settings, making it hard to apply the results to real life. Also, a lot of research focuses on wealthy areas and ignores developing regions where using AR can be more difficult because of money and resources. There is also a lack of diversity in the students studied, with few looking at how AR affects students with disabilities or those from different cultures. This can lead to biased results and unfair outcomes. Additionally, many studies talk about how exciting AR is but often miss issues like mental strain, tiredness, or how well it works for long-term learning.

15. FUTURE RESEARCH DIRECTIONS

Future studies on using AR for environmental education should look at long-term effects. Right now, most studies only check short-term learning or attitude changes after quick use of AR apps. This doesn't show if AR truly helps people develop lasting habits like recycling more or being more eco-friendly. Long-term research should track learners for months or years to see how environmental experiences change their habits and community involvement. These studies should also consider how often people use AR, how it works with regular learning, and how support from others helps change behavior. It's important to work with experts in different fields to create effective AR programs and find good ways to measure long-term effects. This will help make sure AR is not just fun but also a useful tool for encouraging eco-friendly behaviors in the future.

16. CONCLUSION

Augmented reality (AR) is a helpful tool for teaching about the environment and encouraging sustainability. It can engage students in important issues like climate change, deforestation, and the loss of plants and animals. AR makes complicated environmental topics easier to understand, especially in a study area called socio-hydrology. However, there are challenges in using AR in schools, like high costs and the need for training for users and teachers. Working together in virtual spaces can improve teamwork across different areas, which is important for solving environmental issues. AR can also work with big data and artificial intelligence (AI) to provide real-time information about the environment, making learning better. But future AR technology needs to be developed carefully, considering privacy, the environmental effects of servers, and how much time people spend on screens.

REFERENCES

Abad-Segura, E., Gonzalez-Zamar, M. D., Rosa, A. L. D. L., & Morales Cevallos, M. B. (2020). Sustainability of educational technologies: An approach to augmented reality research. *Sustainability (Basel)*, *12*(10), 4091. DOI: 10.3390/su12104091

Acosta Castellanos, P. M., & Queiruga-Dios, A. (2022). From environmental education to education for sustainable development in higher education: A systematic review. *International Journal of Sustainability in Higher Education*, *23*(3), 622–644. DOI: 10.1108/IJSHE-04-2021-0167

Ahvenniemi, H., Huovila, A., Pinto-Seppä, I., & Airaksinen, M. (2017). What are the differences between sustainable and smart cities? *Cities (London, England)*, *60*, 234–245. DOI: 10.1016/j.cities.2016.09.009

Akçayır, M., & Akçayır, G. (2017). Advantages and challenges associated with augmented reality for education: A systematic review of the literature. *Educational Research Review*, *20*, 1–11. DOI: 10.1016/j.edurev.2016.11.002

Al-Ansi, A. M., Jaboob, M., Garad, A., & Al-Ansi, A. (2023). Analyzing augmented reality (AR) and virtual reality (VR) recent development in education. *Social Sciences & Humanities Open*, *8*(1), 100532. DOI: 10.1016/j.ssaho.2023.100532

Alam, M. F., Katsikas, S., Beltramello, O., & Hadjiefthymiades, S. (2017). Augmented and virtual reality based monitoring and safety system: A prototype IoT platform. *Journal of Network and Computer Applications*, *89*, 109–119. DOI: 10.1016/j.jnca.2017.03.022

Altinpulluk, H. (2019). Determining the trends of using augmented reality in education between 2006-2016. *Education and Information Technologies*, *24*(2), 1089–1114. DOI: 10.1007/s10639-018-9806-3

Andújar, J. M., Mejías, A., & Márquez, M. A. (2011). Augmented reality for the improvement of remote laboratories: An augmented remote laboratory. *IEEE Transactions on Education*, *54*(3), 492–500. DOI: 10.1109/TE.2010.2085047

Araujo, A. C., de Oliveira, J. R., & Lima, N. E. (2021). Augmented reality for biodiversity conservation: Challenges and opportunities in education. *Sustainability*, *13*(8), 4103.

Ardyansyah, A., & Rahayu, S. (2023). Development and implementation of augmented reality-based card game learning media with environmental literacy in improving students' understanding of carbon compounds. *Orbital*, *15*(2), 118–126. DOI: 10.17807/orbital.v15i2.17617

Arici, F., Yildirim, P., Caliklar, S., & Yilmaz, R. M. (2019). Research trends in the use of augmented reality in science education: Content and bibliometric mapping analysis. *Computers & Education, 142*, 103647. DOI: 10.1016/j.compedu.2019.103647

Arvanitis, T. N., Petrou, A., Knight, J. F., Savas, S., Sotiriou, S., Gargalakos, M., & Gialouri, E. (2007). Human factors and qualitative pedagogical evaluation of a mobile augmented reality system for science education used by learners with physical disabilities. *Personal and Ubiquitous Computing, 13*(3), 243–250. DOI: 10.1007/s00779-007-0187-7

Azuma, R., Baillot, Y., Behringer, R., Feiner, S., Julier, S., MacIntyre, B., 2001. Recent advances in augmented reality. IEEE Comput. Graph. Appl. 21 (6), 34e47.

Bacca, J., Baldiris, S., Fabregat, R., & Graf, S., & Kinshuk. (2014). Augmented reality trends in education: A systematic review of research and applications. *Journal of Educational Technology & Society, 17*(4), 133–149.

Berger-Tal, O., & Lahoz-Monfort, J. J. (2018). Conservation technology: The next generation. *Conservation Letters, 11*(6), e12458. DOI: 10.1111/conl.12458

Billinghurst, M., Clark, A., & Lee, G. (2015). A survey of augmented reality. *Foundations and Trends® in Human–Computer Interaction, 8*(2-3), 73–272.

Billinghurst, M., & Duenser, A. (2012). Augmented reality in the classroom. *Computer, 45*(7), 56–63. DOI: 10.1109/MC.2012.111

Bronack, S. C. (2011). The role of immersive media in online education. *The Journal of Continuing Higher Education, 59*(2), 113–117. DOI: 10.1080/07377363.2011.583186

Brown, J. R., Doore, S. A., Dimmel, J. K., Giudice, N., & Giudice, N. A. (2023). Comparing Natural Language and Vibro-Audio Modalities for Inclusive STEM learning with blind and low vision users. In *Proceedings of the 25th International ACM SIGACCESS Conference on Computers and Accessibility*, pp. 1-17. DOI: DOI: 10.1145/3597638.3608429

Buchner, J., & Kerres, M. (2023). Media comparison studies dominate comparative research on augmented reality in education. *Computers & Education, 195*, 104711. DOI: 10.1016/j.compedu.2022.104711

Burton, E. P., Frazier, W., Annetta, L., Lamb, R., Cheng, R., & Chmiel, M. (2011). Modeling Augmented Reality Games with Preservice. *Journal of Technology and Teacher Education, 19*(3), 303–329.

Cabero, J., & Barroso, J. (2016). The educational possibilities of Augmented Reality. *Journal of New Approaches in Educational Research*, *5*(1), 44–50. DOI: 10.7821/naer.2016.1.140

Caudell, T. P., & Mizell, D. W. (1992). Augmented reality: An application of heads-up display technology to manual manufacturing processes. *Proceedings of the 25th Hawaii International Conference on System Sciences*, 2, 659–669. DOI: 10.1109/HICSS.1992.183317

Chai, C. S., Koh, J. H. L., & Teo, Y. H. (2019). Fostering collaboration and creativity in education through AR and AI. *Journal of Educational Technology & Society*, *22*(2), 45–56.

Chang, H.Y., Wu, H.K., Hsu, Y.S., 2013. Integrating a mobile augmented reality activity to contextualize student learning of a socioscientific issue. Br. J. Educ. Technol. 44 (3), E95eE99.

Cheng, K. H., & Tsai, C. C. (2020). A case study of immersive virtual field trips in an elementary classroom: Students' learning experience and teacher-student interaction behaviors. *Interactive Learning Environments*, *28*(4), 432–446.

Daniela, L., Visvizi, A., Gutierrez-Braojos, C., & Lytras, M. (2018). Sustainable higher education and technology-enhanced learning (TEL). *Sustainability (Basel)*, *10*(11), 3883. DOI: 10.3390/su10113883

de Moraes Rossetto, A. G., Martins, T. C., Silva, L. A., Leithardt, D. R., Bermejo-Gil, B. M., & Leithardt, V. R. (2023). An analysis of the use of augmented reality and virtual reality as educational resources. *Computer Applications in Engineering Education*, *31*(6), 1761–1775. DOI: 10.1002/cae.22671

Dede, C. (2009). Immersive interfaces for engagement and learning. *Science*, *323*(5910), 66–69. DOI: 10.1126/science.1167311

Dunleavy, M. (2010). Persisient design challenges: Augmeflling reafiry for teaming with wireless mobile devices. San Diego, CA: Invitation Symposia at Society for Information Technology and Teacher Education (SITE).

Dunleavy, M., Dede, C., & Mitchell, R. (2009). Affordances and limitations of immersive participatory augmented reality simulations for teaching and learning. *Journal of Science Education and Technology*, *18*(1), 7–22. DOI: 10.1007/s10956-008-9119-1

El Sayed, N. A. M., Zayed, H. H., & Sharawy, M. I. (2011). ARSC: Augmented reality student card – an augmented reality solution for the education field. *Computers & Education*, *56*(4), 1045–1061. DOI: 10.1016/j.compedu.2010.10.019

Enyedy, N., Danish, J. A., Delacruz, G., & Kumar, M. (2012). Learning physics through play in an augmented reality environment. *International Journal of Computer-Supported Collaborative Learning*, *7*(3), 347–378. DOI: 10.1007/s11412-012-9150-3

Farroni, T., Valori, I., & Carnevali, L. (2022). Multimedia Interventions for Neurodiversity: Leveraging Insights from Developmental Cognitive Neuroscience to Build an Innovative Practice. *Brain Sciences*, *12*(2), 147. DOI: 10.3390/brainsci12020147

Fazel, A., & Izadi, A. (2018). An interactive augmented reality tool for constructing free-form modular surfaces. *Automation in Construction*, *85*, 135–145. DOI: 10.1016/j.autcon.2017.10.015

Fernández-Robles, B. (2018). La utilización de objetos de aprendizaje de realidad aumentada en la enseñanza universitaria de educación primaria. *IJERI Int. J. Educ. Res. Innov.*, *9*, 90–104.

Ferrer-Torregrosa, J., Torralba, J., Jimenez, M. A., García, S., & Barcia, J. M. (2015). ARBOOK: Development and assessment of a tool based on augmented reality for anatomy. *Journal of Science Education and Technology*, *24*(1), 119–124. DOI: 10.1007/s10956-014-9526-4

Garzón, J. (2021). An overview of twenty-five years of augmented reality in education. *Multimodal Technologies and Interaction*, *5*(7), 37. DOI: 10.3390/mti5070037

Garzón, J., & Acevedo, J. (2019). Meta-analysis of the impact of augmented reality on students' learning gains. *Educational Research Review, 27*, 244–260. 1

Geroimenko, V. (2020). *Augmented reality in education. Springer Series on Cultural Computing*. Springer International Publishing. DOI: 10.1007/978-3-030-42156-4

Geroimenko, V. (2020). *Augmented reality in tourism, museums and heritage: Emerging applications and future trends*. Springer.

Gkini, E., Voyiatzis, I., & Sgouropoulou, C. 2021. Head-Mounted Display Systems as Visual Aids for the Visually Impaired: A Survey. In *25th Pan-Hellenic Conference on Informatics (PCI 2021)*, November 26–28, 2021, Volos, Greece. ACM, New York, NY, USA 5 Pages. DOI: 10.1145/3503823.3503883

Hewitt, N., Wood, S., & Wilson, B. (2022). Ecosystem education with augmented reality: A flexible tool for in-field learning. *The Professional Geographer*, *75*(4), 577–590. DOI: 10.1080/00330124.2022.2134151

Huang, C., Zhao, Y., & Zhang, M. (2022). Smart urban sustainability: Integrating AR, AI, and IoT for environmental awareness. *Sustainable Cities and Society*, *84*, 104030.

Ibáñez, M. B., & Delgado-Kloos, C. (2018). Augmented reality for STEM learning: A systematic review. *Computers & Education, 123*, 109–123. DOI: 10.1016/j.compedu.2018.05.002

Issa, T., Isaias, P., & Issa, T. (2017). *Sustainability, Green IT and Education Strategies in the Twenty-First Century*. Springer International Publishing. DOI: 10.1007/978-3-319-57070-9

Itamiya, T., Tohara, H., & Nasuda, Y. 2019, March. Augmented Reality Floods and Smoke Smartphone App Disaster Scope utilizing Real-time Occlusion. In 2019 IEEE Conference on Virtual Reality and 3D User Interfaces (VR)(pp. 1397-1397). IEEE. DOI: 10.1109/VR.2019.8798269

Jewell, C. 2014. Augmented Reality Could Have Sustainability Applications. https:// www.thefifthestate.com.au/innovation/design/augmented-reality-could -havesustainability-applications/61208. (Accessed 13 May 2018).

Kamarainen, A. M., Metcalf, S., Grotzer, T., Browne, A., Mazzuca, D., Tutwiler, M. S., & Dede, C. (2013). EcoMOBILE: Integrating augmented reality and probeware with environmental education field trips. *Computers & Education*, *68*, 545–556. DOI: 10.1016/j.compedu.2013.02.018

Kato, H., & Billinghurst, M. (1999). Marker tracking and HMD calibration for a video-based augmented reality conferencing system. *Proceedings of the 2nd International Workshop on Augmented Reality*, 85–94. DOI: 10.1109/IWAR.1999.803809

Kerawalla, L., Luckin, R., Seljeflot, S., & Woolard, A. (2006). "Making it real": Exploring the potential of augmented reality for teaching primary school science. *Virtual Reality (Waltham Cross)*, *10*(3), 163–174. DOI: 10.1007/s10055-006-0036-4

Klopfer, E., & Squire, K. (2008). Environmental detectives: The development of an augmented reality platform for environmental simulations. *Educational Technology Research and Development*, *56*(2), 203–228. DOI: 10.1007/s11423-007-9037-6

Kolb, D. A. (1984). *Experiential learning: Experience as the source of learning and development*. Prentice Hall.

Krueger, M. W. (1977). Responsive environments. *Proceedings of the National Computer Conference*, 267–277.

Kuhlen, T., & Dohle, C. (1995). Virtual reality for physically disabled people. *Computers in Biology and Medicine*, *25*(2), 205–211. DOI: 10.1016/0010-4825(94)00039-S

Lee, S. W.-Y., Tsai, C.-C., Wu, Y.-T., Tsai, M.-J., Liu, T.-C., Hwang, F.-K., Lai, C.-H., Liang, J.-C., Wu, H.-C., & Chang, C.-Y. (2011). Internet-based science learning: A review of journal publications. *International Journal of Science Education*, *33*(14), 1893–1925. DOI: 10.1080/09500693.2010.536998

Lopez-Medina, I. M., Álvarez-Nieto, C., Grose, J., Elsbernd, A., Huss, N., Huynen, M., & Richardson, J. (2019). Competencies on environmental health and pedagogical approaches in the nursing curriculum: A systematic review of the litera- ture. *Nurse Education in Practice*, *37*, 1–8. DOI: 10.1016/j.nepr.2019.04.004

Macchione, F., Costabile, P., Costanzo, C., & De Santis, R. (2019). Moving to 3-D flood hazard maps for enhancing risk communication. *Environmental Modelling & Software*, *111*, 510–522. DOI: 10.1016/j.envsoft.2018.11.005

Macdonald, S. J., & Clayton, J. (2013). Back to the Future, Disability and the Digital Divide. *Disability & Society*, *28*(5), 702–718. DOI: 10.1080/09687599.2012.732538

Majeed, Z. H., & Ali, H. A. (2020). A review of augmented reality in educational applications. *International Journal of Advanced Technology and Engineering Exploration*, *7*(62), 20–27. DOI: 10.19101/IJATEE.2019.650068

Mappin, M. J., & Johnson, E. A. (2005). Changing perspectives of ecology and education in environmental education. In Johnson, E. A., & Mappin, M. J. (Eds.), *Environmental education and advocacy* (pp. 1–27). Cambridge University Press.

Martin, S., Diaz, G., Sancristobal, E., Gil, R., Castro, M., & Peire, J. (2011). New technology trends in education: Seven years of forecasts and convergence. *Computers & Education*, *57*(3), 1893–1906. DOI: 10.1016/j.compedu.2011.04.003

Martin-Gutierrez, J., Guinters, E., Perez-Lopez, D., 2012. Improving strategy of selflearning in engineering: laboratories with augmented reality. Procedia e Soc. Behav. Sci. 51, 832e839.

McElwee, P., Turnout, E., Chiroleu-Assouline, M., Clapp, J., Isenhour, C., Jackson, T., Kelemen, E., Miller, D. C., Rusch, G., Spangenberg, J. H., Waldron, A., Baumgartner, R. J., Bleys, B., Howard, M. W., Mungatana, E., Ngo, H., Ring, I., & Santos, R. (2020). Ensuring a Post-COVID Economic Agenda Tackles Global Biodiversity Loss. *One Earth*, *3*(4), 448–461. DOI: 10.1016/j.oneear.2020.09.011

Miller, D.. (2020). Real-time waste monitoring using AR and IoT: Applications in smart waste management. *Journal of Cleaner Production*, *258*, 120758.

Mills, K.. (2020). Geospatial AR in climate education: Real-world data visualization for secondary learners. *Environmental Science Education*, *45*(2), 103–118.

Mistry P, Maes P, Chang L (2009) WUW – Wear Ur World – A Wearable Gestural Interface", ACM, CHI 2009, Boston, April 4-9.

Munoz-Cristobal, J. A., Jorrin-Abellan, I. M., Asensio-Perez, J. I., Martinez-Mones, A., Prieto, L. P., & Dimitriadis, Y. (2015). Supporting teacher orchestration in ubiquitous learning environments: A study in primary education. Learning Technologies. *IEEE Transactions on Learning Technologies*, *8*(1), 83–97. DOI: 10.1109/TLT.2014.2370634

Muñoz-Saavedra, L., Miró-Amarante, L., & Domínguez-Morales, M. (2020). Augmented and Virtual Reality Evolution and Future Tendency. *Applied Sciences (Basel, Switzerland)*, *10*(1), 322. DOI: 10.3390/app10010322

Nisi, V., Paredes, H., Jardim, M. A., & Nunes, N. (2019). Exploring AR for urban sustainability engagement. *International Journal of Human-Computer Studies*, *130*, 52–65.

Perry, J., Klopfer, E., Norton, M., Sutch, D., Sandford, R., & Facer, K. (2008). AR gone wild: Two approaches to using augmented reality learning games in zoos. Proceedings of the 8th International Conference on International C01iference for the Learning Sciences, The Netherlands, (pp. 322-329).

Pierson, M. E. (2001). Technology Integration Practice as a Function of Pedagogical Expertise. *Journal of Research on Computing in Education*, *33*(4), 413–430. DOI: 10.1080/08886504.2001.10782325

Preece, J., Sharp, H., & Rogers, Y. (2021). *Interaction design: Beyond human-computer interaction* (5th ed.). Wiley.

Radu, I. (2014). Augmented reality in education: A meta-review and cross-media analysis. *Personal and Ubiquitous Computing*, *18*(6), 1533–1543. DOI: 10.1007/s00779-013-0747-y

Rahman, A.. (2022). AR-based tools for inclusive climate change education in urban classrooms. *Journal of Environmental Studies (Northborough, Mass.)*, *54*(4), 209–222.

Rosenbaum, E., Klopfer, E., & Perry, J. (2007). On location learning: Authentic applied science with networked augmented realities. *Journal of Science Education and Technology*, *16*(1), 31–45. DOI: 10.1007/s10956-006-9036-0

Sermet, Y., & Demir, I. (2020). Virtual and augmented reality applications for environmental science education and training. In *New perspectives on virtual and augmented reality* (pp. 261–275). Routledge. DOI: 10.4324/9781003001874-17

Shakirova, N., Berechikidze, I., & Gafiyatullina, E. (2023). The effects of immersive AR technology on the environmental literacy, intrinsic motivation, and cognitive load of high school students. *Education and Information Technologies*, *29*, 6121–9138.

Shankar, A. U., Tewari, V., Rahman, M., Mishra, A., & Bajaj, K. K. (2023). Impact of Virtual Reality (VR) and Augmented Reality (AR) in education. Tuijin Jishu/ Journal of Propulsion Technology, 44(4), 1310-1318. DOI: DOI: 10.52783/tjjpt. v44.i4.1014

Sivapalan, M., Savenije, H. H., & Blöschl, G. (2012). Socio-hydrology: A new science of people and water. *Hydrological Processes*, *26*(8), 1270–1276. DOI: 10.1002/hyp.8426

Smith, J.. (2020). Augmented reality applications for enhancing recycling behavior: A pilot study. *Journal of Environmental Psychology*, *68*, 101382.

Sutherland, I. (1968). A head-mounted three-dimensional display. *Proceedings of the AFIPS Fall Joint Computer Conference, 33*, 757–764.

Teófilo, M., Lucena, V. F., Nascimento, J., Miyagawa, T., & Maciel, F. "Evaluating accessibility features designed for virtual reality context," *2018 IEEE International Conference on Consumer Electronics (ICCE)*, 2018, pp. 1-6. DOI: 10.1109/ ICCE.2018.8326167

Thew H, Graves C, Reay D, Smith S, Petersen K, et al. Mainstreaming climate change education in UK higher education institutions 2021.

UNESCO. (2014). *Roadmap for implementing the global 760 action programme on education for sustainable development*. UNESCO.

Veas, E., Grasset, R., Ferencik, I., Grünewald, T., & Schmalstieg, D. (2013). Mobile augmented reality for environmental monitoring. *Personal and Ubiquitous Computing*, *17*(7), 1515–1531. DOI: 10.1007/s00779-012-0597-z

Wu, H. K., Lee, S. W. Y., Chang, H. Y., & Liang, J. C. (2013). Current status, opportunities and challenges of augmented reality in education. *Computers & Education*, *62*, 41–49. DOI: 10.1016/j.compedu.2012.10.024

Wu, P. H.. (2020). Gamification of environmental education with augmented reality applications: Enhancing critical thinking. *Interactive Learning Environments*, *28*(7), 895–911.

Wu, W., Lee, S. W., Chang, H., & Liang, J. (2020). Current status, opportunities, and challenges of augmented reality in education. *Computers & Education*, *62*(1), 41–49.

Yilmaz, S., & Goktas, Y. (2021). The role of gamified augmented reality in climate change education. *Journal of Interactive Learning Research*, *32*(1), 91–108.

Yoon, S. A., Anderson, E., Lin, J., & Elinich, K. (2017). How augmented reality enables conceptual understanding of challenging science content. *Educational Technology Research and Development*, *65*(4), 611–633.

Chapter 6
IoT and Smart City Solutions for Sustainable Waste Management:
Empowering Green Governance Through Technology in Urban Bangladesh

Imran Hossain
https://orcid.org/0000-0002-7602-053X
Varendra University, Bangladesh

A. K. M. Mahmudul Haque
University of Rajshahi, Bangladesh

Moinuddin Zahangir
Varendra University, Bangladesh

Babasaheb Jadhav
https://orcid.org/0000-0002-1933-0558
Dr. D.Y. Patil Vidyapeeth, India

ABSTRACT

The rapid urbanization in Bangladesh has led to significant challenges in solid waste management, with traditional systems struggling to meet the growing demands of urban centers. This study explores the potential of Internet of Things (IoT) technologies in revolutionizing waste management practices, focusing on their role in enhancing efficiency, sustainability, and citizen engagement. By analyzing global case studies from cities such as Barcelona, Singapore, and New York, this research demonstrates

DOI: 10.4018/979-8-3373-1892-9.ch006

how IoT solutions, such as smart bins, real-time data analytics, and waste-to-energy technologies, can optimize waste collection, reduce operational costs, and promote recycling efforts. However, the adoption of these technologies in Bangladesh faces several barriers, including financial constraints, inadequate infrastructure, a lack of technical expertise, and limited public awareness. The study identifies these challenges and provides recommendations for overcoming them, emphasizing the need for pilot projects, cost-effective solutions, and capacity-building initiatives.

1. INTRODUCTION

In recent years, urbanization in Bangladesh has rapidly expanded, drastically changing the social, economic, and environmental scene of the nation (Hassan, 2017). Cities like Dhaka, Chittagong, and Rajshahi are struggling greatly in properly managing solid waste as the nation's urban population is growing at a startling rate (Hossain et al., 2024a; Roy et al., 2022). With more than 160 million residents, Bangladesh is seeing an annual urban growth rate of roughly 3.5%. Dhaka is among the fastest-growing megacities worldwide (Swapan et al., 2017). This population increase has caused solid waste creation in metropolitan areas to skyrocket, surpassing the already taxed waste management capacity (Hossain et al., 2024b). According to estimates, Dhaka alone produces almost 5,000 tons of waste every day, much of which is incorrectly handled because of logistical and budgetary restrictions (Islam, 2016). In terms of waste collecting, transportation, disposal, and recycling, this mounting waste load presents serious difficulties; it also significantly contributes to environmental contamination, bad public health, and the declining urban livability (Hossen et al., 2024).

The customary methods of waste management in Bangladesh generally rely on inefficient, antiquated, and labor-intensive processes (Hossain et al., 2024a). Manual waste collection and transportation, coupled with a lack of adequate segregation at the source, result in huge amounts of recyclable and organic waste being mixed with non-recyclable waste, further complicating disposal operations (Hossain et al., 2025; Bernat, 2023). Along with being resource-intensive, these conventional methods are prone to inefficiencies, including uneven collecting schedules, poor waste disposal sites, and contamination from open dumpsites. Bangladesh thus urgently needs creative ideas that can solve these flaws and guide the direction toward more environmentally friendly urban waste management techniques.

Smart technologies—especially the Internet of Things (IoT)—have lately become rather popular in waste management systems worldwide. IoT, or the network of physical objects fitted with sensors, software, and other technologies enabling data interchange and process automation (Abdul-Qawy et al. 2015). In the context of waste management, IoT has been employed to construct smart waste management systems that improve waste collection, cut costs, enhance recycling rates, and offer real-time data to authorities (Lakhouit, 2025). The use of IoT-based solutions allows cities to monitor waste levels, detect inefficiencies, and dynamically change collection routes and schedules, contributing to more effective waste management and a reduction in operational costs (Sosunova & Porras, 2022). However, the application of such technologies in Bangladesh confronts various hurdles, many of which come from the country's existing infrastructure, financial constraints, and inadequate technological capacity (Rahman & Parvin, 2024). Although the promise for IoT in waste management is evident, the adoption of these technologies in Bangladesh remains delayed due to numerous obstacles. The country's waste management infrastructure is antiquated, and the lack of suitable waste collection and disposal facilities makes it impossible to incorporate innovative technologies (Roy et al. 2022).

The role of green governance in supporting sustainable waste management is equally crucial. Green governance emphasizes environmental sustainability, citizen participation, and the effective use of resources in policy-making and government (Hawkins & Wang, 2012; Hossain et al., 2023). In the context of urban waste management, green governance means implementing policies and technology that not only solve waste challenges but also promote environmental stewardship and public responsibility. IoT technology can act as a significant enabler of green governance by delivering transparent, real-time data that allows individuals, local governments, and private players to engage in waste management efforts.

This review intends to evaluate the potential of IoT and smart city solutions for revolutionizing waste management in urban Bangladesh. It will assess the existing landscape of waste management in the country, emphasize the benefits and obstacles of adopting IoT technologies, and discuss the role of green governance in promoting the integration of these technologies into waste management practices. By studying worldwide case studies and examining local pilot programs, this article will provide insights into the future direction of waste management in Bangladesh and propose ideas for overcoming present impediments.

2. IOT AND SMART CITY SOLUTIONS FOR WASTE MANAGEMENT

The integration of IoT technologies in waste management has emerged as a fundamental technique for boosting the efficiency and sustainability of municipal waste systems internationally. In a smart city environment, IoT applications offer a solid foundation for real-time monitoring, data collecting, and the automation of various waste management activities (Sosunova & Porras, 2022). This technical breakthrough promises to change traditional waste systems by enhancing the collecting, sorting, recycling, and disposal of waste in ways that were previously impracticable. The implementation of IoT in waste management aligns with the aims of sustainable development, contributing to resource efficiency, cost savings, and environmental preservation (Table 1). In the instance of Bangladesh, where urbanization is quickly developing, the implementation of IoT and smart city solutions has tremendous potential to alleviate the growing waste management crisis (Roy et al., 2023). One of the key applications of IoT in waste management is the deployment of smart bins. These bins are equipped with sensors that detect the fill levels of waste containers and provide real-time data to waste management operators (Afolalu et al., 2021). This technology helps optimize waste collection by scheduling pickups only when bins are full, minimizing wasted trips and boosting fuel efficiency. In Bangladesh, where waste collection services often face challenges due to limited resources, smart bins could significantly reduce inefficiencies in urban waste management, particularly in highly populated areas like Dhaka, where congestion and limited road access can delay waste collection services. Moreover, the adoption of smart bins can be coupled with mobile applications, allowing users to watch the status of waste bins in their locality and report any difficulties, providing a more participative and responsive waste management system.

In addition to smart bins, IoT technologies enable the adoption of sensor-based systems that monitor the waste generation and pickup operations across a whole city (Vishnu et al., 2022). These systems, which might incorporate waste monitoring sensors placed in various locations, provide a comprehensive overview of waste production trends (Ali et al., 2020). By collecting and evaluating this data, city officials may identify high-density waste locations, predict future waste generation trends, and allocate resources more effectively (Ali et al., 2020). In the case of Bangladesh, a comparable IoT-based system might assist in addressing the uneven distribution of waste generation and collection across the city, ensuring that resources are employed where they are most needed.

Table 1. IoT and Smart City Solutions for Waste Management

Technology	Description	Key Benefits	Challenges
Smart Bins	Bins equipped with sensors to monitor fill levels and optimize waste collection.	Reduced fuel consumption, optimized collection schedules, real-time data monitoring.	High initial cost of sensors, integration with existing bins.
Waste-to-Energy	Conversion of waste into renewable energy using IoT-enabled monitoring systems.	Reduced waste sent to landfills, generation of renewable energy, decreased environmental impact.	High infrastructure cost, need for continuous maintenance.
Data Analytics	Use of real-time data tracking to forecast waste generation and optimize management.	Improved resource allocation, better decision-making, predictive waste management.	Data privacy concerns, reliance on data accuracy.
Citizen Engagement	Mobile apps and notifications for public participation in waste management.	Increased public participation, awareness, and recycling rates.	Low public awareness, technology accessibility challenges.
Waste Segregation	IoT-enabled automated sorting of recyclable and non-recyclable waste.	Increased recycling rates, reduced landfill usage.	High cost of sorting technology, need for public behavior change.

IoT also plays a crucial role in enhancing recycling rates through automated sorting systems. In many underdeveloped nations, including Bangladesh, recycling rates remain low due to inefficient sorting processes, lack of public awareness, and inadequate recycling infrastructure. IoT-based waste sorting systems, linked with machine learning algorithms, can identify and separate recyclable items such as plastics, paper, and metals from non-recyclables, thus boosting the total efficiency of recycling processes (Ahmed et al., 2025). These methods not only decrease human error but also speed up the sorting process, increasing the volume of material that may be recycled (Middha et al., 2024). In Bangladesh, where there is minimal recycling infrastructure and many recyclables are still dumped in landfills, IoT-enabled sorting systems could offer a major improvement. By automating waste segregation at the time of collection, the system might ensure that valuable materials are diverted from landfills and delivered to recycling factories, contributing to a circular economy and minimizing environmental damage.

Another significant area where IoT technologies are boosting waste management is in waste transportation. In previous systems, waste collection vehicles followed pre-determined routes regardless of the actual fill levels in bins, leading to inefficiencies and increased expenses. With IoT-enabled waste management, waste pickup vans are integrated with GPS and sensor technologies that allow for real-time route optimization (Medvedev et al., 2015). Data from smart bins can be utilized to guide trucks to only the regions where waste needs to be collected, thus decreasing fuel usage and reducing congestion. Furthermore, these cars can be monitored remotely, allowing fleet managers to check vehicle performance, fuel consumption, and maintenance needs, ensuring that the fleet is functioning at optimal efficiency. In Dhaka and other urban regions in Bangladesh, where traffic congestion sometimes disrupts waste collection schedules, IoT-based route optimization could help eliminate delays and assure timely waste pickup.

While the promise of IoT in waste management is immense, the adoption of these technologies in Bangladesh confronts various hurdles. One of the significant challenges is the expensive initial expenditure required to construct IoT infrastructure, which may be difficult to secure in a developing country with low financial resources. Moreover, the lack of technological competence and capacity for managing such systems poses a considerable issue for local administrations. To overcome these challenges, there needs to be a concerted effort from both the government and the business sector to invest in the necessary infrastructure, develop technical skills, and create public awareness about the benefits of IoT-enabled waste management systems.

3. GREEN GOVERNANCE THROUGH INTERNET OF THINGS (IOT)

The concept of green governance is crucial to the achievement of sustainable urban development, particularly in rapidly urbanizing countries like Bangladesh. Green governance refers to a governance model that promotes environmental sustainability, resource conservation, and equitable development through the involvement of numerous stakeholders, including citizens, corporations, and government authorities (Li et al., 2018). It underlines the need for policy frameworks and technology that not only safeguard the environment but also promote social inclusion and economic progress. In the context of urban waste management, green governance argues for the efficient management of resources, decrease of waste output, and increased recycling and recovery rates (Liu et al., 2021). The incorporation of IoT technology into waste management systems has the potential to dramatically advance green governance by boosting transparency, efficiency, and citizen participation, ultimately contributing to the overarching objective of sustainable urban development (Figure 1).

Figure 1. Green Governance through IoT

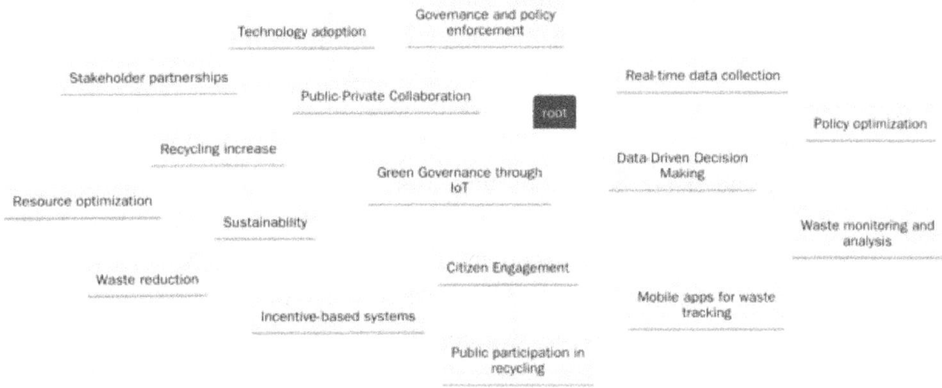

IoT-based waste management systems give a wealth of data that can inform decision-making at multiple levels of governance. One of the fundamental parts of green governance is the use of data to promote informed policy-making and resource allocation. IoT technologies, by embedding sensors in waste bins, vehicles, and processing plants, provide massive volumes of real-time data on waste generation, collection, and disposal trends (Vishnu et al., 2022). This data can be evaluated to identify waste generation hotspots, track recycling progress, and evaluate the success of waste management programs. By employing this data, local governments may make more informed decisions on waste collection schedules, resource allocation, and infrastructure development.

In Bangladesh, where the lack of reliable data on waste generation and disposal is a serious concern, IoT technologies can play a revolutionary role in enhancing the accuracy of waste management data. Real-time data collected from smart waste bins and other IoT-enabled devices can provide municipalities with granular insights into the waste management system, such as tracking the amount of recyclable materials created in different sections of the city. This data can inform the creation of targeted policies and initiatives, such as optimizing waste collection routes, boosting the availability of recycling bins in high-density regions, and enhancing waste segregation at the source. The capacity to monitor waste flows in real time guarantees that waste management policies are responsive and adaptable and that resources are allocated where they are most required, thus boosting the efficiency and sustainability of waste management systems.

Another crucial feature of green governance is citizen participation, which is essential for guaranteeing the success of sustainable waste management projects. IoT technologies promote improved public engagement by giving real-time information on waste collection schedules, waste diversion rates, and recycling activities

(Lakhouit, 2025). For instance, mobile applications that are integrated with IoT waste management systems can tell residents about the status of waste pickup in their areas and remind them of recycling days. These applications can also provide citizens with feedback on their waste disposal behaviors, such as whether they are recycling correctly or whether their waste bins are overfilled, promoting more responsible behavior.

In addition to increasing citizen participation, IoT-enabled waste management systems can increase the role of local governments in green governance by offering tools for better monitoring and enforcement of waste management policies (Hossain et al., 2025). In many cities in Bangladesh, the absence of effective implementation of waste segregation laws and regulations leads to the inefficiency of waste management systems. IoT solutions can assist in addressing this issue by providing authorities with real-time data on waste disposal methods, enabling them to detect places where policies are not being implemented. For example, smart bins equipped with RFID tags can track the origin of waste, ensuring that only correctly segregated items are collected for recycling. This data can be utilized by local governments to fine individuals or enterprises that fail to comply with waste segregation requirements, thus enhancing policy enforcement and encouraging greater adherence to environmental regulations.

While the incorporation of IoT in waste management has enormous promise for developing green governance, the successful implementation of these systems in Bangladesh involves overcoming many difficulties. The absence of technological infrastructure, financial constraints, and low public understanding of IoT technology are key impediments to the mainstream implementation of smart waste management solutions. Addressing these difficulties will need a coordinated effort involving government agencies, commercial organizations, and citizens. Policymakers must invest in the necessary infrastructure, create supporting regulatory frameworks, and promote public awareness of the benefits of IoT-enabled waste management systems.

4. METHODOLOGY OF THE STUDY

This study employs a systematic review methodology to explore the integration of IoT technologies in waste management systems and its potential to advance green governance in urban Bangladesh. The review focuses on synthesizing existing literature, case studies, and pilot projects to assess how IoT-based solutions can improve the efficiency and sustainability of waste management practices, as well as foster citizen engagement and data-driven decision-making in the context of Bangladesh's urban centers. By reviewing a wide range of scholarly articles, government reports, and industry publications, this study aims to provide a comprehensive understanding of the current state of IoT applications in waste management and the challenges and opportunities they present in the specific context of Bangladesh.

To begin, the study undertakes an extensive search of academic databases, including Google Scholar, ScienceDirect, and SpringerLink, using key terms such as “IoT waste management,” “smart cities,” “green governance,” and “sustainable urban development.” The search is limited to studies published in the last decade to ensure that the findings reflect the most current developments in IoT and smart city technologies. Articles that were found to be highly relevant to the scope of the study were selected based on their contributions to understanding the role of IoT in waste management, particularly in urban settings similar to Bangladesh. Studies discussing global examples of IoT-driven waste management systems, such as those in cities like Barcelona, Singapore, and New York, are included to provide a comparative perspective. The inclusion of these international case studies offers valuable insights into how IoT technologies can be adapted to meet the specific challenges faced by Bangladesh’s urban waste management systems.

The review also explores local studies and reports to examine the existing infrastructure and policies surrounding waste management in Bangladesh. This includes government publications, reports from non-governmental organizations (NGOs), and industry insights into the current status of waste management in major cities like Dhaka and Chittagong. Particular attention is given to pilot projects that have implemented IoT-based solutions in waste management in Bangladesh, focusing on their design, implementation, outcomes, and lessons learned. By incorporating both international and local sources, the study provides a balanced view of how IoT technologies can be tailored to the specific needs and constraints of Bangladesh.

5. CURRENT LANDSCAPE OF SMART WASTE MANAGEMENT IN BANGLADESH

The urbanization of Bangladesh has given birth to considerable issues in waste management, especially in large cities like Dhaka, Chittagong, and Khulna, where fast population development has placed great pressure on current waste management systems. Despite the crucial necessity of efficient waste management for public health, environmental sustainability, and urban livability, Bangladesh continues to experience issues in managing solid waste properly. The conventional waste management systems in Bangladesh are often inefficient, underfunded, and inadequate to fulfill the demands of rapidly developing urban areas (Ahmed, 2024). Poor waste collection, insufficient recycling infrastructure, and poor waste disposal procedures have resulted in environmental deterioration, increased pollution, and a decline in the quality of life for urban people (Abubakar et al., 2022). In response to these issues, smart waste management solutions, including the deployment of IoT technologies, have been recommended as a novel method to increase waste management efficiency and sustainability in Bangladesh's metropolitan centers.

While the notion of smart waste management is relatively new in Bangladesh, there have been initiatives to deploy IoT-based solutions to optimize waste collection and disposal. The government and numerous corporate sector partners have recognized the potential of smart city solutions to address urban waste management concerns, particularly in major metropolitan regions. However, the adoption of IoT technologies in waste management remains in its fledgling phases. Several pilot projects have been started in places like Dhaka, but their scale and impact have been limited due to various challenges, including financial constraints, lack of technical skills, and insufficient infrastructure (Roy, 2009). Despite these limitations, there have been positive developments in certain areas, such as the introduction of smart waste bins, waste tracking systems, and waste-to-energy projects, all of which provide a glimpse into the potential of IoT technologies to revolutionize waste management practices in Bangladesh.

One of the biggest problems to adopting smart waste management systems in Bangladesh is the absence of reliable and complete data on waste generation and disposal. Traditional waste management systems in Bangladesh rely on inefficient human techniques for rubbish collection, frequently based on guesses rather than actual data. As a result, waste collection routes are not optimized, and the collection schedules are sometimes inconsistent, leading to overflowing bins and delays in waste removal. IoT-enabled waste management systems, by comparison, offer a data-driven strategy that can monitor the fill levels of waste bins in real time, enabling waste collection vehicles to follow optimum routes and timetables based on actual demand. This technology has been deployed in experimental programs in Dhaka, where sensors placed in rubbish bins allow waste management authorities to monitor waste levels remotely and alter pickup schedules accordingly. Although these systems are still in the early phases of deployment, they have shown the ability to increase collection efficiency, cut fuel consumption, and improve waste management services in urban settings.

In addition to smart bins, the adoption of waste-to-energy (WTE) technology has been considered as a means to address the increasing volume of waste in Bangladesh. Waste-to-energy plants transform non-recyclable waste into electricity or heat, lowering the quantity of waste sent to landfills and providing a renewable source of energy (Roy et al., 2022). Several WTE projects have been suggested in Bangladesh, but their implementation has been difficult because of high starting costs, lack of technical skills, and worries about environmental pollution. However, the potential for IoT to boost the efficiency of WTE plants is enormous. IoT technology can be integrated into WTE systems to monitor the functioning of incineration and gasification units, discover inefficiencies, and assure compliance with environmental regulations. Real-time monitoring of WTE facilities can help improve waste sorting, lowering the environmental impact of waste disposal and boosting the overall sustainability of waste management systems. For Bangladesh, the implementation of IoT in WTE facilities might be a game-changer, particularly given the country's limited capacity for waste disposal and the growing need for renewable energy sources.

Another interesting development in the current landscape of smart waste management in Bangladesh is the steady deployment of mobile applications and public interaction platforms. These platforms allow residents to report difficulties relating to waste collection, track waste collection schedules, and receive information about recycling activities (Hossain et al., 2024a). The use of mobile technology in waste management is a fundamental element of green governance, as it stimulates active citizen participation in waste management and creates more openness in waste disposal procedures (Hossain et al., 2024a). In metropolitan Bangladesh, a pilot mobile application has been created to give citizens real-time updates on waste collection and to promote waste segregation at the household level (Hossain et al., 2024a). By allowing citizens to track waste collection schedules and identify problems with waste collection, these applications assist in establishing a more responsive and responsible waste management system (Hossain et al., 2024b). Furthermore, IoT-enabled mobile systems can provide feedback to citizens on their waste disposal patterns, motivating responsible conduct and promoting environmental consciousness.

Despite these developments, significant challenges to the broad application of smart waste management systems remain (Table 2). One of the biggest challenges is the financial cost of implementing IoT technologies, which can be expensive for local governments in Bangladesh, particularly given the country's limited public sector funding for infrastructure development. The initial expenditure required for smart bins, waste tracking systems, and IoT-based waste management software is expensive, and without appropriate funding, these projects are generally limited to small-scale experimental programs. Additionally, the lack of technical capability and competence in administering IoT systems creates another issue. While there is a growing interest in IoT technology, there is a need for training and capacity-building activities to educate local authorities and waste management experts with the essential skills to run and maintain these systems.

Table 2. Challenges and Barriers to IoT Waste Management Implementation

Challenges/Barriers	Description	Impact on IoT Implementation	Recommendations for Overcoming Barriers
High Initial Cost	The cost of setting up IoT infrastructure, including smart bins, sensors, and data analytics systems, is high, making it difficult for local governments to allocate resources for large-scale deployment.	High upfront costs limit the ability of municipalities to invest in large-scale smart waste management systems.	Explore international funding and public-private partnerships to share the financial burden. Consider phased rollouts to reduce initial costs.
Inadequate Infrastructure	Many urban areas in Bangladesh, especially smaller towns and low-income neighborhoods, lack the infrastructure necessary to support IoT-based waste management systems. This includes inadequate waste bins, outdated vehicles, and poor road conditions.	Without the necessary physical infrastructure, IoT systems cannot be effectively implemented, reducing their operational impact.	Invest in basic infrastructure upgrades and ensure that waste management systems are compatible with IoT solutions. Improve energy access to support smart systems.
Technical Capacity and Expertise	There is a shortage of skilled professionals in Bangladesh who can design, implement, and maintain IoT-based waste management systems. This skills gap limits the effectiveness of the systems.	Lack of technical expertise limits the ability to maintain and operate IoT systems, affecting long-term functionality and efficiency.	Focus on training and capacity-building initiatives to develop local expertise in IoT system integration and maintenance. Encourage knowledge transfer from international experts.

Furthermore, public awareness of the benefits of IoT-based waste management systems is still rather low in Bangladesh. Many citizens are unaware of the relevance of waste segregation and the role of smart technologies in enhancing waste management. Public education campaigns and community outreach programs are vital to promoting better awareness and encouraging participation in smart waste management initiatives. Without active engagement from the public, the effectiveness of IoT-based technologies in waste management may be limited.

In terms of legislative and regulatory frameworks, Bangladesh is still in the process of establishing and implementing comprehensive waste management policies that combine smart city solutions. While there have been some regulatory measures aimed at improving waste management, such as the National Waste Management Regulatory, the lack of a clear and coordinated strategy for the implementation of IoT technology in waste management remains a significant hurdle. The creation of supportive laws and regulations that stimulate the adoption of smart waste management technologies, provide financial support for pilot projects, and foster public-private partnerships is vital for the wider application of IoT in waste management.

6. CASE STUDIES AND GLOBAL PERSPECTIVES

The integration of IoT technologies into waste management systems is not only an emerging trend in Bangladesh but also a developing global phenomenon. Across the world, cities have begun to implement smart waste management technologies as part of broader smart city projects aimed at increasing urban sustainability, operational efficiency, and quality of life (Table 3). These case studies offer significant insights into the potential of IoT to transform waste management methods and provide useful lessons that could shape the future of waste management in Bangladesh. By examining successful implementations in cities such as Barcelona, Singapore, and New York, as well as some emerging smart waste management initiatives in developing countries, this section explores the global perspectives on IoT-based waste management and the lessons that can be drawn from these examples.

Table 3. Global Case Studies on Smart Waste Management

City	Technology Solutions	Objectives	Key Achievements	Challenges Faced	Lessons Learned	Relevance to Bangladesh
Barcelona	Smart bins with sensors, real-time data analytics, optimized waste collection routes	To improve waste collection efficiency, reduce fuel consumption, and minimize environmental impact through data-driven decision making.	Improved waste collection efficiency by 20%, reduced fuel consumption by 10%, and enhanced recycling rates in select districts.	High initial cost of sensor technology, difficulty in integration with existing infrastructure, and the need for constant system upgrades.	Data-driven waste management systems can significantly enhance operational efficiency and reduce environmental impact, but long-term investments in infrastructure are essential.	Barcelona's focus on data-driven efficiency and optimized waste collection routes is highly relevant for Dhaka, where traffic congestion and inefficiencies in collection are prevalent.
Singapore	Smart bins with sensors, automated collection systems, waste tracking, real-time data analytics	To optimize waste collection schedules, increase recycling rates, and reduce waste generation by using smart technologies.	Achieved significant reductions in operational costs, improved waste segregation, and increased recycling rates in key districts.	High cost of implementation, integration challenges with the smart city infrastructure, and public reluctance to adopt new waste sorting practices.	Combining waste tracking with real-time analytics and recycling initiatives can drastically improve waste management, but public participation and awareness are key.	Singapore's experience with automated systems and real-time data analytics for waste management provides an insightful model for improving waste segregation and increasing recycling rates in Bangladesh.
New York	Smart bins with sensors, waste tracking, data analytics for optimized waste collection	To reduce operational costs, improve efficiency, and ensure timely waste collection in a large metropolitan area.	Optimized collection routes leading to a 15% reduction in truck fleet requirements and a significant decrease in carbon emissions.	Integration with existing systems, high operational costs, and the complexity of scaling the system in a diverse metropolitan environment.	Even in large cities with logistical challenges, IoT solutions can optimize waste management, but integration with existing infrastructure requires careful planning.	New York's smart bin and route optimization systems could be useful for Dhaka's urban areas, reducing waste collection costs and emissions while improving service efficiency.

continued on following page

Table 3. Continued

City	Technology Solutions	Objectives	Key Achievements	Challenges Faced	Lessons Learned	Relevance to Bangladesh
Kigali (Rwanda)	Sensor-equipped waste bins, mobile apps for citizen engagement, real-time waste tracking	To engage citizens in waste management via mobile applications and optimize waste collection routes using real-time data.	Increased citizen engagement and participation by 25% through the mobile app and optimized collection routes, contributing to cleaner neighborhoods.	Limited financial resources for technology adoption in lower-income neighborhoods, infrastructure constraints in rural areas, and scalability challenges.	Citizen engagement through mobile apps and IoT technologies is crucial for success, but the integration of these systems requires targeted outreach and capacity-building programs.	Kigali's model of integrating mobile apps for public participation can help increase citizen engagement in waste management in Bangladesh, fostering responsible waste disposal behavior.
Curitiba (Brazil)	Smart bins with sensors, waste-to-energy technologies, waste composition monitoring	To enhance waste collection efficiency, integrate renewable energy generation, and improve recycling rates using smart technologies.	Enhanced waste-to-energy operations, reducing landfill use by 30%, and generated renewable energy contributing to the cityâ€™s grid.	The cost of implementing waste-to-energy infrastructure, regulatory barriers, and the complexity of integrating with other sustainability programs.	Combining IoT technologies with waste-to-energy can be a viable solution for waste management and energy production, but high costs and infrastructure needs must be addressed.	Curitiba's integration of waste-to-energy with IoT systems can provide a model for Bangladesh to address waste disposal and energy generation, particularly in cities with limited landfill space.

Barcelona, a major smart city in Europe, has long been at the forefront of embracing IoT for waste management (Bibri & Krogstie, 2020; Szpilko et al., 2023). The city established a comprehensive smart waste management system that comprises sensor-equipped waste bins, real-time data analytics, and optimized waste collection routes (Szpilko et al., 2023). The smart bins, which are connected with fill-level sensors, communicate data to a central system that analyzes the waste levels and tells waste management authorities when the bins need to be emptied. This system has considerably increased the efficiency of waste collection, decreasing superfluous waste collection trips and minimizing fuel usage and emissions. Furthermore, the system enables the optimization of waste collection routes based on real-time data, which helps ensure that waste collection is carried out in a timely and cost-effective manner. The success of Barcelona's smart waste management system demonstrates the potential of IoT to streamline waste collection operations, reduce costs, and improve environmental sustainability, providing a useful model for urban areas in Bangladesh, where similar issues of inefficient waste management and high operational costs are prevalent.

Another noteworthy example of IoT-based waste management comes from Singapore, one of the world's top smart cities (Wong et al., 2021). Singapore has integrated IoT technologies into its waste management system using a combination of smart bins, automated collection systems, and waste tracking systems (Vishnu et al., 2021). For example, Singapore has introduced an innovative system known as the "Smart Waste Bin" initiative, which uses sensors to monitor the fill levels of waste bins and optimize waste collection schedules. The city also utilizes a waste tracking system that offers real-time data on the waste generated by different sectors, helping the authorities to manage waste more efficiently and reduce overall waste creation. The success of Singapore's smart waste management initiative highlights the importance of using IoT not only to improve waste collection but also to foster sustainable consumption and waste reduction, lessons that are crucial for cities like Dhaka, which face challenges with waste segregation and recycling.

New York City is another important case study, particularly in terms of scaling up IoT-enabled waste management systems in a broad metropolitan region. In 2018, New York City initiated a pilot study to assess the usefulness of smart bins fitted with sensors that monitor waste levels in real time (Fang et al., 2023). The project was targeted at reducing waste collection costs and boosting operating efficiency. The results of the pilot project were positive, since the sensors allowed for optimal collection schedules, which in turn reduced the number of vehicles needed to collect waste and minimized the environmental impact of the waste management system. New York's project also focused on integrating data analytics with waste management technologies, allowing the city to estimate future waste amounts and alter the system accordingly. The experiment revealed that even in a city with substantial logistical challenges, IoT technologies might lead to more efficient waste management operations. However, the project also encountered issues relating to the cost of implementation and the integration of new technologies into existing infrastructure, which are lessons that Bangladesh must consider when evaluating the practicality of IoT solutions in its own metropolitan areas.

In addition to these established countries, emerging economies in Africa and Latin America have also begun to examine the potential of IoT for waste management, bringing useful insights for Bangladesh's situation. For example, the city of Kigali in Rwanda, one of the fastest-growing cities in Africa, has begun using smart waste management solutions to address its rapid urbanization and waste management concerns (Manirakiza et al., 2019; Burns, 2021). Kigali has deployed sensor-equipped rubbish bins to monitor waste levels and optimize waste pickup routes. The project, inaugurated in 2019, also combines mobile applications to educate individuals in waste segregation and encourage more sustainable waste habits (Burns, 2021). Although still in the early stages, the Kigali project demonstrates the potential for IoT in waste management in low- and middle-income nations, where infrastructure restrictions and limited financial resources create distinct problems. The usage of mobile applications for citizen engagement and real-time waste tracking in Kigali also mimics the efforts seen in Bangladesh, where mobile technology might be utilized to promote public participation in waste management activities.

In Brazil, the city of Curitiba, noted for its creative urban planning solutions, has integrated smart waste management technologies as part of its broader sustainability strategy (Spinosa & Costa, 2020). Curitiba's waste management system features smart bins with sensors that check waste levels, optimize waste collection routes, and analyze waste composition to boost recycling rates (Sosunova & Porras, 2022). The city has also employed waste-to-energy systems to reduce the volume of rubbish going to landfills. Curitiba's experience highlights the value of merging IoT technologies with other sustainable municipal practices, such as waste-to-energy, to create a comprehensive waste management system. The city's success is a promising model for Bangladesh, where waste-to-energy technologies have yet to be completely adopted but show tremendous promise for solving both waste disposal and electricity generation demands.

In the context of Bangladesh, these global case studies give useful insights in both the potential benefits and obstacles involved with using IoT in waste management systems. The successful implementation of smart waste management in cities like Barcelona, Singapore, New York, Kigali, and Curitiba highlights several key factors that contribute to the success of IoT-based waste management systems, including the importance of data-driven decision-making, real-time monitoring, citizen engagement, and the integration of waste management with broader smart city infrastructures. For Bangladesh, the most essential takeaway is the need to invest in both technology infrastructure and public awareness to guarantee the effective implementation of smart waste management technologies. Moreover, the lessons acquired from developing nations such as Kigali and Curitiba illustrate that IoT solutions can be implemented even in resource-constrained situations, providing that there is political will, community participation, and successful public-private partnerships.

7. RECOMMENDATIONS FOR FUTURE RESEARCH AND IMPLEMENTATION

The integration of IoT technologies into waste management systems in Bangladesh is a promising approach for enhancing efficiency, sustainability, and environmental effect. However, problems such as infrastructure, financial limits, technical capacity, and public participation remain. To fully achieve the potential of IoT in improving waste management, future research and practical implementation initiatives must address these limitations and investigate novel approaches to optimize the system for local conditions.

To verify the feasibility and effectiveness of IoT-based waste management solutions in different urban and rural environments, thorough pilot projects should be done. These should examine the outcomes of these pilot projects, discuss the problems faced during implementation, and identify significant success factors. Understanding local impediments to implementation, such as road infrastructure concerns in densely populated areas or electrical access in rural regions, is vital for building systems that are adaptable to diverse municipalities.

Underdeveloped cost-effective IoT solutions adapted to the financial constraints of underdeveloped countries like Bangladesh is another key subject for future research. This could involve studying cheaper, more durable sensors, low-cost data storage options, and open-source software platforms that can minimize the cost of adopting and maintaining IoT-based waste management systems. Public-private partnerships (PPP) can harness private sector investment and innovation while ensuring that the public sector's demands are addressed.

Capacity building is also vital for tackling the huge technical and human resource limitations in Bangladesh. Research should focus on training and educating local engineers, waste management specialists, and government officials in the skills needed to build, install, and maintain IoT systems. Partnerships with universities, research institutions, and international organizations can play a significant role in facilitating information exchange and offering technical expertise.

In addition, data privacy and security are key concerns when adopting IoT devices. Future research should study the establishment of secure and privacy-preserving data management frameworks for IoT-based waste management systems, ensuring that data flows easily across multiple systems and contributes to the overall sustainability of the city.

Finally, policy development and regulatory support are required for the successful integration of IoT technology into waste management systems in Bangladesh. Future research should focus on developing policy recommendations that support the adoption of smart waste management systems, including incentives for municipalities to implement IoT solutions and collaboration between national and local governments to create a regulatory framework that supports the development of IoT-based solutions while ensuring public interests.

8. CONCLUSION

The integration of IoT technologies into waste management systems is a transformative strategy to address the expanding waste management challenges encountered by rapidly urbanizing areas, particularly those in Bangladesh. As urban populations continue to expand, the inefficiencies of traditional waste management systems are becoming increasingly obvious, leading to greater environmental degradation, resource depletion, and public health threats. In Bangladesh, where waste management methods are generally underdeveloped and underfunded, IoT-based solutions offer a promising pathway for boosting the efficiency, sustainability, and efficacy of waste management systems, particularly in large urban centers like Dhaka and

Chittagong. This study has studied the potential benefits of IoT in waste management, concentrating on its capacity to optimize waste collection, boost recycling rates, and promote citizen engagement through real-time data and smart technologies. Global case studies from cities like Barcelona, Singapore, and New York have proven that IoT may greatly enhance waste management efficiency by enabling data-driven decision-making, improving waste collection routes, decreasing costs, and creating more sustainable urban settings. The lessons learnt from these worldwide examples provide useful insights that can be tailored to the specific demands and problems of Bangladesh's urban context. Despite the promise that IoT-based solutions hold, there are major impediments to their mainstream adoption in Bangladesh. These include budgetary limits, poor infrastructure, lack of technical skills, insufficient regulatory frameworks, and limited public awareness. These difficulties need to be addressed by combined efforts from the government, commercial sector, and civil society. The success of IoT adoption will depend on overcoming these constraints and ensuring that the technologies are scalable, cost-effective, and linked with the country's broader sustainability goals. Additionally, the integration of IoT with other smart city technologies, such as traffic management and environmental monitoring, can further enhance the overall sustainability of urban environments. Future research and practical implementation efforts should focus on piloting IoT-based waste management systems in varied urban and rural environments inside Bangladesh. By evaluating the outcomes of these pilot projects, finding best practices, and scaling successful models, Bangladesh can create the framework for a comprehensive national waste management strategy that incorporates smart technologies. Moreover, research should study cost-effective solutions adapted to local financial restrictions, as well as emphasis on capacity building to ensure that local governments and waste management professionals have the skills and knowledge to implement and operate these systems efficiently.

ACKNOWLEDGMENTS

The research was funded by the Center for Interdisciplinary Research (CIR), Varendra University, Bangladesh. The memo number is VU/CIR/53-02-04, dated: February, 2024. The authors would like to give special thanks to the funding authority for supporting research grants.

REFERENCES

Abdul-Qawy, A. S., Pramod, P. J., Magesh, E., & Srinivasulu, T. (2015). The internet of things (iot): An overview. *International Journal of Engineering Research and Applications*, *5*(12), 71–82.

Abubakar, I. R., Maniruzzaman, K. M., Dano, U. L., AlShihri, F. S., AlShammari, M. S., Ahmed, S. M. S., Al-Gehlani, W. A. G., & Alrawaf, T. I. (2022). Environmental sustainability impacts of solid waste management practices in the global South. *International Journal of Environmental Research and Public Health*, *19*(19), 12717. DOI: 10.3390/ijerph191912717 PMID: 36232017

Afolalu, S. A., Noiki, A. A., Ikumapayi, O. M., Ogundipe, A. T., & Oloyede, O. R. (2021). Development of smart waste bin for solid waste management. *International Journal of Sustainable Development and Planning*, *16*(8), 1449–1454. DOI: 10.18280/ijsdp.160805

Ahmed, R. (2024). Innovative Waste Management Solutions: A Global Perspective Challenges and Opportunities and the Bangladesh Context.

Ahmed, S. T., Reddy, R. V., & KM, K. R. (2025). Artificial Intelligence for Sustainable Waste Management: A Literature Review. *Artificial Intelligence for Sustainable Waste Management: A Literature Review (April 17, 2025).*

Ali, T., Irfan, M., Alwadie, A. S., & Glowacz, A. (2020). IoT-based smart waste bin monitoring and municipal solid waste management system for smart cities. *Arabian Journal for Science and Engineering*, *45*(12), 10185–10198. DOI: 10.1007/s13369-020-04637-w

Bernat, K. (2023). Post-consumer plastic waste management: From collection and sortation to mechanical recycling. *Energies*, *16*(8), 3504. DOI: 10.3390/en16083504

Bibri, S. E., & Krogstie, J. (2020). The emerging data–driven Smart City and its innovative applied solutions for sustainability: The cases of London and Barcelona. *Energy Informatics*, *3*(1), 5. DOI: 10.1186/s42162-020-00108-6

Burns, H. (2021). A smart city masterplan, Kigali. *Urban planning for transitions*, 153-169.

Fang, B., Yu, J., Chen, Z., Osman, A. I., Farghali, M., Ihara, I., Hamza, E. H., Rooney, D. W., & Yap, P. S. (2023). Artificial intelligence for waste management in smart cities: A review. *Environmental Chemistry Letters*, *21*(4), 1959–1989. DOI: 10.1007/s10311-023-01604-3 PMID: 37362015

Hassan, M. M. (2017). Monitoring land use/land cover change, urban growth dynamics and landscape pattern analysis in five fastest urbanized cities in Bangladesh. *Remote Sensing Applications: Society and Environment*, *7*, 69–83. DOI: 10.1016/j.rsase.2017.07.001

Hawkins, C. V., & Wang, X. (2012). Sustainable development governance: Citizen participation and support networks in local sustainability initiatives. *Public Works Management & Policy*, *17*(1), 7–29. DOI: 10.1177/1087724X11429045

Hossain, I., Haque, A. M., Kılıç, Z., Ullah, S. A., Azrour, M., & Mabrouki, J. (2024). Exploring Household Waste Management Practices and IoT Adoption in Barisal City Corporation. In *Smart Internet of Things for Environment and Healthcare* (pp. 27–45). Springer Nature Switzerland. DOI: 10.1007/978-3-031-70102-3_2

Hossain, I., Haque, A. M., & Ullah, S. A. (2024). Assessing sustainable waste management practices in Rajshahi City Corporation: An analysis for local government enhancement using IoT, AI, and Android technology. *Environmental Science and Pollution Research International*, •••, 1–19. DOI: 10.1007/s11356-024-33171-7 PMID: 38581631

Islam, F. S. (2016). Solid waste management system in Dhaka City of Bangladesh. *Journal of Modern Science and Technology*, *4*(1), 192–209.

Lakhouit, A. (2025). Revolutionizing Urban Solid Waste Management with AI and IoT: A review of smart solutions for waste collection, sorting, and recycling. *Results in Engineering*, *25*, 104018. DOI: 10.1016/j.rineng.2025.104018

Li, W., Xu, J., & Zheng, M. (2018). Green governance: New perspective from open innovation. *Sustainability (Basel)*, *10*(11), 3845. DOI: 10.3390/su10113845

Liu, Z., Liu, T., Liu, X., Wei, A., Wang, X., Yin, Y., & Li, Y. (2021). Research on optimization of healthcare waste management system based on green governance principle in the COVID-19 pandemic. *International Journal of Environmental Research and Public Health*, *18*(10), 5316. DOI: 10.3390/ijerph18105316 PMID: 34067709

Manirakiza, V., Mugabe, L., Nsabimana, A., & Nzayirambaho, M. (2019). City Profile: Kigali, Rwanda. *Environment & Urbanization Asia, 10*(2), 290–307. DOI: 10.1177/0975425319867485

Medvedev, A., Fedchenkov, P., Zaslavsky, A., Anagnostopoulos, T., & Khoruzhnikov, S. (2015). Waste management as an IoT-enabled service in smart cities. In *Internet of Things, Smart Spaces, and Next Generation Networks and Systems: 15th International Conference, NEW2AN 2015, and 8th Conference, ruSMART 2015, St. Petersburg, Russia, August 26-28, 2015* [Springer International Publishing.]. *Proceedings, 15*, 104–115.

Middha, R., Srivastava, N., & Saxena, N. (2024). Chemical Principles in Waste Segregation and Recycling. In *Waste Management for Smart Cities* (pp. 1–46). Springer Nature Singapore.

Rahman, M., & Parvin, R. (2024). Advancing Bangladesh's Digital Transition: From Digital Foundations to Technological Leadership.

Roy, H., Alam, S. R., Bin-Masud, R., Prantika, T. R., Pervez, M. N., Islam, M. S., & Naddeo, V. (2022). A review on characteristics, techniques, and waste-to-energy aspects of municipal solid waste management: Bangladesh perspective. *Sustainability (Basel), 14*(16), 10265. DOI: 10.3390/su141610265

Roy, H., Alam, S. R., Bin-Masud, R., Prantika, T. R., Pervez, M. N., Islam, M. S., & Naddeo, V. (2022). A review on characteristics, techniques, and waste-to-energy aspects of municipal solid waste management: Bangladesh perspective. *Sustainability (Basel), 14*(16), 10265. DOI: 10.3390/su141610265

Roy, M. (2009). Planning for sustainable urbanisation in fast growing cities: Mitigation and adaptation issues addressed in Dhaka, Bangladesh. *Habitat International, 33*(3), 276–286. DOI: 10.1016/j.habitatint.2008.10.022

Roy, S., Rautela, R., & Kumar, S. (2023). Towards a sustainable future: Nexus between the sustainable development goals and waste management in the built environment. *Journal of Cleaner Production, 415*, 137865. DOI: 10.1016/j.jclepro.2023.137865

Sosunova, I., & Porras, J. (2022). IoT-enabled smart waste management systems for smart cities: A systematic review. *IEEE Access : Practical Innovations, Open Solutions, 10*, 73326–73363. DOI: 10.1109/ACCESS.2022.3188308

Sosunova, I., & Porras, J. (2022). IoT-enabled smart waste management systems for smart cities: A systematic review. *IEEE Access : Practical Innovations, Open Solutions, 10*, 73326–73363. DOI: 10.1109/ACCESS.2022.3188308

Spinosa, L. M., & Costa, E. M. (2020). Urban innovation ecosystem & humane and sustainable smart city: A balanced approach in Curitiba. *Handbook of smart cities*, 1-23.

Swapan, M. S. H., Zaman, A. U., Ahsan, T., & Ahmed, F. (2017). Transforming urban dichotomies and challenges of South Asian megacities: Rethinking sustainable growth of Dhaka, Bangladesh. *Urban Science (Basel, Switzerland)*, *1*(4), 31. DOI: 10.3390/urbansci1040031

Szpilko, D., de la Torre Gallegos, A., Jimenez Naharro, F., Rzepka, A., & Remiszewska, A. (2023). Waste management in the smart city: Current practices and future directions. *Resources*, *12*(10), 115. DOI: 10.3390/resources12100115

Vishnu, S., Ramson, S. J., Rukmini, M. S. S., & Abu-Mahfouz, A. M. (2022). Sensor-based solid waste handling systems: A survey. *Sensors (Basel)*, *22*(6), 2340. DOI: 10.3390/s22062340 PMID: 35336511

Vishnu, S., Ramson, S. J., Senith, S., Anagnostopoulos, T., Abu-Mahfouz, A. M., Fan, X., Srinivasan, S., & Kirubaraj, A. A. (2021). IoT-enabled solid waste management in smart cities. *Smart Cities*, *4*(3), 1004–1017. DOI: 10.3390/smartcities4030053

Wong, C., Wood, J., & Paturi, S. (2021). Managing waste in the smart city of Singapore. *Tropical Constrained Environments and Sustainable Adaptations: Businesses and Communities*, 225-241.

Chapter 7
AI and Robotics in Urban Infrastructure Maintenance

S. Mohana Saranya
https://orcid.org/0000-0002-6004-924X
Kongu Engineering College, India

R. Mohemmed Yousuf
Bannari Amman Institute of Technology, India

M. R. Prasanndh Raaju
Kongu Engineering College, India

Dinesh Komarasamy
Kongu Engineering College, India

R. Prabavathi
Sri Sai Ram Institute of Technology, India

M. Kavyadarshini
Kongu Engineering College, India

ABSTRACT

The integration of Artificial Intelligence (AI) and robotics is transforming urban infrastructure maintenance, addressing challenges associated with traditional methods. Predictive maintenance, real-time monitoring, and collaborative frameworks enhance efficiency and resource management. Robotics, including cobots and swarm technologies, play a pivotal role in large-scale inspection and repair tasks, ensuring precision and scalability. The chapter explores the collation of AI and robotics, emphasizing data-driven approaches and real-world case studies to highlight their potential in achieving sustainable and inclusive urban development. It also discusses environmental, economic, and social benefits while addressing challenges like technical constraints, ethical concerns, and safety. Future directions envision smarter cities with innovative AI-driven solutions.

DOI: 10.4018/979-8-3373-1892-9.ch007

1. INTRODUCTION

Urban infrastructure serves as the backbone of modern society, encompassing transportation systems, water supply networks, energy grids, waste management systems, and public facilities (Hossain et al., 2024; Hossain et al., 2025). Maintaining this critical element is essential for ensuring the seamless functioning of cities, fostering economic growth, and enhancing the quality of life for urban populations.

However, as cities expand and evolve, the combined challenges of aging infrastructure, increasing environmental concerns, and budgetary constraints are reshaping professional and public priorities (Hossen et al., 2024). Traditional maintenance methods—relying on periodic inspections and reactive repairs—are proving inadequate to address the growing complexities of urban systems. This chapter challenges the conventional notion that infrastructure maturity signifies a phase of urban and infrastructural stability. Instead, it argues that we are entering an "age of infrastructure maintenance," where the focus has shifted from expansion to repair and upkeep.

This shift, evident in an increasing number of countries, especially in the global North, marks a departure from the modern infrastructural ideal of relentless progress and growth. Drawing insights from practices such as the maintenance and asset management of French water networks, this chapter highlights two critical challenges of adopting a maintenance-cantered perspective: knowledge production practices and forms of infrastructure valuation.

The integration of Artificial Intelligence (AI) and robotics into urban infrastructure maintenance offers transformative opportunities to address these challenges. By enabling predictive maintenance, real-time monitoring, and the automation of labour-intensive processes, these technologies redefine how infrastructure systems are managed. They promise significant gains in cost efficiency, operational effectiveness, and sustainability, facilitating a data-driven approach to ensure infrastructure longevity and adaptability (Denis, 2024).

This chapter explores how AI and robotics are reshaping urban infrastructure maintenance by examining existing practices, introducing innovative AI-driven approaches, and integrating robotics into asset management. Furthermore, it delves into the implications of sustainability and the challenges associated with adopting advanced technologies, paving the way for smarter, more resilient, and inclusive urban environments. Ultimately, it envisions a future where the age of maintenance aligns with the ideals of smart cities, transforming infrastructure management into a cornerstone of urban resilience and sustainability.

2. CURRENT STATE OF URBAN INDRASTRUCTURE MAINTENANCE

Maintaining the Urban Infrastructure is very important for ensuring the effective functioning by the metrics shown in the Figure 2.1, is essential for systems like transportation, water supply, energy grids. Aging assets, environmental challenges, and financial constraints will significantly affect the effectiveness of the traditional approaches that include periodic inspections and reactive repairs. Such limitations will result in inefficiencies of a system operation and expose people to risks associated with public safety (Hossain et al., 2025). As cities become more complex, there is an increasing need to move towards more sustainable and adaptive maintenance strategies beyond the conventional practices.

2.1 Traditional Maintenance Methods

Traditional maintenance methods in urban infrastructure typically rely on routine inspections, reactive repairs, and manual monitoring. These conventional approaches have become increasingly ineffective as urban systems grow more complex and infrastructures age. Periodic inspections involve manually checking roads, pipelines, electricity grids, and other essential urban structures for damages, often missing smaller, critical issues that could later lead to system failures (Denis, 2024). Furthermore, maintenance tends to be reactive, with repairs initiated only when problems arise, leading to higher costs and delays in addressing critical failures (Venigandla et al., 2024). This approach, while commonplace, involves labor-intensive processes that demand significant human resources for monitoring and repair tasks, which adds to the operational inefficiencies (Padhiary et al., 2025). Additionally, these methods often fail to leverage modern data technologies, relying on manual assessments rather than sophisticated digital tools or real-time monitoring systems. As a result, the limited use of data impedes the ability to predict or prevent potential infrastructure failures, thus exacerbating the challenges faced by urban maintenance systems (Gkontzis et al., 2024).

Figure 1. Urban Maintenance Metrix

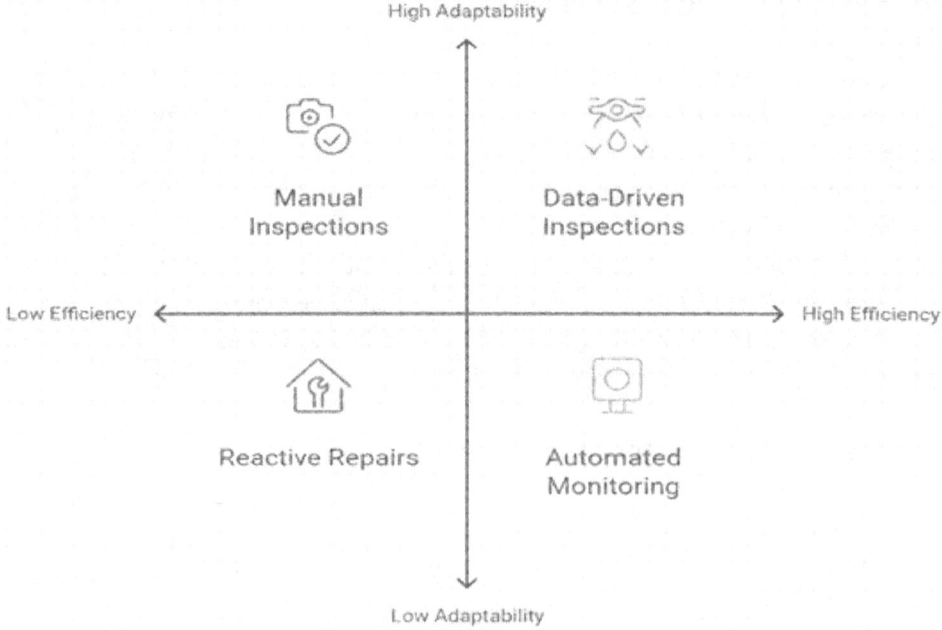

2.2 Limitations in Existing Practicecs

With the increasing complexity of urban infrastructure, the existing maintenance system, primarily based on traditional approaches such as manual inspections and reactive repairs, is becoming increasingly ineffective. These methods tend to result in significant inefficiencies, as manual inspections often fail to detect minor yet critical damage, causing delays in identifying and addressing issues (Venigandla et al., 2024). Additionally, the reliance on reactive repairs, where maintenance is only performed after a failure occurs, leads to higher operational costs due to the need for emergency interventions and the labor-intensive nature of these processes (Fathy et al., 2024). Such approaches also contribute to extended downtime, as reactive maintenance can cause prolonged system interruptions, severely affecting urban services and disrupting daily operations (Padhiary et al., 2025). Furthermore, traditional maintenance methods struggle with prioritizing repairs effectively, often resulting in delays in addressing the most urgent needs, which compromises the overall efficiency and reliability of urban systems (Gkontzis et al., 2024).

Figure 2. Disadvantages of Traditional Urban Maintenance Practices

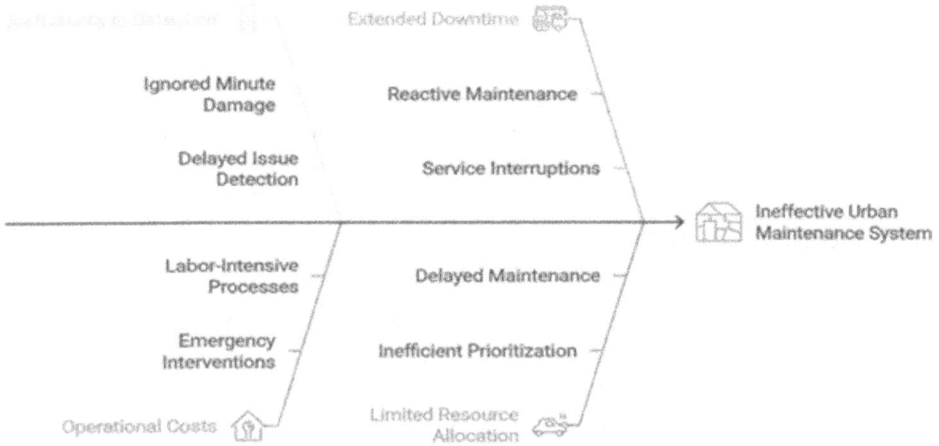

2.3 Opportunities for Automation and AI Integration

Urban infrastructure maintenance can be made more efficient through automation. This is through inspection and data collection by using drones, robots, and autonomous vehicles. They help eliminate labour work and identify needs more efficiently. Risks are also mitigated in dangerous or inaccessible places.

The AI integration further enhance maintenance by providing predictive models that predict probable failures and optimize resource allocation. Machine learning algorithms can analyse data to make proactive decisions, which reduces downtime and repair costs, thereby improving the sustainability and long-term resilience of urban systems.

Automation and AI integration in urban infrastructure maintenance enhance efficiency. Automation uses drones, robots, and autonomous vehicles for inspections and data collection, replacing manual labour and identifying maintenance needs more effectively. It also reduces risks in dangerous or hard-to-reach areas. With AI, predictive models are created to forecast potential failures, optimize resource allocation, and proactively manage urban systems. Machine learning algorithms analyse real-time data to make proactive decisions, reduce downtime, save on repairs, and contribute to the long-term sustainability and resilience of city infrastructure [Venigandla, 2024].

3. AI – DRIVEN STRATEGIES FOR MAINTENANCE

Application of AI-driven strategies in urban infrastructure maintenance aims to overcome the disadvantages of the traditional methods. This approach utilizes advanced data analytical techniques along with machine learning in order to maximize reliability and efficiency of urban systems. The three major AI driven strategies are predictive maintenance, real-time monitoring, and data analytics.

3.1 Predictive Maintenance

Predictive maintenance requires condition-monitoring tools and techniques that capture performance trends of the urban infrastructure with time. While ordinarily, maintenance may occur according to a preplanned schedule, in predictive terms, a maintenance depends upon the actual conditions of the urban infrastructure. It relies on real-time data obtained from sensors placed within the infrastructure. Sensors include those sensing the structural integrity; vibrations or movements which might point to potential failure. The process of predictive maintenance as shown in the Figure 3.1, firstly starts with the data collection from the Internet-of-Things (IoT) devises then followed by data analysis through which we can identify the anomalies and prioritize the maintenance to take the timely intervention.

Figure 3. Process flow of predictive maintenance

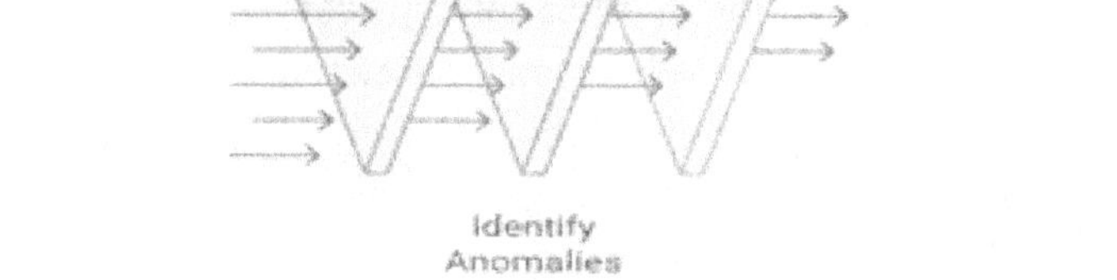

One of the key advantage of predictive maintenance is that it lowers the operational costs. Unlike conventional methods which includes scheduled checks and fixes after breakdown, predictive maintenance involves almost no downtime and eliminates a lot of expenditure. Early detection of failure enables the municipality to deploy its financial resources on specific repairs, ensuring public safety and prolonging the service life of critical assets.

An example of predictive maintenance is the Pavement Rehabilitation Predictive Model (PRPM), which allows long-term forecasting of road network's condition. It facilitates planning by identifying the most cost-effective rehabilitation policies for a road asset, both from the technical and the budgetary points of view. An empirical study comparing three rehabilitation variants found that PRPM reduces maintenance costs by as much as 20% when compared to traditional methods. Such models emphasize the effectiveness of predictive maintenance as saving costs and lengthening the life cycle of infrastructure assets [Fathy, 2024].

3.2 Real – Time Monitoring and Data Analytics

The process flow of the Real-Time Monitoring and Data Analytics is a cyclic flow as depicted in the Figure 3.2. The Real-time monitoring depends on the deployment of advanced IoT sensors and the use of digital twin techniques that continuously monitor the status of urban infrastructures. Sensors can get real-time measurements of temperature, pressure, and other forms of vibration and traffic flows etc., AI algorithms are used to analyse these collected data to detect anomalies as listed below.

AI Algorithms in Urban Maintenance

Artificial Intelligence (AI) plays a crucial role in modernizing urban maintenance systems by enabling automation, pattern recognition, and intelligent decision-making. By leveraging AI models, cities can enhance their ability to predict and manage infrastructure needs, making systems more responsive and efficient. Various AI algorithms are employed depending on the specific task, each offering distinct advantages. For example, Support Vector Machines (SVM) are effective for binary classification tasks, such as determining whether a structural anomaly is critical or benign. Random Forests and Decision Trees are commonly used in predictive analytics, especially for energy consumption and waste management, due to their ability to interpret and manage large datasets. Artificial Neural Networks (ANN) are valuable for forecasting equipment degradation and predicting failures like water pipeline bursts. Convolutional Neural Networks (CNN) are employed in image-based inspections, such as pothole detection from street camera feeds, while Long Short-Term Memory (LSTM) Networks are applied to predict time-series data like urban utility usage. Additionally, Reinforcement Learning (RL) optimizes robotic paths and adaptive traffic control, improving traffic flow and system management.

An example of AI's application can be seen in Singapore's Smart Nation initiative, where deep learning models power smart cameras to detect illegal dumping. This system enhances response times and reduces the need for manual oversight. However, successful AI deployment in urban systems requires careful consideration of several factors. Model transparency is crucial to ensure public trust, as citizens need to understand how decisions are made. Moreover, training data must accurately represent the diverse conditions of urban environments to ensure reliable AI predictions. Lastly, the use of Edge AI can reduce latency, making real-time applications, such as monitoring and decision-making, more effective in time-sensitive situations.

Digital twin technology enables the creation of virtual replicas of urban environments, facilitating simulations and scenario studies to fine-tune city performance and identify system bottlenecks. AI-powered predictive models significantly enhance these simulations by enabling scenario-based decision-making and optimization. Through digital twin technology, real-time monitoring allows for improved forecasting and analysis of urban system responses. This capability not only enhances resilience but also enables proactive management of urban challenges. By leveraging AI, forecasting is further refined through the identification of trends and patterns from real-time data. This integration ensures timely corrective actions while contributing to long-term urban planning, ultimately improving city functionality and enhancing the quality of life for residents. [Gkontzis, 2024].

Figure 4. Cyclic Process Flow of Real Time Monitoring & Data Analytics

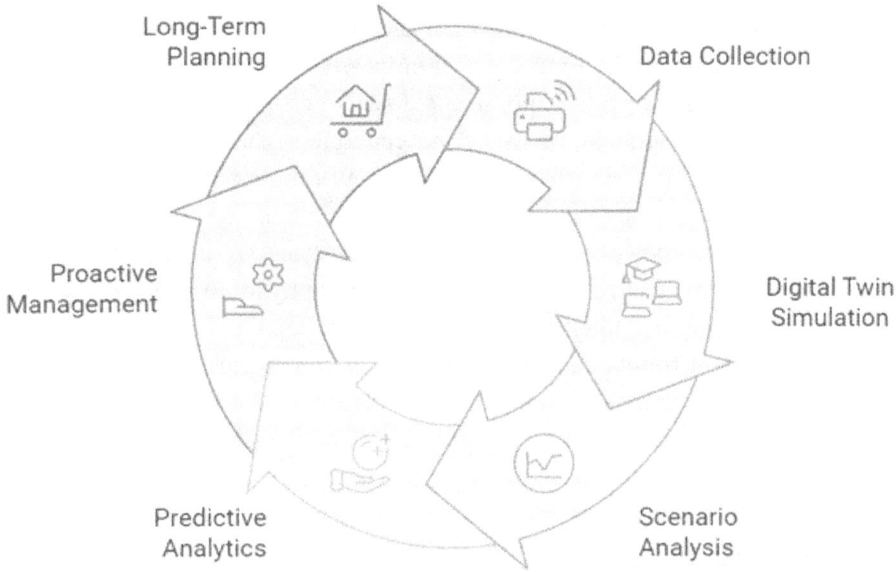

Data analytics involves processing vast amounts of data from real-time monitoring systems to identify patterns and trends that guide maintenance decisions. Advanced techniques such as machine learning and predictive modelling are used to analyse historical and real-time data to forecast future infrastructure needs and identify areas for improvement. The municipalities can optimize resource allocation, make effective planning of repairs, and allocate data-driven decisions to prolong the life of urban infrastructure without increased costs through the leverage of data analytics.

4. ROBOTICS AND URBAN MAINTENANCE

The integration of robotics into urban framework maintenance implies a world-view move in managing and maintaining basic urban resources. Robotics improves accuracy, operational productivity, and the capacity to perform tasks in hazardous or remote locations. By addressing challenges over the lifecycle of infrastructure maintenance, robotics drives advancement and ensures long-term maintainability for urban frameworks [Padhiary, 2025]. The development of Artificial Intelligence (AI) and Internet of Things (IoT) technologies has further empowered groundbreaking urban network, contributing essentially to smart city activities.

4.1 Role of Inception and Repair

Robotic frameworks equipped with advanced sensors, cameras, and imaging technologies as shown in the Figure 4.1, it empowers comprehensive inspections. These frameworks give detailed appraisals of structural keenness, detect potential issues such as cracks, erosion, or wear, and generate high-resolution information for arranging and intervention. Drones and ground robots are commonly utilized for studying hard-to-reach zones, such as bridges, tunnels, or high-rise buildings. Moreover, robotic frameworks can perform non-destructive testing (NDT) strategies such as ultrasonic testing and infrared thermography to distinguish subsurface defects, guaranteeing that potential weaknesses are recognized before they become critical. Independent robots can explore hazardous territories, such as crumbling streets or unsteady structures, guaranteeing that the inspection process is both secure and careful. The information collected is then nourished into predictive maintenance models, permitting urban planners and engineers to make informed decisions around necessary repairs or reinforcements.

In the repair stage, robotics exceeds expectations in executing precise and often hazardous tasks. Examples incorporate automated welding of damaged structures, applying protective coatings to amplify fabric lifespans, and robotic arms repairing underground pipelines. Robotic frameworks can moreover replace or reestablish components in dangerous situations, such as high-voltage electrical frameworks or gas pipelines, where human intervention would posture significant dangers. Drones prepared with specialized tools such as, cutting and drilling tools, coating and spraying tools, cable management tools, fastening and assembling tools etc., can perform high-altitude repairs, such as supplanting defective components on power lines or communication towers. By minimizing human association in risky conditions, robotics guarantees higher security and consistency in repairs whereas decreasing operational downtime. The integration of IoT gadgets enhances real-time observing and predictive analytics during these phases, guaranteeing convenient maintenance and amplifying the lifespan of urban framework. Moreover, the use of robotic exoskeletons or teleoperated frameworks can help human labourers by expanding their quality and accuracy, empowering them to perform errands that would otherwise be as well strenuous or dangerous. These advances also guarantee that repairs are conducted with minimal disturbance to ongoing urban activities, such as traffic or business operations.

Figure 5. Robotics in urban maintenance

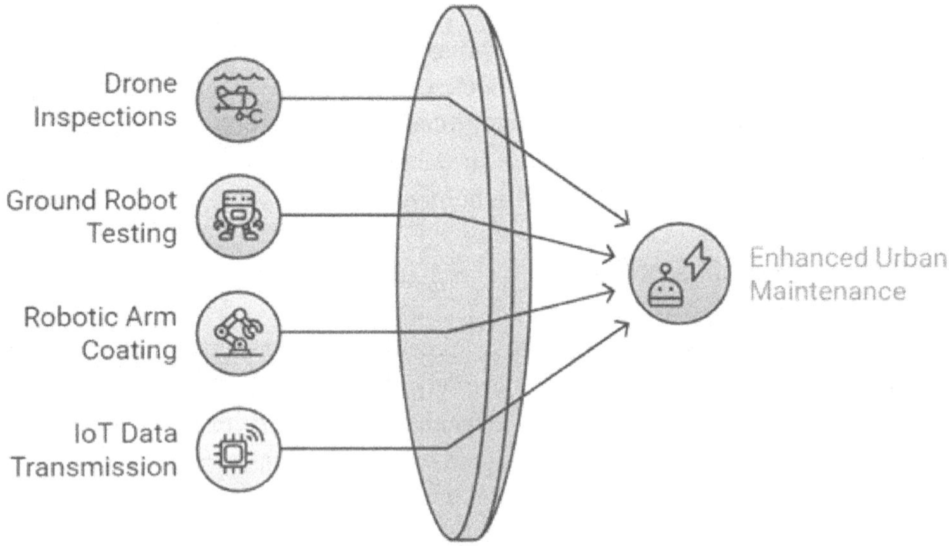

Drones prepared with specialized tools such as, cutting and drilling tools, coating and spraying tools, cable management tools, fastening and assembling tools etc., can perform high-altitude repairs, such as supplanting defective components on power lines or communication towers. By minimizing human association in risky conditions, robotics guarantees higher security and consistency in repairs whereas decreasing operational downtime. The integration of IoT gadgets enhances real-time observing and predictive analytics during these phases, guaranteeing convenient maintenance and amplifying the lifespan of urban framework. Moreover, the use of robotic exoskeletons or teleoperated frameworks can help human labourers by expanding their quality and accuracy, empowering them to perform errands that would otherwise be as well strenuous or dangerous. These advances also guarantee that repairs are conducted with minimal disturbance to ongoing urban activities, such as traffic or business operations.

Beyond the quick stages of inspection and repair, robotics contributes to long-term foundation maintainability. Independent robots, prepared with machine learning capabilities, can persistently screen basic wellbeing, independently identify wear and tear, and adjust to evolving maintenance needs. These robots, coordinates with smart sensors, are capable of performing continuous assessments and issuing alarms when conditions change, permitting for prescient maintenance cycles that minimize cost and maximize the life expectancy of foundation. By robotizing tedious tasks such as cleaning, painting, or joint sealing, robots not only enhance the efficiency of maintenance but moreover guarantee consistency in quality over time [Caleb, 2025].

4.2 COBOTS (Collaborative Robots)

Collaborative robots, commonly known as cobots, are redefining the future of urban maintenance by enabling safe, effective, and human-friendly collaboration between robots and people. Unlike traditional industrial robots that operate in isolation, cobots are designed with built-in safety features and intelligent systems, allowing them to work side-by-side with human operators.

Core Characteristics of Cobots:

Collaborative robots, or cobots, are designed with several core characteristics that enable them to work safely and effectively alongside humans. One of the key features of cobots is their emphasis on safety and human interaction. Equipped with proximity sensors, force-limiting joints, and emergency stop mechanisms, cobots can detect and respond to human presence, ensuring that they operate in a safe and co-working environment. This safety-oriented design allows for seamless collaboration between robots and human workers, making them ideal for tasks where human intervention is necessary, yet safety cannot be compromised.

In addition to their safety features, cobots are equipped with AI-driven flexibility. The integration of AI capabilities allows cobots to understand the context of their environment, adapt to different tasks, and even learn from previous operations. This adaptability significantly enhances their usability across a wide range of urban maintenance scenarios, making them versatile tools that can perform multiple tasks efficiently. Whether it's repairing infrastructure, cleaning, or assisting in routine maintenance, cobots can adjust to evolving needs and improve over time.

Cobots also feature intuitive interfaces that make them accessible to non-specialist workers. With easy-to-use programming options like drag-and-drop apps or voice commands, cobots can be deployed and managed with minimal training, allowing workers without specialized technical skills to operate them effectively. This ease of use empowers a broader range of workers to harness the potential of robotics, reducing the learning curve and increasing the efficiency of urban maintenance tasks..

Urban Maintenance Applications

Cobots are highly effective in urban maintenance tasks that are repetitive, ergonomically challenging, or require precision and flexibility. They are particularly useful for applications like graffiti removal and wall cleaning, where they utilize automated spray or scrubbing systems, guided by vision systems and AI to detect affected areas. Cobots also assist in building facade maintenance, where their compact design allows them to climb scaffolding or work on aerial platforms, cleaning or repairing building exteriors while minimizing risks to human workers and ensuring consistency. Additionally, cobots play a crucial role in component assembly within confined spaces, such as metro stations, tunnels, or maintenance ducts, where they work alongside technicians to assemble or repair mechanical and electrical components, enhancing productivity in environments with limited human mobility.

Advantages of Deploying Cobots:

Deploying cobots in urban maintenance offers several advantages. Their lightweight and compact design make them highly space-efficient, making them ideal for dense urban environments where spatial constraints are a challenge. Cobots also enhance worker safety and comfort by taking over physically strenuous or repetitive tasks, reducing the risk of injuries and fatigue. Additionally, they improve accuracy and reduce errors, thanks to AI-powered sensors and real-time feedback loops, which help minimize human mistakes during routine maintenance operations. Furthermore, cobots contribute to resource optimization through AI analytics by collecting operational data that can be analyzed to optimize maintenance schedules, predict resource needs, and enhance task planning and logistics, making urban maintenance more efficient and cost-effective.

Cobots not only streamline urban maintenance but also empower human workers to shift toward more strategic, decision-oriented roles, thus promoting a more intelligent and collaborative workforce model in smart cities [Sanchez (2024)]. Cobots represent a balanced integration of human intuition and robotic precision, making urban maintenance more safe, efficient, and intelligent. Their deployment contributes to workforce augmentation rather than replacement, positioning them as key assets in building sustainable and adaptive urban environments.

Figure 6. Work of cobots

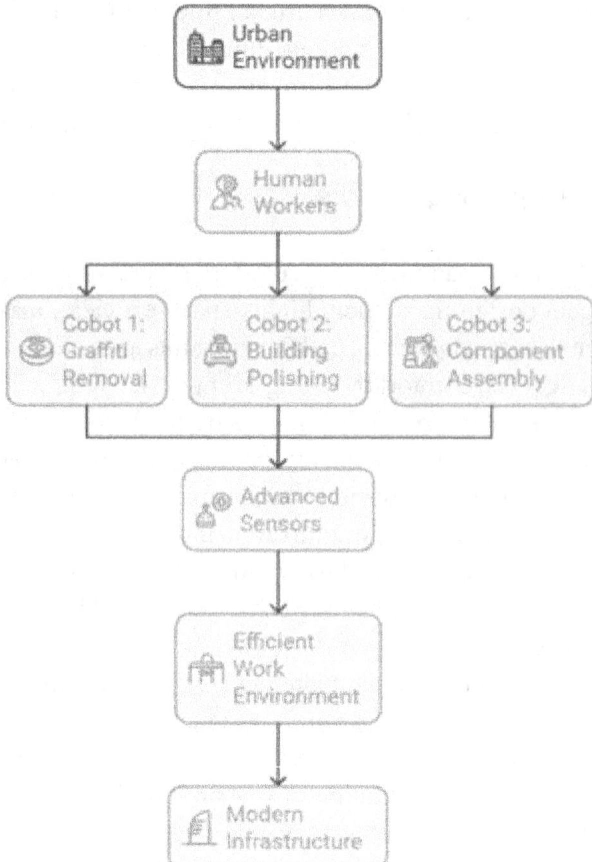

4.3 SWARN Robotics (Large – Scale Tasks)

Swarm robotics is an emerging field inspired by the behavior of social insects like ants, bees, and termites. It leverages decentralized control, local interactions, and collective intelligence to perform complex tasks using a large group of relatively simple robots. This model offers a scalable, fault-tolerant, and adaptive approach to solving large-scale urban infrastructure problems.

Principles behind SWARN Robotics:

- **Decentralized Control:** Each robot operates independently using local data, without relying on a central coordinator. This reduces system vulnerabilities and enables autonomous operation even in dynamic environments.

- **Self-Organization:** Swarm members follow simple rules, but through local interactions, they collectively achieve complex objectives.
- **Scalability and Redundancy:** Adding more robots improves performance without the need for reprogramming, and individual robot failure does not compromise the entire system.

Key Applications in Urban Maintenance

Swarm robotics is particularly effective in scenarios that require coverage over wide areas, parallel operations, or quick adaptation to changing environments. Major use cases include:

1. **Urban Space Cleaning:** Swarm robots equipped with sensors and cleaning modules can autonomously clean large public spaces such as parks, stadiums, or streets. Their distributed nature ensures even coverage and quick adaptation to foot traffic patterns.
2. **Infrastructure Inspection (e.g., Highways, Bridges, Pipelines):** Equipped with cameras, ultrasonic, or thermal sensors, swarm robots can be deployed to inspect extensive linear infrastructure for cracks, corrosion, or deformation. They coordinate to split coverage areas, avoid redundancy, and compile data into centralized AI systems for analysis.
3. **Disaster Response and Search & Rescue:** In post-disaster scenarios (earthquakes, floods, industrial accidents), swarm robots can enter hazardous or collapsed zones to:
    - Map the area with onboard LiDAR or SLAM systems.
    - Detect human presence using thermal sensors.
    - Communicate findings back in real time to rescue teams via mesh networking.

Functionality and Technical Strengths:

- **Dynamic Task Redistribution:** Swarm members communicate locally (via short-range wireless links like Bluetooth or Zigbee), enabling them to reassign tasks on the fly when one unit fails or encounters obstacles.
- **Environmental Adaptability:** Robots can adapt their paths and roles based on immediate surroundings and sensor data, making them well-suited for unpredictable urban conditions.
- **Real-Time Coordination via IoT Integration:** Swarm systems linked with IoT infrastructures can:
- Access and share sensor data across the city.

- Interact with smart infrastructure (e.g., traffic systems, environmental sensors).
- Enable real-time updates to optimize tasks and routes.

The integration of swarm robotics with AI and IoT infrastructure positions it as a transformative tool for handling complex, large-scale urban maintenance challenges in a highly autonomous, resilient, and efficient manner [Maffettone (2024)].

Swarm robotics presents a decentralized, flexible, and intelligent method for addressing the growing complexity of urban maintenance tasks. As cities become more connected and data-driven, the fusion of swarm intelligence with real-time IoT systems will be essential for scalable, smart, and sustainable urban management.

5. COLLATION OF AI AND ROBOTICS IN URBAN MAINTENANCE

The cooperative energy between AI and robotics represents a transformative approach to urban maintenance as shown in the Figure 5.1. By joining these progressed innovations, cities can accomplish a consistent and data-driven framework that improves productivity, decreases costs, and advances supportability. In this topic we shall investigate about the collaborative frameworks, data-driven robotics, and real-world case studies that highlight the potential of this integration.

Figure 7. Vision of AI and robotics in urban maintenance

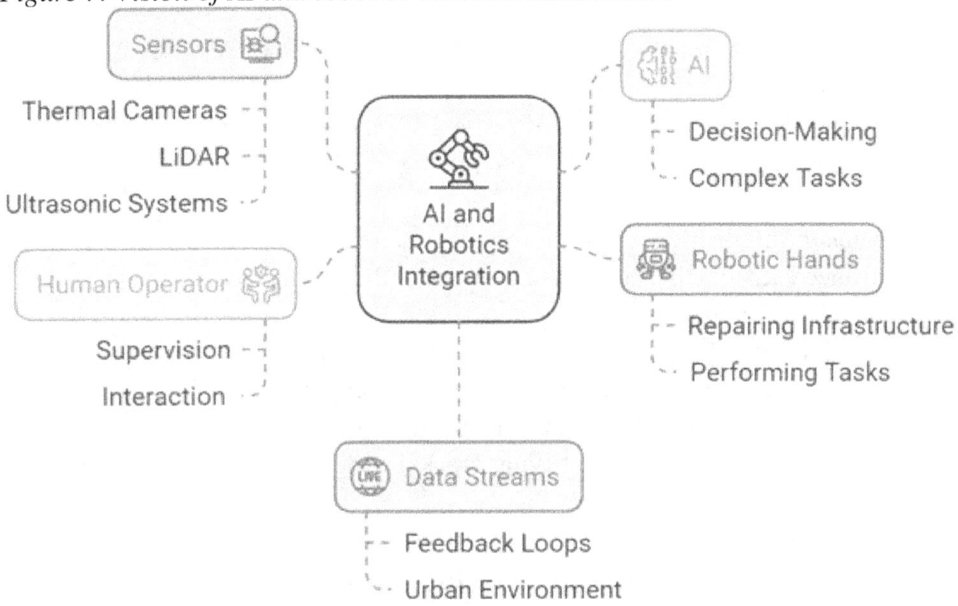

5.1 Collaborative Framework

The integration of AI and robotics requires a well-defined collaborative system to guarantee their consistent interaction and viability:

Coordination Between AI and Robotics: AI serves as the brain, preparing information and creating insights, whereas robotics acts as the hands, executing errands based on AI-driven choices. This collaboration empowers prescient analytics, versatile support plans, and robotized task execution.

Human-Robot Collaboration: In urban maintenance, human skill remains crucial. AI and robotics complement human endeavors by robotizing repetitive and hazardous tasks, whereas people direct complex decision-making and give vital direction.

Interoperability: A collaborative system emphasizes interoperability between AI calculations, robotic platforms, and existing urban framework systems. Open guidelines and secluded plans encourage integration and versatility.

5.2 Data – Driven Robotics

Data is the foundational element that enables AI and robotics to work in synergy, allowing systems to make informed decisions, detect issues early, and streamline urban maintenance operations. The effective collection, analysis, and utilization of data is what turns automated systems into intelligent, adaptive tools for smart cities.

Sensor Integration: Modern robotic systems are equipped with a suite of advanced sensors that continuously monitor the condition of urban infrastructure. These include:

- **Thermal Cameras** – Detect overheating or temperature anomalies in electrical systems, pipes, or structural components.
- **LiDAR (Light Detection and Ranging)** – Creates high-resolution 3D maps of surfaces, identifying cracks, surface deformations, and structural misalignments.
- **Ultrasonic Sensors** – Penetrate materials to detect internal faults like voids, leaks, or corrosion in walls, pipes, and foundations.

These sensors generate large volumes of real-time data, which is immediately fed into AI algorithms for interpretation and anomaly detection. Real-time data enables quicker detection of critical issues, supports preventive action, and drastically reduces the risk of infrastructure failure.

Machine Learning and Analytics: AI models, particularly those powered by machine learning, process and analyze the sensor data to extract meaningful patterns. This involves:

- **Training AI models on historical maintenance and failure data**, enabling systems to recognize early warning signs.
- **Deploying predictive algorithms** that assess the likelihood of system degradation or breakdowns based on current trends and environmental factors.
- **Adaptive learning models** that evolve over time, improving their accuracy and reliability with each data cycle.

Predictive maintenance strategies informed by machine learning significantly cut down on unexpected infrastructure failures and reduce long-term operational costs. Maintenance becomes pre-emptive rather than reactive, leading to better resource allocation, fewer emergency repairs, and extended asset lifecycles (Wang, 2024).

Feedback Circles: A hallmark of AI-robotics integration is the **closed feedback loop**, where robotic systems and AI algorithms interact in real time to refine operations. The process includes:

- **Robots collecting operational and environmental data** during maintenance tasks.
- **AI models using this feedback** to fine-tune control commands, navigation, and task execution strategies.
- **Reinforcement learning mechanisms**, where robots learn from previous actions to improve future performance.

This dynamic learning environment ensures that maintenance operations are not only automated but intelligently adaptive. Systems become more precise, efficient, and context-aware, enhancing both the quality and speed of urban maintenance tasks.

The successful integration of data, AI, and robotics is transforming urban maintenance from a manual, time-consuming process into an intelligent, predictive, and efficient system. While the benefits are substantial, the implementation also introduces complexities in terms of data privacy, system interoperability, and model transparency. These opportunities and challenges are explored in greater detail in Figure 5.2 and in the subsequent sections of this chapter.

Figure 8. Pros & cons due to integration of AI and robotics

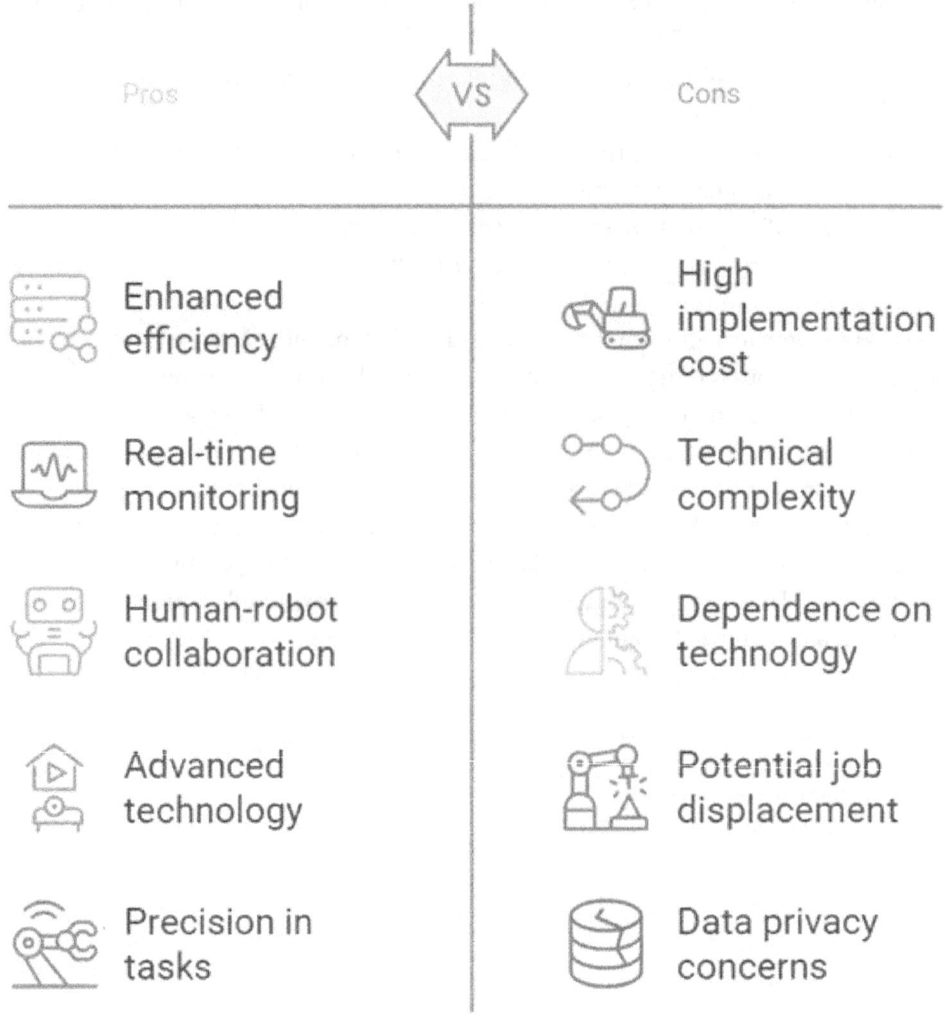

5.3 Case Studies

The integration of Artificial Intelligence (AI) and robotics is revolutionizing urban maintenance and infrastructure management, offering enhanced efficiency, safety, and sustainability. Several case studies highlight the real-world applications of these technologies. In urban centers, autonomous drones equipped with AI-driven computer vision systems are transforming bridge inspections. Traditionally, these inspections required human workers, scaffolding, and traffic disruption, but AI-powered drones with high-resolution cameras, LiDAR, and thermal sensors now detect structural defects, such as cracks and corrosion, and compare current data with historical maintenance records. This approach reduces inspection time by over 50%, minimizes human safety risks, and lowers the costs associated with manual inspections and equipment rentals.

In pipeline maintenance, robotic crawlers integrated with AI navigate complex underground networks, equipped with HD cameras, ultrasonic sensors, and chemical detectors to monitor pipeline integrity. AI-based analytics process this data to detect leaks, rust, and blockages while predictive maintenance models recommend timely repairs, reducing downtime and extending the lifespan of pipeline infrastructure. This system also enhances environmental safety by quickly identifying hazardous leaks.

Cities like Singapore and Amsterdam are implementing smart waste management systems that combine AI with IoT technologies. IoT-enabled smart bins measure fill levels and movement, while AI algorithms analyze waste generation patterns across city zones. These systems optimize garbage truck routes and collection schedules, reducing fuel consumption, lowering carbon emissions, and cutting operational costs by 30-40%. They also ensure cleaner public spaces and more timely waste collection.

AI-robotics systems are also crucial in urban disaster response and recovery. During events such as earthquakes or floods, robots equipped with AI-powered vision and mapping systems assess structural damage, identify inaccessible areas using thermal imaging and 3D scanning, and support search-and-rescue missions. Japan's use of robotic systems after the tsunami exemplifies how these technologies provide situational awareness, speed recovery efforts, and reduce risks for human responders.

By combining AI and robotics, urban maintenance becomes more proactive, adaptable, and resilient. These case studies illustrate how the synergy between these technologies is reshaping urban infrastructure maintenance, improving efficiency, enhancing safety, and fostering smarter, more sustainable cities.

6. SUSTAINABILITY AND INCLUSIVITY THROUGH COLLABORATION

Sustainability and inclusivity are basic pillars of urban advancement, requiring a collective approach where governments, private organizations, non-profits, and communities work together. This collaboration guarantees that cities evolve into versatile, resilient ecosystems capable of tending to natural, financial, and social challenges. Through shared responsibility, innovative solutions, and vital partnerships, collaborative endeavors are reshaping urban maintenance practices to prioritize natural preservation, financial development, and social value. The deployment of

AI and robotic technologies in city maintenance contributes significantly to environmental sustainability. These systems enable better resource management, reduce emissions, and minimize waste.

Quantifiable Impacts:

Smart waste bins equipped with fill-level sensors have reduced garbage collection frequency by up to 40%, lowering fuel usage and CO_2 emissions.

AI-driven energy optimization in buildings has led to energy savings of 15–25% in pilot smart city programs.

Drones used for monitoring air and water quality have improved regulatory compliance and enabled quicker identification of pollution sources.

These technologies contribute to sustainable urban development by optimizing operations, promoting circular economy principles, and ensuring environmental compliance with less human intervention.

6.1 Environmental Benefits

Collaboration among different stakeholders is key to achieving environmental supportability in urban maintenance. By pooling resources and skills, cities can actualize arrangements that minimize natural harm whereas advancing ecological conservation. For instance, organizations between neighborhood governments and private ventures have driven to the selection of renewable energy in urban frameworks, decreasing the carbon impression of city operations. Electrification of public transportation and civil fleets, supported by public-private subsidizing, has moreover gotten to be a foundation of economical urban planning. As depicted in the Figure 6.1, this approach aligns with the principles of the circular economic model, emphasizing resource optimization and waste minimization.

Figure 9. Circular economic model

Urban greening activities, such as tree planting drives, vertical gardens, and the rebuilding of urban wetlands, benefit enormously from collaborative endeavors including environmental organizations, neighborhood authorities, and citizen groups. These projects not only upgrade air quality and biodiversity but moreover mitigate urban heat island impacts. Moreover, progressions in waste administration frameworks, such as smart waste containers associated to IoT systems, have been made possible through innovative organizations. These frameworks optimize waste collection routes, decrease energy utilization, and empower proficient reusing, demonstrating how collaboration drives natural progress [Yigitcanlar, 2020].

6.2 Economic Advantages

Collaboration is a driving force behind the financial supportability of urban systems. Associations between governments and private organizations encourage the improvement and scaling of inventive urban technologies, such as AI-powered monitoring systems and robotics for foundation support. These progressions not only diminish operational costs but moreover attract ventures that fuel assist development. Collaborative research activities empower the discovery of cost-effective arrangements, such as self-repairing materials for foundation, which essentially lower long-term maintenance costs.

Economic advantages moreover arise from the shared utilization of assets. For occurrence, joint procurement agreements for sustainable materials permit cities to benefit from economies of scale, whereas the integration of prescient support systems minimizes costly crisis repairs. Collaboration moreover makes business openings, especially in segments like green technology and feasible urban administration. Workforce upskilling programs, supported by associations between educational institutions and businesses, prepare inhabitants for rising work parts, cultivating financial versatility.

6.3 Social Impact

The social impact of collaborative urban maintenance expands far beyond the immediate benefits of improved infrastructure. By including diverse partners, cities can guarantee that urban frameworks cater to the needs of all inhabitants, cultivating inclusivity and value. Citizen-centric arrangements created through community engagement sessions permit individuals to voice their concerns and contribute to urban planning forms. For example, portable applications created through public-private associations give inhabitants with real-time updates on maintenance exercises, guaranteeing straightforwardness and responsibility.

Collaboration too engages marginalized communities by offering inclusive opportunities for cooperation in urban ventures. Activities like urban farming, driven by community groups with bolster from non-profits and local governments, improve nourishment security and financial strengthening. Cleaner open spaces improved sanitation, and equitable access to fundamental administrations like healthcare contribute to upgraded social well-being and security, cultivating a sense of having a place among inhabitants.

Besides, collaborative efforts play a vital part in disaster flexibility. Community-led readiness programs, backed by non-governmental organizations and government organizations, equip inhabitants with the tools and information to reply successfully to normal disasters. Advanced technologies, such as AI-robotics integration for search-and-rescue missions, encourage support urban strength, minimizing the effect of unanticipated occasions.

7. FUTURE DIRECTION

The future of AI and robotics in urban foundation maintenance is set to be a cornerstone within the advancement of smart cities. This change aligns with the broader objectives of making interconnected, maintainable, and flexible urban ecosystems that are responsive to both current and future challenges. The integration of AI and robotics in urban maintenance frameworks guarantees significant advancements in operational effectiveness, exactness, and versatility, tending to the complexities and demands of advanced urban environments.

7.1 Smart Cities Vision

Smart cities are characterized by their use of interconnected framework and intelligent systems that improve the quality of life for urban inhabitants whereas promoting maintainability. The vision of a smart city intensely depends on innovations such as Robotics and AI, which give the establishment for robotizing and optimizing urban administrations. IoT connects physical objects utilizing sensors, hardware, and communication systems, changing city foundation into energetic frameworks capable of reacting to real-time information. These interconnecting permits cities to oversee assets more effectively and improve service delivery.

AI plays a vital part by analysing tremendous amounts of IoT-generated information, empowering more brilliant decision-making for urban administration. Through progressed AI calculations, cities can optimize traffic flow, manage waste, and screen environmental conditions. Robotics, when coordinates into this ecosystem, can handle maintenance errands such as road repairs, waste administration, and basic inspections. For example, autonomous drones and robots can monitor infrastructure wellbeing in real-time, essentially improving the speed and precision of maintenance work whereas decreasing costs and improving security.

In addition, AI and robotics can bolster green activities in smart cities. They contribute to energy-efficient urban administration by computerizing systems that monitor energy use, waste disposal, and air quality, in this way adjusting with worldwide sustainability goals. AI-enabled frameworks can distinguish pollutants, estimate natural conditions, and suggest policies to mitigate negative impacts on urban environments. These advances empower cities to progress their supportability and efficiency, contributing to a much better quality of life for their inhabitants [Alahi, 2023].

As 5G systems continue to advance, they will assist back the smart city vision by giving high-speed, low-latency communication, vital for real-time information exchange and independent system operation. By coupling 5G with AI, cities can upgrade their ability to monitor framework and react rapidly to emerging issues, hence guaranteeing more prominent productivity in urban administration.

7.2 Research and Development

Advancements in AI and robotics for urban foundation maintenance require strong research and development (R&D) to address specialized, ethical, and security challenges. Proceeded innovation in areas such as AI-driven prescient analytics, independent navigation, swarm intelligence, and progressed sensors will be basic for guaranteeing that these advances can work consistently in complex urban situations.

One critical region of R&D is the improvement of progressed materials and energy-efficient robotics. As cities develop and face modern challenges, it is pivotal that automated frameworks are outlined to resist harsh urban conditions whereas remaining energy-efficient. Research in materials science will focus on making lightweight, strong, and cost-effective components that can expand the lifespan of robots and diminish the require for frequent support.

Predictive analytics, powered by AI, is anticipated to play a key part within the future of urban maintenance. Machine learning models will be able to analyze information from IoT sensors deployed all through the city to estimate when maintenance is required, in this way avoiding expensive repairs and downtime. By predicting infrastructure failures before they happen, cities can decrease the risk of accidents and progress the in general security and reliability of urban frameworks.

Swarm intelligence, where multiple robots collaborate to perform errands, is another zone of great promise. In urban maintenance, a swarm of drones or robots may well be deployed at the same time to perform inspections, repairs, or observation errands, moving forward effectiveness and diminishing downtime. These frameworks will be empowered by AI calculations that permit robots to communicate and work together, learning from each other to complete tasks quicker and more viably.

Moreover, the integration of 5G systems will be vital for supporting these progressions. 5G's high-speed network and low-latency capabilities will empower real-time communication between robots and IoT gadgets, permitting them to function independently and make moment choices based on approaching information. This will be especially critical in keen cities, where robots and AI frameworks will be responsible for keeping up a wide range of urban framework, from roads and bridges to utilities and open services.

Collaboration between academia, industry, and government will be fundamental for overcoming these challenges and driving advancement in AI and robotics. Tending to the moral suggestions, such as security concerns and work displacement, will require thoughtful policy improvement. Moreover, preparing programs focused on upskilling workers in robotics, AI, and IoT innovations will be crucial for guaranteeing the smooth integration of these innovations into existing frameworks [Padhiary, 2025].

The integration of IoT and AI in smart city foundations moreover presents openings to optimize communication innovations. Research into remote communication systems, such as machine-to-machine (M2M) communication and sensor systems, will be key to enhancing the adequacy of IoT deployments. By determining the most suitable communication technologies for particular urban use cases, cities can guarantee consistent network and interoperability among different smart devices and frameworks.

8. CONCLUSION

The integration of AI and robotics in urban maintenance is no longer a futuristic ideal—it is rapidly becoming a practical necessity. From predictive maintenance to autonomous cleaning systems and environmental monitoring, these technologies are reshaping how cities function, offering improved efficiency, safety, and sustainability.

This chapter has highlighted the transformative potential of AI-powered solutions and robotic systems, while also acknowledging the pressing challenges such as cost, interoperability, and ethical considerations. However, with strategic planning, collaborative governance, and citizen engagement, these barriers can be effectively addressed.

Looking ahead, several exciting research directions are worth exploring:

AI model transparency and explainability, especially in critical urban operations
Adaptive learning systems that evolve based on city-specific data patterns
Edge-AI for real-time decision-making in constrained urban environments
Cross-sector collaborations that unify public, private, and academic efforts in urban AI deployment

Ultimately, the goal should not merely be technological advancement, but the creation of smarter, safer, and more inclusive urban spaces. Continued innovation, paired with responsible implementation, will be key to realizing the full potential of AI and robotics in urban maintenance and beyond. The vision of smart cities, driven by interconnected IoT frameworks, AI, and autonomous robots, is rapidly getting to be a reality. These technologies will not only streamline urban maintenance but moreover contribute to green activities, vitality productivity, and the achievement of worldwide sustainability objectives. By harnessing the control of AI to anticipate framework failures and optimize maintenance plans, cities can move from responsive to proactive administration, amplifying the lifespan of basic framework and decreasing the event of expensive emergencies.

However, the effective execution of these advances requires critical progressions in research and advancement, focusing on energy-efficient robotics, advanced materials, AI-driven analytics, and collaborative robotic frameworks. As these advances evolve, continuous venture in R&D, beside cross-sector collaboration, will guarantee that AI and robotics can be consistently coordinates into urban frameworks. Additionally, tending to moral considerations, guaranteeing security, and upskilling the workforce will be fundamental in supporting the transition to an AI-driven urban landscape.

In conclusion, long run of AI and robotics in urban framework support is bright, promising a more effective, feasible, and flexible future for cities around the world. By progressing these innovations and fostering advancement over academia, industry, and government, we will construct smarter, greener cities that not only meet the needs of today but are also prepared to handle the challenges of tomorrow.

REFERENCES

Alahi, M. E. E., Sukkuea, A., Tina, F. W., Nag, A., Kurdthongmee, W., Suwannarat, K., & Mukhopadhyay, S. C. (2023). Integration of IoT-enabled technologies and artificial intelligence (AI) for smart city scenario: Recent advancements and future trends. *Sensors (Basel), 23*(11), 5206. DOI: 10.3390/s23115206 PMID: 37299934

Denis, J., & Florentin, D. (2024). *"Urban infrastructures' maturity and the age (s) of maintenance," in Handbook of Infrastructures and Cities*. Edward Elgar Publishing.

Fathy, A. E. A. E., Abdel-Hamid, M., & Abdelhaleem, H. M. (2024). Developing a predictive maintenance model for the rehabilitation of roads' assets. *International Journal of Construction Management*, •••, 1–11.

Gkontzis, A. F., Kotsiantis, S., Feretzakis, G., & Verykios, V. S. (2024). Enhancing urban resilience: Smart city data analyses, forecasts, and digital twin techniques at the neighborhood level. *Future Internet, 16*(2), 47. DOI: 10.3390/fi16020047

Hossain, I., Haque, A. K. M., Ullah, S. M., & Salehin, A. (2025). Assessing the Role of Urban Governance in Promoting Sustainable Cities and Communities: Evidence from Four City Corporations in Bangladesh. *International Journal of Community Well-being*, •••, 1–28. DOI: 10.1007/s42413-025-00247-2

Hossain, I., Haque, A. M., & Ullah, S. A. (2024). Policy evaluation and performance assessment for sustainable urbanization: A study of selected city corporations in Bangladesh. *Frontiers in Sustainable Cities, 6*, 1377310. DOI: 10.3389/frsc.2024.1377310

Hossen, M. S., Haque, A. M., Hossain, I., Haque, M. N., & Hossain, M. K. (2024). Towards comprehensive urban sustainability: Navigating predominant urban challenges and assessing their severity differential in Bangladeshi city corporations. *Urbanization. Sustainability Science, 1*(1), 1–17. DOI: 10.1108/USS-10-2023-0009

Macrorie, R., Marvin, S., & While, A. (2021). Robotics and automation in the city: A research agenda. *Urban Geography, 42*(2), 197–217. DOI: 10.1080/02723638.2019.1698868

Maffettone, G. C., Liguori, L., Palermo, E., Di Bernardo, M., & Porfiri, M. (2024). Mixed reality environment and high-dimensional continuification control for swarm robotics. *IEEE Transactions on Control Systems Technology.*

Padhiary, M., Roy, P., & Roy, D. (2025). *"The Future of Urban Connectivity: AI and IoT in Smart Cities," in Sustainable Smart Cities and the Future of Urban Development.* IGI Global Scientific Publishing.

Pollak, A., Gupta, A., & Göhlich, D. (2024). Optimized Operation Management With Predicted Filling Levels of the Litter Bins for a Fleet of Autonomous Urban Service Robots. *IEEE Access : Practical Innovations, Open Solutions*, *12*, 7689–7703. DOI: 10.1109/ACCESS.2024.3352436

Sanchez-Calleja, I., Martínez-Gutiérrez, A., Ferrero-Guillén, R., Díez-González, J., & Perez, H. (2024). Contact system method for the precise interaction between cobots and mobile robots in smart manufacturing. *International Journal of Precision Engineering and Manufacturing*, *25*(2), 303–318. DOI: 10.1007/s12541-023-00907-3

Venigandla, K., Vemuri, N., & Aneke, E. N. (2024). Empowering Smart Cities with AI and RPA: Strategies for Intelligent Urban Management and Sustainable Development. *Valley Int. J. Digit. Libr.*, *2024*(4), 1117–1125. DOI: 10.18535/ijsrm/v12i04.ec02

Wang, H. (2024). Application of new features based on artificial intelligent robot technology in medium-scale urban design pedigree and intelligent management and control. *Intelligent Systems with Applications*, *22*, 200379. DOI: 10.1016/j.iswa.2024.200379

Yigitcanlar, T., & Cugurullo, F. (2020). The sustainability of artificial intelligence: An urbanistic viewpoint from the lens of smart and sustainable cities. *Sustainability (Basel)*, *12*(20), 8548. DOI: 10.3390/su12208548

Chapter 8
Key Barriers to the Extensive Adoption of Robotics in the Nigerian Built Environment

Oluwadamilare Olamide Ilesanmi
https://orcid.org/0009-0002-7916-8932
Federal University of Technology, Akure, Nigeria

Ayodeji Emmanuel Oke
Federal University of Technology, Akure, Nigeria

Dorcas T. Moyanga
Federal University of Technology, Akure, Nigeria

John Aliu
University of Georgia, Athens, USA

Clinton Aigbavboa
University of Johannesburg, South Africa

ABSTRACT

Robotics has revolutionized global construction, boosting productivity and safety while advancing sustainable cities. However, its adoption in Nigeria faces major obstacles. This study identifies key barriers hindering robotics adoption in the Nigerian built environment. A quantitative approach was employed, using a structured questionnaire based on a comprehensive literature review. Data analysis involved mean item score, exploratory factor analysis, the Mann-Whitney U test, and frequency

DOI: 10.4018/979-8-3373-1892-9.ch008

distribution. Results highlight electricity supply limitations, high initial costs, and workforce displacement concerns as primary barriers. Four critical barrier groups emerged: investment in innovation, environmental impediments, technological gaps, and resource and risk concerns. The Mann-Whitney U test revealed significant differences in perceptions between indigenous and foreign firms. Addressing these challenges can drive robotics adoption, enhancing efficiency, safety, and innovation while fostering sustainable development in Nigeria's construction sector.

INTRODUCTION

Achieving the Sustainable Development Goals (SDGs) in the Age of Industry 4.0 requires innovative technological integration across all sectors, including construction (Wei et al., 2023). However, the Nigerian construction industry faces significant challenges in adopting robotics, including high costs, inadequate infrastructure, and workforce limitations (Oke et al., 2023). Labor shortages, low productivity, material waste, and unsafe working conditions persist, largely due to the sector's dependence on manual and repetitive tasks (Haung et al., 2022; Mohapatra 2024). These challenges endure despite efforts to incorporate prefabrication, automation, and robotics, which remain underutilized (Pradhananga et al., 2021). Robotics and automation are revolutionizing construction methods, enhancing sustainability, and presenting opportunities for improving infrastructure systems (Gharbia et al., 2020). While robotics can enhance safety, efficiency, and precision, it also introduces ethical and regulatory concerns that require cooperation across sectors (Chea et al., 2020).

In the Nigerian context, several key barriers hinder the widespread adoption of robotics in the construction industry. The high initial cost of procuring and implementing robotic systems is one of the primary obstacles, compounded by the country's unreliable electricity supply, which is essential for operating advanced technology (Amaifeobu et al., 2023). Additionally, the Nigerian construction workforce is predominantly unskilled, and there are significant concerns about job displacement, which creates resistance to automation (Oke et al., 2023). Moreover, the lack of government incentives and policies promoting technological adoption further stifles progress (Adedeji et al., 2021). These factors create a complex environment where the benefits of robotics, such as improved safety, efficiency, and reduced project delays, are overshadowed by economic, technical, and social challenges (Gharbia et al., 2020). Addressing these issues requires a coordinated effort among policymakers, industry leaders, and educational institutions to build capacity, improve infrastructure, and provide incentives for robotics adoption (Ilesanmi et al., 2024).

The construction sector, traditionally labor-intensive, is seeing gradual improvements through the integration of robotics, with an increasing number of robots working alongside human workers (Wei et al., 2023). However, the sector's continued reliance on outdated construction methods hinders progress. As the fourth industrial revolution (4IR) unfolds, embracing new technologies is imperative to meet the evolving demands of construction projects (Ikuabe et al., 2022). Ebekozien and Aigbavboa (2021), caution that while the integration of technology in construction offers potential benefits, it could also introduce uncertainties. Despite these concerns, the use of robotics promises several advantages, such as faster project delivery, reduced injuries, improved task accuracy, cost-efficiency, reduced waste, and enhanced site security (Ikuabe et al., 2023). Hoeft et al., (2022), argue that for sustainable cities and communities to emerge, civil engineers, construction companies, universities, suppliers, clients, and SMEs must collaborate on a strategic roadmap to integrate robotics into the construction industry. While various sectors have increasingly adopted robotics, its application in the built environment remains limited.

Previous studies have focused on robotics adoption in construction across different countries. Mahbub (2008), explored barriers to robotics adoption in Malaysia, Australia, and Japan, while Ojha et al., (2022), examined obstacles in the United States, offering suggestions for eliminating these impediments. Boya et al. (2022) assessed challenges in the South African construction sector, and Pradhananga et al. (2021) provided a thorough understanding of barriers in the US construction industry. However, these studies have not addressed the specific barriers in Nigeria, underscoring the need for context-specific research (Oluseye et al., 2020). In Nigeria, Amaifeobu et al., (2023), examined the limitations of robotics in construction, while Oke et al. (2023) introduced innovative ideas to further global discussions on robotics adoption. However, these studies have not fully explored the unique barriers faced by Nigerian construction industries in urban centers like Lagos and Abuja. These cities, with their dynamic environments and extensive infrastructure projects, provide a rich setting for understanding how robotics can be applied more effectively categorizing these barriers into manageable sizes and offering practical solutions. This study fills these gaps by offering a comprehensive evaluation of the barriers to robotics adoption in Nigeria's built environment.

(a) It addresses key obstacles hindering widespread robotics implementation in developing countries, focusing on the Nigerian context.
(b) The research provides a comparative analysis with other regions, identifying unique challenges and proposing solutions relevant to the Nigerian construction industry.

The study's contribution is critical for promoting economic growth, innovation, and sustainable development in Nigeria, particularly in advancing the construction sector's competitiveness in a global context. By providing an all-encompassing strategy, this research aims to not only address barriers but also propose a roadmap for wider robotics integration in both Nigeria and other international settings, ensuring long-term safety, efficiency, and sustainability in the built environment.

LITERATURE REVIEW

The Current Situation of Robotic Application in the Built Environment of Nigeria

The Nigerian built environment must deepen its understanding of mechanization and technological advancements, with future trends and opportunities for robotization and autonomous applications rapidly emerging (Adedeji et al., 2021). Significant advancements in robotics and artificial intelligence (AI) have intersected, transforming automation capabilities (Olaoye et al., 2024). However, despite these breakthroughs, the construction sector in developing countries like Nigeria has received far less attention, resulting in challenges to optimal performance and productivity due to late technology adoption (John et al., 2022). Although the construction industry is utilizing technologies like BIM, automation, and wearables to boost productivity, limited research exists regarding their awareness and adoption (Ibrahim et al., 2024). These emerging technologies are revolutionizing industries and production techniques globally to meet the growing demand for a variety of goods and services (Mali 2022). The Nigerian built environment, however, faces significant challenges in adopting robotic and automation technologies due to various human, site, financial, technical, material, governmental, ethical, and cultural barriers (Amaifeobu et al., 2023). A study by Oke et al., (2023), provides a roadmap for improving the understanding of robotics in developing economies like Nigeria, yet the rate of adoption in Nigeria's construction sector remains alarmingly low, raising concerns. The perception that construction in Nigeria is labor-intensive presents a key obstacle to robotics adoption (Oluseye and Aigbavboa, 2020). While microphones, speakers, cameras, and global positioning systems (GPSs) are commonly used cyber-physical systems (CPSs) in the construction industry, most professionals are primarily aware of HVAC systems (Adeosun and Oke, 2024). Inhibiting factors include high costs, resistance to new technologies, lack of standardized protocols, insufficient maintenance capabilities, and inadequate government support (Ilesanmi et al., 2024). For Nigeria to fully benefit from this industrial revolution, there is a pressing need to combine the best aspects of human intelligence with machine capabilities (Ejohwomu et al., 2021).

Key Challenges Hindering the Adoption of Robotics in Construction

Industries worldwide are integrating robotics and digital manufacturing to improve construction efficiency, yet the sector lags in technological adoption. While robotic intelligence enhances productivity and safety, construction firms still struggle to identify the most beneficial technologies. In Nigeria, several obstacles hinder robotics adoption despite growing interest in prefabrication and BIM technologies. Electricity supply limitations remain a critical challenge, as only a fraction of the population has consistent access to power (Nkollo et al., 2023; Owolabi et al., 2022). High initial investment costs make robotics unaffordable for most firms, while limited access to financing prevents small and medium-sized enterprises from integrating automation (Ilesanmi et al., 2024; Nachiappan et al., 2024; Chan et al., 2024; Pradhananga et al., 2021). Moreover, economic instability, driven by inflation and currency fluctuations, discourages long-term investments (Asif et al., 2024; Thirumal et al., 2024).

Beyond financial barriers, the industry faces displacement of the construction workforce, leading to employee resistance and cultural norms that favor traditional labor-intensive methods (Leesakul et al., 2022; Pradhananga et al., 2021; Waqar et al., 2023; Nachiappan et al., 2024). Limited awareness and education, alongside employees' restricted access to advanced technology, further slow adoption, as many stakeholders lack the necessary ICT expertise to leverage automation (Gharbia et al., 2020; Abioye et al., 2021; Javaid et al., 2021; Regona et al., 2022). The lack of technical expertise (12) and limited investment in research and development (15) exacerbate this gap, preventing innovation and local adaptation of robotic solutions. Institutional challenges also play a role (Huang et al., 2022; Cai et al., 2020; Oke et al., 2023; Ngcobo and Akinradewo, 2023). The lack of regulatory and policy frameworks creates uncertainty for construction firms, while support challenges from the government reduce incentives for private-sector adoption (Pan et al., 2020; Bademosi and Issa, 2021; Ilesanmi et al., 2024; Kissi et al., 2023). Corruption rates and supply chain challenges further complicate procurement and integration processes (Ikuabe et al., 2023; Faltein and Sekdeo, 2024). Additionally, insurance and liability issues deter firms from adopting robotics, as legal protections remain weak (Nachiappan et al., 2024; Chan and Muralidharan, 2024).

The construction industry's structure and workflow systems pose another challenge (Guma, 2021; Tijani and Feng, 2021; Hoeft et al., 2022). Robotics must integrate with fragmented design processes, making project management complexity a significant concern (Hoeft et al., 2022; Ejohwomu et al., 2021). Standardization issues also hinder widespread adoption, as firms struggle with compatibility across different systems (Musarat et al., 2024; Mohapatra et al., 2024). Environmental factors, including sustainability regulations and environmental concerns, limit the scalability of robotic technologies (Oluseye and Aigbavboa 2020; Olaoye et al., 2024) Additionally, cybersecurity concerns pose risks, as interconnected robotic systems are vulnerable to hacking and data breaches (Zafar et al., 2024; Van Tam et al., 2024). Finally, maintenance and technical support issues make long-term sustainability difficult, as local expertise is scarce, and reliance on foreign suppliers increases operational costs (Faltein amd Sukdeo, 2024; Wei et al., 2023). Addressing these barriers requires a strong policy framework, increased government incentives, and investment in education and workforce training (Yi and Hiroatsu, 2022; Shojaei and Burgess, 2022). By overcoming these challenges, Nigeria's construction sector can align with global trends, fostering innovation, efficiency, and sustainable development in the age of Industry 4.0.

This study used various journals, conference proceedings, and book chapters published in respectable journals and academic databases (see Table 1 below).

Table 1. Summary of the Evaluated Literature on Barriers to the Widespread Adoption of Robotics

S/N	Crucial Barriers	References
1	Electricity Supply Limitations	(Nkollo et al., 2023; Owolabi et al., 2022).
2	High Initial Investment Cost	(Ilesanmi et al., 2024;Nachiappan et al., 2024)
3	Displacement of Construction Workforce	(Leesakul et al., 2022; Pradhananga et al., 2021).
4	Lack of Regulatory and Policy Framework	(Pan et al., 2020; Bademosi and Issa, 2021).
5	Limited Awareness and Education	(Gharbia et al., 2020;Abioye et al., 2021)
6	Corruption Rate	(Odulana and Oyewobi, 2019;
7	Cultural Norms and Human Element	(Waqar et al., 2023; Nachiappan et al., 2024)
8	Structure and Organization of the Construction Industry	(Guma, 2021; Tijani and Feng, 2021)
9	Support Challenges from the Government	(Ilesanmi et al., 2024; Kissi et al., 2023)
10	Employees' Limited Access to Advanced Technology	(Javaid et al., 2021; Regona et al., 2022)
11	Construction Product and Workflow Systems	(Guma 2021; Hoeft et al., 2022)
12	Lack of Technical Expertise	[3;42]
13	Lack of Incentives	(Yi and Hiroatsu, 2022; Shojaei and Burgess, 2022).
14	Limited Access to Financing	(Chan et al., 2024; Pradhananga et al., 2021)
15	Limited Investment in Research and Development	(Oke et al., 2023; Ngcobo and Akinradewo, 2023)
16	Economic Instability	(Asif et al., 2024; Thirumal et al., 2024)

continued on following page

Table 1. Continued

S/N	Crucial Barriers	References
17	Environmental Concern	Oluseye and Aigbavboa 2020; Olaoye et al., 2024
18	Cybersecurity Concern	(Zafar et al., 2024; Van Tam et al., 2024)
19	Maintenance and Technical Support Issues	(Faltein amd Sukdeo, 2024; Wei et al., 2023).
20	Insurance and Liability Issues	(Nachiappan et al., 2024; Chan and Muralidharan, 2024)
21	Project Management Complexity	(Hoeft et al., 2022; Ejohwomu et al., 2021)
22	Supply Chain Challenges	(Ikuabe et al., 2023; Faltein and Sekdeo, 2024)
23	Standardization Issues	(Musarat et al., 2024; Mohapatra et al., 2024)

RESEARCH METHODS

The research method employed for this study was quantitative, aimed at collecting and analyzing numerical data to evaluate the barriers hindering the adoption of robotics in the Nigerian built environment. The choice of Lagos and Abuja as the study locations is justified by their prominence in Nigeria's construction industry. Lagos, being the commercial capital, hosts a significant portion of construction activities and innovations, while Abuja, as the federal capital, is home to many high-profile government and private sector construction projects. Both cities represent diverse perspectives and scales of construction, making them ideal for capturing the range of factors affecting robotics adoption across the Nigerian-built environment (Olanipekun et al., 2015).

The research utilized structured and self-administered questionnaires targeting construction experts in active firms operating in Lagos and the Federal Capital Territory (FCT) of Abuja. The sample size was calculated using the Yamane formula, which provided a statistically representative sample of the population (Yamane, 1967). The study population consisted of 193 registered construction firms in Lagos and 102 in Abuja, sourced from NIIQS, NIA, NSE, NIOB, and PMI (2024). A total of 195 questionnaires were retrieved, with 123 from Lagos and 72 from Abuja. This sample was deemed sufficient to capture a range of opinions and insights from key industry players.

A pilot test was conducted to ensure the validity and reliability of the questionnaire, aligning with the recommendation by Moores et al., (2012), which suggests pilot testing a limited number of questionnaires to ensure the instrument's effectiveness before broader deployment. Convenience sampling, a variant of random sampling, was employed to select respondents based on criteria such as availability, accessibility, and relevance, following similar methodologies in previous studies (Creswell 2014). To analyze the data, frequency distribution and percentage were used to describe the demographic attributes of the sampled firms. The study employed mean item scores to rank the level of agreement or disagreement with the identified barriers on a 5-point Likert scale (Taherdoost 2019). The mean item score is a widely recognized method for determining the average level of response to a given set of items, providing an effective way to gauge the relative importance of each barrier. Exploratory Factor Analysis (EFA) was applied to identify underlying dimensions within the data, enabling the reduction of numerous barriers into smaller, more manageable factors. EFA helps uncover latent variables that contribute to observed responses, adding depth to the interpretation of the barriers (Babalola and Harinarain, 2024). This statistical technique is vital in identifying patterns and relationships within complex datasets, which in turn aids in making informed recommendations. Additionally, the Mann-Whitney U Test was used to compare perceptions between indigenous and foreign construction firms regarding the barriers to robotics adoption. This non-parametric test is suitable for comparing two independent groups, particularly when the data is not normally distributed (Field, 2018). The test provided insights into whether significant differences exist between these two groups, informing the study's conclusions and recommendations for overcoming identified barriers. In summary, this quantitative approach, underpinned by robust statistical methods such as mean item scores, exploratory factor analysis, and the Mann-Whitney U Test, provided detailed insights into the barriers limiting robotics adoption in the Nigerian-built environment. Ethical considerations, including informed consent and the confidentiality of respondents, were duly observed, ensuring the integrity of the research process (Sales, 2024)

RESULTS

Respondent Demographics

Table 2 provides an overview of the types of firms that participated in the survey. Among the respondents, 32.3% work as contractors, 26.2% are consultants or provide professional services, 23.1% are subcontractors, and 18.5% are developers. The table also highlights the qualifications of the respondents, reflecting their expertise in various sectors. Specifically, 28.2% of respondents specialize in building construction, 22.1% in non-residential construction, 20.5% in civil engineering, and 15.9% in infrastructure, with 13.3% involved in other areas. Additionally, 21% of the respondents represent foreign firms, while 79% are from indigenous companies.

Table 2. Respondent's Demographics

Factor	Variables	Frequency	Percent
Type of Firm	Contractor	63	32.3
	Consultant or Professional services	51	26.2
	Sub-Contractor	45	23.1
	Developer	36	18.5
	Total	**195**	**100**
Area of Specialization	Building	55	28.2
	Non-Residential	43	22.1
	Civil Engineering	40	20.5
	Infrastructure	31	15.9
	Others Specify	26	13.3
	Total	**195**	**100**
Category of Firm	Indigenous Firm	120	61.5
	Foreign Firm	75	38.5
	Total	**195**	**100**

Mean Score

The barriers to the extensive adoption of robotics in the Nigerian built environment were identified through a comprehensive literature review. Of the 12 variables considered in this study, those with a median value of ≥3.5 were classified as "significant factors." The results indicate that all components above this threshold validate the relevance and distinctiveness of the barriers identified (Wuni et al., 2020; Khoshfetrat et al., 2022). Table 3 details these barriers. The highest-ranked barrier, with a mean score of 4.34, is the limitation in electricity supply. This is followed by high initial investment costs, with a mean score of 4.33. The third-ranked barrier is the displacement of the construction workforce, with a mean score of 4.05. At the bottom, standardization issues were rated lowest, with a mean score of 3.21.

Table 3. Barriers to the Extensive Adoption of Robotics

Barriers	Mean	Std. Deviation	Rank
Electricity Supply Limitations	4.34	0.795	1
High Initial Investment Cost	4.33	0.714	2
Displacement of Construction Workforce	4.05	0.754	3
Lack of Regulatory and Policy Framework	4.04	0.698	4
Limited Awareness and Education	3.98	0.845	5
Corruption Rate	3.92	0.954	6
Cultural Norms and Human Element	3.88	0.969	7
Structure and Organization of the Construction Industry	3.87	0.720	8
Support Challenges from the Government	3.84	1.056	9
Employees' Limited Access to Advanced Technology	3.81	0.732	10
Construction product and workflow system	3.69	0.859	11
Lack of Technical Skill	3.66	0.946	12
Lack of Incentives	3.63	0.607	13
Limited Access to Financing	3.61	0.712	14
Limited Investment in Research and Development	3.57	0.690	15
Economic Instability	3.49	0.693	16
Environmental Concern	3.42	0.802	17
Cybersecurity Concern	3.39	0.911	18
Maintenance and Technical Support Issues	3.36	0.832	19
Insurance and Liability Issues	3.32	0.866	20
Project Management Complexity	3.27	0.816	21
Supply Chain Challenges	3.23	0.921	22
Standardization issues	3.21	0.829	23

Exploratory Factor Analysis

To further analyze the study, exploratory factor analysis (EFA) was employed. After the initial extraction, a rotation approach was applied to refine the factor structure and ensure clearer interpretability using the Varimax rotation method. The significant result from Bartlett's test of sphericity (1251.003, $p < 0.001$) and a Kaiser-Meyer-Olkin measure of 0.811 confirmed the appropriateness of the sampled data, validating the factor analysis (Field, 2018). The scree plot and Kaiser's criterion were used to determine the optimal number of factors to be regrouped (Shrestha, 2021). A total of four factors were retained, accounting for 65.67% of the variance in the dataset. Table 4 below illustrates the items grouped within each component. The study categorized the technological barriers to the extensive adoption of robotics and automation into four clusters: "Innovation Investment" (25.246%), "Environmental Impediments" (16.362%), "Technological Gaps" (13.729%), and "Resource and Risk Concerns" (10.334%). These renamed clusters provide concise, comprehensive terms to represent the bundled barriers to the widespread adoption of robotics in the built environment.

Table 4. Exploratory Factor Analysis Showing the Three Distinct Factors, Description, and Factor Loading

Factor	Description	Factor loading
KMO= 0.811	Sig=0.000	
Innovation Investment	Limited Awareness and Education	0.871
	Lack of Incentives	0.854
	Limited investment in Research and Development	0.776
	Standardization Issues	0.719
	Lack of Regulatory and Policy Framework	0.629
	Displacement of Construction Workforce	0.616
	High Initial Investment Cost	0.614
	Electricity Supply Limitations	0.587
Environmental Impediments	Support Challenges from the Government	0.738
	Economic Instability	0.709
	Environmental Concerns	0.675
	Structure and Organization of the Construction Industry	0.664
	Cultural Norms and Human Element	0.657
	Corruption Rate	0.633
Technological Gaps	Lack of Technical Expertise	0.759
	Cybersecurity Concern	0.645
	Maintenance and Technical Support Issues	0.633
	Construction Product and Workflow Systems	0.613
	Employee's Limited Access to Advanced Technology	0.593
Resource and Risk Concerns	Limited Access to Financing	0.712
	Supply Chain Challenges	0.665
	Project Management Complexity	0.602
	Insurance and Liability Issues	0.567

Mann-Whitney U Test

The Mann-Whitney U test was used to determine whether there are significant differences in the perceptions or attitudes between indigenous and foreign firms regarding the adoption of robotics in the Nigerian built environment. This analysis provides valuable insights into the factors that may influence these perceptions and helps inform strategies to overcome barriers to robotics adoption. The mean rank, sum of ranks, and median values are shown in Table 5. The mean rank represents the average position of the observations within each group, while the sum of ranks reflects the total rank assigned to each group's observations. For skewed distributions, the median serves as the measure of central tendency. The results in Table 6 indicate strong statistical evidence of a difference between the perceptions of indigenous and foreign firms concerning the barriers to robotics adoption. This is supported by a significance level of 0.000, which is below the 0.05 threshold ($p < 0.05$).

Table 5. Mann-Whitney U Test (Indigenous and Foreign Firms)

Category of firm		N	Mean Rank	Sum of Ranks	median
Crucial Barrier to the widespread implementation of robotics	**Indigenous firms**	**120**	**47.60**	**3522.50**	**3.00**
	Foreign firms	75	12.05	132.50	1.00
	Total	195			

Table 6. Test Statistics (Indigenous and Foreign Firms)

	Crucial Barrier to the Widespread Implementation of Robotics
Mann-Whitney U	66.500
Wilcoxon W	132.500
Z	-4.828
Asymp. Sig. (2-tailed)	.000

- Grouping Variable: Category of firm

Justification of Rankings with Selected Regions

To enhance the robustness and applicability of the key barriers to the widespread adoption of robotics in the built environment, the findings of this study were further compared with data from other countries and contexts. These include

South Africa (Aghimien et al., 2024), India (Huang et al., 2022), China and the UK (Li et al., 2024), Malaysia (Waqar et al., 2023), and the US (Vora et al., 2024), alongside the current study from Nigeria, as shown in Table 7. The barriers were categorized and analyzed based on their frequency and distinct characteristics. The most common barrier across all countries is the "High Initial Investment Cost." This universal barrier represents the substantial financial expenditure required to integrate robotics into the construction industry. It suggests that cost-cutting measures, such as government incentives, subsidies, or cooperative funding schemes, are necessary to mitigate this challenge. Another significant barrier is the "Lack of Skilled Labor," which is evident in South Africa, the US, China, and the UK (where it manifests as a "lack of training"). This reflects the shortage of qualified workers to operate and maintain robotics, especially in nations with advanced construction industries. Funding programs for education and training could help reduce this barrier. "Resistance to Change/Job Security Concerns" is a notable barrier in South Africa, Nigeria, and the US. This highlights the human and social dimensions of robotics implementation, including resistance from labor unions and workers due to concerns about job displacement. Addressing these issues will require retraining programs, policies, and effective change management strategies to safeguard jobs. Additionally, "Regulatory and Policy Framework Issues" are prominent in Nigeria and India. The absence of clear standards and regulations in these countries presents one of the greatest challenges, especially in regions where policy frameworks have not yet adapted to accommodate robotic technologies. Establishing clear standards and streamlining regulatory processes are essential steps.

"Nigeria's Electricity Supply Limitation" is a distinct barrier unique to Nigeria. Infrastructure issues, such as inconsistent electricity supply, hinder the adoption of robotics. Solutions include developing robotics that can operate on multiple power sources or upgrading the nation's power infrastructure. In India, "Limited Design and Research Input" is another unique barrier. India's construction sector would benefit from localized research and development to create robotics solutions tailored to the region's specific needs. Promoting investment in local research institutions could drive innovation. "Aversion to Change in the Construction Industry" in South Africa highlights the need for awareness campaigns that showcase the long-term benefits of robotics while identifying culturally specific barriers to adopting new technologies. India's "Unstructured Construction Environment" presents further challenges due to the lack of standardization in construction practices. Encouraging the use of organized project management methods could help address this issue. Malaysia faces a unique concern with "Privacy Violations." As the only country where privacy concerns are noted, this likely stems from the data collection capabilities of robotic systems. Ensuring strict privacy and data protection protocols is crucial in overcoming this barrier. Finally, "Management Issues" in the US and

Malaysia reflect challenges related to the integration of robotics in construction management. Effective leadership and change management strategies are critical to overcoming these obstacles.

In conclusion, a combination of financial strategies, educational and training initiatives, regulatory reforms, and cultural change programs is required to effectively address these barriers. This comprehensive approach can facilitate the broader adoption of robotics in the built environment across various international contexts.

Table 7. Juxtaposition of Findings

Barriers Rankings	Current Nigerian Context	South African Context (Aghimien et al., 2024)	Indian Context (Huang et al., 2022)	China and UK Context (Li et al., 2024)	Malaysian Context (Waqar et al., 2023)	US Context (Vora et al., 2024)
1	Electricity supply limitation	High investment cost	Limited research and design input	High Cost	Inadequate technology	High Capital Investment
2	High Initial Investment cost	Aversion to change in the industry	High purchase cost	Lack of Training	Management issues	Lack of Skilled Labour
3	Displacement of the construction workforce	Resistance by workers and Unions	Unstructured construction environment	Limited Awareness	Lack of Privacy	Job security
4	Lack of regulatory and policy framework	Insufficient skilled personnel	Regulatory and Standardization issues	Industry fragmentation	High Cost	Management over time

DISCUSSION OF FINDINGS

This study identified key barriers to the widespread adoption of robotics in Nigeria's built environment, grouping them into four main categories: Innovation Investment, Environmental Impediments, Technological Gaps, and Resource and Risk Concerns.

Innovation Investment

The first group, "Innovation Investment", encompasses eight variables: limited awareness and education, lack of incentives, underinvestment in research and development (R&D), standardization issues, lack of regulatory and policy frameworks, concerns about workforce displacement, high initial investment costs, and electricity supply limitations. In Nigeria's built environment, a lack of awareness and education about robotics, coupled with misconceptions regarding workforce displacement and inadequate training, impedes the industry's embrace of automation (Abioye et al., 2021). Stakeholders remain hesitant to invest in robotics due to the absence of incentives and regulatory support, which affects their confidence in the stability and profitability of these technologies (Shojaei and Burgess, 2022). Furthermore, the insufficient allocation of funds for R&D stifles innovation and forces reliance on expensive, imported solutions (Ngcobo and Akinradewo, 2023). The absence of standardized regulations complicates the deployment of robotics, as inconsistencies in quality, interoperability, and safety deter investment (Musurat et al., 2024). A robust legal and regulatory framework is essential to drive the adoption of robotics in Nigeria's construction sector (Rane, 2023). Moreover, fears of job loss and inadequate reskilling options exacerbate resistance to automation, especially in a country with high unemployment rates (Pradhananga et al., 2021). Addressing this will require comprehensive retraining programs and policies aimed at equitable economic growth (Leesakul et al., 2022). High initial investment and maintenance costs, coupled with reliance on costly foreign technical services, further limit adoption (Bademosi and Issa, 2021; Boya et al.,2022). Overcoming these barriers demands innovative funding strategies, partnerships with financial institutions, and local technical capacity building (Ilesanmi et al., 2024; Ojha et al., 2022). Additionally, the inconsistency of Nigeria's electricity supply discourages investment in robotics, while reliance on backup power sources like generators adds complexity and cost (Owolabi et al., 2022). Thus, the development of renewable energy and power infrastructure is vital (Nkollo and Saidu 2023).

Environmental Impediments

The second group, "Environmental Impediments", consists of six variables: lack of government support, economic instability, environmental concerns, the structure of the construction industry, cultural norms, and corruption. Government support, such as infrastructure development, grants, and subsidies, could significantly reduce the cost of adopting automation technologies (Ibekwe et al., 2024). However, Nigeria's economic instability, marked by inflation and currency fluctuations, discourages investment in expensive technologies like robotics (Asif et al., 2024). Environmental concerns about the potential clash between robotics and sustainability goals, as well as public pressure to adopt greener practices, also present significant barriers (Olaoye et al., 2024). The fragmented nature of the construction industry, characterized by a lack of collaboration among stakeholders, further complicates the adoption of robotics, particularly for smaller firms that cannot afford the high cost of equipment (Thirumal et al., 2024). Cultural resistance to modern technology and a preference for traditional methods inhibit the spread of automation (Van Tam et al., 2024). Moreover, corruption exacerbates these issues, leading to unfair practices, inefficiencies, and an uneven playing field for technology deployment (Gomaa et al., 2023; Odulana and Oyewobi, 2019).

Technological Gaps

The third group, Technological Gaps, includes five variables: lack of technical expertise, cybersecurity concerns, maintenance and support issues, traditional construction systems, and limited access to advanced technology. Nigeria's construction industry struggles with a shortage of skilled personnel capable of programming and maintaining robotic systems, severely hindering the adoption of these technologies (Akande et al., 2023). Cybersecurity threats, including data breaches and hacking, make companies wary of investing in robotics, due to potential revenue loss and brand damage (Zafar et al., 2024). Maintenance issues, compounded by slow internet access, import taxes, and reliance on costly offshore services, further complicate adoption (Gonsalves et al., 2024). Traditional methods and workflow systems, resistant to change, also obstruct the integration of new technologies in Nigeria's construction organization (Seyman-Guray and Kismet, 2023). Limited access to advanced technologies, along with insufficient training, results in fear and resistance among employees, necessitating a stronger focus on technological infrastructure and educational programs (.Oke et al., 2023).

Resource and Risk Concerns

The final group, "Resource and Risk Concerns", comprises four variables: limited access to financing, supply chain challenges, project management complexity, and insurance and liability issues. Financial limitations remain a significant barrier to robotics adoption, as high initial costs and limited funding options curtail innovation (Chen et al., 2022). Insurance and liability concerns, coupled with unclear legal frameworks, deter firms from adopting robotics due to potential legal disputes and the inherent risks of using such systems (Chan et al., 2024). The complexity of managing robotics-related projects, perceived risks, and potential delays also hinder widespread adoption, as stakeholders are reluctant to take on the perceived burden (Abutaha et al., 2024). Additionally, supply chain issues, such as high import taxes, logistical delays, and geopolitical unrest, raise the cost and complexity of acquiring necessary components, further discouraging investment in robotics (Shakur et al., 2024).

Implication of Findings

This study aims to bridge the knowledge gap between robotics theory and its practical application in the Nigerian built environment by identifying the key barriers to robotics adoption. The findings have significant implications for research, practice, public policy, and society. By classifying the barriers into four main components; "Innovation Investment", "Environmental Impediments", "Technological Gaps", and "Resource and Risk Concerns"; this study provides a comprehensive understanding of the obstacles limiting robotics integration in Nigeria's built environment.

This research offers crucial insights for industry practitioners, policymakers, and researchers, helping them to understand and navigate the challenges that inhibit the adoption of robotics in construction projects. In doing so, the study contributes to advancing sustainable and innovative practices within the construction industry, which aligns with the global push for technological transformation and sustainability. Furthermore, the study emphasizes the importance of addressing these barriers to enable the Nigerian construction sector to evolve alongside global trends in automation and digital transformation. The study's identification of the key barriers to robotics adoption enriches the body of knowledge in construction automation and offers a basis for future research. By pinpointing factors such as limited innovation investment and technological gaps, the research lays the groundwork for scholars to explore targeted solutions that address these barriers. It also bridges the gap between theory and practice by providing empirical evidence of the practical challenges faced in implementing robotics technologies in the Nigerian context. This will guide future studies on how to overcome these barriers through focused technological, regulatory, and educational interventions.

For practitioners, the findings highlight actionable areas where intervention is needed. The study points to the importance of investment in research and development, standardization, and infrastructure improvement as critical enablers of robotics adoption. Construction firms and stakeholders are encouraged to focus on upskilling the workforce, fostering collaboration, and investing in advanced technologies. Addressing Technological Gaps through training and capacity building will ensure that employees are equipped with the skills necessary to operate and maintain robotic systems, which in turn improves efficiency, reduces project costs, and enhances the overall quality of construction projects.

From a policy perspective, the findings underscore the need for government support to create a favorable environment for robotics adoption. Policymakers can draw from this research to establish robust legal frameworks, offer financial incentives, and promote collaboration between private and public sectors. Addressing Environmental Impediments such as economic instability and government support is essential to creating policies that foster innovation while considering local economic and social conditions. Moreover, this study provides insights that can guide the Nigerian government in crafting regulations that promote sustainable automation practices, which align with the country's broader development goals. The adoption of robotics in the Nigerian built environment holds considerable potential for societal benefits. As highlighted by the study, overcoming these barriers can lead to improved infrastructure quality, faster project delivery, and enhanced safety, ultimately improving the quality of life for Nigerian citizens. Furthermore, by fostering a culture of innovation, the widespread use of robotics can reduce construction waste, energy consumption, and emissions, thus contributing to the global agenda for sustainability. This shift can influence public attitudes towards embracing technology in construction, demonstrating that automation can coexist with job creation through retraining and reskilling initiatives.

In conclusion, this study not only identifies the barriers to robotics adoption in Nigeria's construction industry but also provides clear implications for overcoming these challenges through targeted research, policy, and practice. By aligning with Nigeria's developmental goals, the study has the potential to influence public policy, improve industry competitiveness, and promote sustainable development, thereby ensuring that the benefits of automation and robotics are fully realized in the built environment.

CONCLUSION

To foster sustainable cities and communities, this study investigates the barriers to the broad adoption of robotics in the Nigerian construction industry and raises awareness of these challenges. To achieve this goal, a well-designed questionnaire was utilized to gather information from experts in the Nigerian construction sector. Based on the data analyzed, four major obstacles to robotics adoption in the construction sector were identified: innovation investment,' 'environmental impediments,' 'technological gaps,' and 'resource and risk concerns.' While this research has evaluated and highlighted the numerous barriers to the widespread use of robotics, the successful application of these technologies could significantly advance and enhance the growth of Nigeria's built environment. For robotics to be adopted, it must be broadly acknowledged within the industry. This necessitates that practitioners and stakeholders remain receptive to this innovation in their efforts to construct sustainable cities and communities.

Furthermore, the study recommends that the construction sector view itself beyond traditional processes and actively embrace technological advancements, including robotics. Additionally, governments and senior construction professionals should take the lead in implementing this innovation. This can be achieved by forming innovation teams, educating the industry about the advantages of robotics, and raising public awareness. A limitation of this study is its focus on Nigeria; however, further research should be conducted in other developing nations. It would also be beneficial to carry out case studies in similar countries to identify best practices and lessons learned regarding the use of robots in the built environment. Future research endeavors may concentrate on examining the potential financial benefits associated with the extensive implementation of these technologies within the sector.

REFERENCES

Abioye, S. O., Oyedele, L. O., Akanbi, L., Ajayi, A., Delgado, J. M. D., Bilal, M., & Ahmed, A. (2021). Artificial intelligence in the construction industry: A review of present status, opportunities, and future challenges. *Journal of Building Engineering*, *44*, 10–32. DOI: 10.1016/j.jobe.2021.103299

Abutaha, F., & Dinler, C. (2024). Adopting Robotic Process Automation (RPA) in the Construction Industry. In *Complex AI Dynamics and Interactions in Management; 111* (pp. 273–310). IGI Global. DOI: 10.4018/979-8-3693-0712-0.ch013

Adedeji, A., Ayodele, E., & Falaye, T. (2021). Barriers to the Adoption of Innovative Technologies in the Nigerian Construction Industry. *Journal of Construction Engineering and Management*, *147*(8), 1–12. DOI: 10.1061/(ASCE)CO.1943-7862.0002084

Adeosun, F. E., & Oke, A. E. (2024). Examining the awareness and usage of cyber-physical systems for construction projects in Nigeria. *Journal of Engineering. Defence Technology*, *22*(1), 281–294.

Aghimien, D., Ikuabe, M., Aghimien, L. M., Aigbavboa, C., Ngcobo, N., & Yankah, J. (2024). PLS-SEM assessment of the impediments of robotics and automation deployment for effective construction health and safety. *Journal of Facilities Management*, *22*(3), 458–478. DOI: 10.1108/JFM-04-2022-0037

Akande, I., Alagbe, O. T., Adebayo, O. A., Olaosun, S., & Oyedele, I. O. (2023). Prospects And Challenges Of Emerging Technologies In Architectural Design Practice Within The Nigerian Context. Journal of Namibian Studies, 33.

Amaifeobu, O., Ogunbiyi, O., & Ayeni, O. (2023). The Challenges of Robotics in the Nigerian Construction Industry. *International Journal of Construction Research and Technology*, *15*(3), 456–471. DOI: 10.1016/j.jcrt.2023.05.007

Asif, M., Naeem, G., & Khalid, M. (2024). Digitalization for sustainable buildings: Technologies, applications, potential, and challenges. *Journal of Cleaner Production*, *450*, 141814. Advance online publication. DOI: 10.1016/j.jclepro.2024.141814

Babalola, O., & Harinarain, N. (2024). Challenges in Adopting Construction Robotics in Sub-Saharan Africa. *Journal of Construction Engineering and Management*, *145*(2), 215–228.

Bademosi, F., & Issa, R. R. (2021) Factors influencing adoption and integration of construction robotics technology in the US. *Journal of Construction Engineering and Management*; 147(8), pp.04-21.

Boya, A., Akinradewo, O., Aigbavboa, C., Ebekozien, A., & Ramabodu, M. (2022) Bottlenecks to the Implementation of Automation and Robotics in the Construction Industry; In International Conference on Computing in Civil and Building Engineering pp. 139-149. Cham: Springer Nature Switzerland.

Cai, S., Ma, Z., Skibniewski, M. J., Bao, S., & Wang, H.; (2020). Construction Robotics for high-rise buildings: Development priorities and key challenges. *Journal of Construction Engineering and Management,* 146(8), pp. 04-20.

Chan, H. Y., & Muralidharan, A. (2024). Care Robots for the Elderly: Legal, Ethical Considerations and Regulatory Strategies. In *Developments in Intellectual Property Strategy: The Impact of Artificial Intelligence, Robotics and New Technologies Cham* (pp. 129–156). Springer International Publishing. DOI: 10.1007/978-3-031-42576-9_6

Chea, C. P., Bai, Y., Pan, X., Arashpour, M., & Xie, Y. (2020). An integrated review of automation and robotic technologies for structural prefabrication and construction. *Transportation Safety and Environment*, *2*(2), 81–96. DOI: 10.1093/tse/tdaa007

Chen, X., Chang-Richards, A. Y., Pelosi, A., Jia, Y., Shen, X., Siddiqui, M. K., & Yang, N. (2022). Implementation of technologies in the construction industry: A systematic review. *Engineering, Construction, and Architectural Management*, *29*(8), 3181–3209. DOI: 10.1108/ECAM-02-2021-0172

Creswell, J. W. (2014). *Research Design: Qualitative, Quantitative, and Mixed Methods Approaches* (4th ed.). SAGE Publications.

de Soto, B. G., & Skibniewski, M. J. (2020). Future of robotics in construction. In *Construction 4.0* (pp. 289–306). Routledge. DOI: 10.1201/9780429398100-15

Delgado, J. M. D., Oyedele, L., Ajayi, A., Akanbi, L., Akinade, O., Bilal, M., & Owolabi, H. (2019). Robotics and automated systems in construction: Understanding industry-specific challenges for adoption. *Journal of Building Engineering*, *26*, 10–18. DOI: 10.1016/j.jobe.2019.100868

Ebekozien, A., & Aigbavboa, C. (2021). Technological Integration in the Construction Industry: Opportunities and Uncertainties. *Journal of Building Engineering*, *40*, 1–10. DOI: 10.1016/j.jobe.2021.102773

Ejohwomu, O., Adekunle, S. A., Aigbavboa, C. O., & Bukoye, O. T. (2021). *Construction Industry and the Fourth Industrial Revolution: Issues and Strategies. The Construction Industry: Global Trends, Job Burnout, and Safety Issues*. Nova Science Publishers.

Faltein, S. A., & Sukdeo, N. I. (2024). Culture-driven quality enhancement: Uncovering the impact of robotics integration in the South African construction sector. *Ain Shams Engineering Journal, 102*(6), 728. DOI: 10.1016/j.asej.2024.102728

Field, A. (2018). *Discovering Statistics Using IBM SPSS Statistics* (5th ed.). SAGE Publications.

García de Soto, B., Agustí-Juan, I., Joss, S., & Hunhevicz, J. (2022). Implications of Construction 4.0 to the workforce and organizational structures. *International Journal of Construction Management, 22*(2), 205–217. DOI: 10.1080/15623599.2019.1616414

Gharbia, M., Chang-Richards, A., Lu, Y., Zhong, R. Y., & Li, H. (2020). Robotic technologies for on-site building construction: A systematic review. *Journal of Building Engineering, 32*, 101–115. DOI: 10.1016/j.jobe.2020.101584

Gomaa, M., Schade, S., Bao, D. W., & Xie, Y. M. (2023). Automation in rammed earth construction for industry 4.0: Precedent work, current progress, and prospect. *Journal of Cleaner Production, 398*, 13–25. DOI: 10.1016/j.jclepro.2023.136569

Gonsalves, N., Akanmu, A., Shojaei, A., & Agee, P. (2024). Factors influencing the adoption of passive exoskeletons in the construction industry: Industry perspectives. *International Journal of Industrial Ergonomics, 100*, 10–35. DOI: 10.1016/j.ergon.2024.103549

Guma, T. N. (2021). An Overview of Industrial Automation and its Implementation Approaches for Boosting Nigerian and Other Developing Economies. Federal University Wukari. *Trends in Science & Technology Journal, 6*(1), 154–160.

Hoeft, M., Kronsell, S., Manzoor, S., Johansson, F., Gustafson, A., von Haslingen, T., & Eriksson, K. (2022).. . *Construction Robotics in Infrastructure., 12*, 23–34.

Huang, Z., Mao, C., Wang, J., & Sadick, A. M. (2022);Huang, Z., Mao, C., Wang, J., & Sadick, A.-M. (2022, November 24). Understanding the key takeaway of construction robots towards construction automation. *Engineering, Construction, and Architectural Management, 29*(9), 3664–3688. DOI: 10.1108/ECAM-03-2021-0267

Ibekwe, K. I., Umoh, A. A., Nwokediegwu, Z. Q. S., Etukudoh, E. A., Ilojianya, V. I., & Adefemi, A. (2024). Energy efficiency in industrial sectors: A review of technologies and policy measures. *Engineering Science & Technology Journal, 5*(1), 169–184.

Ibrahim, K., Simpeh, F., & Adebowale, O. J. (2024). Awareness and adoption of wearable technologies for health and safety management in the Nigerian construction industry. *Frontiers in Engineering and Built Environment*, *4*(1), 15–28. DOI: 10.1108/FEBE-11-2022-0041

Ikuabe, M., Obikwelu, O., & Ebekozien, A. (2022). Robotic Advancements in the Fourth Industrial Revolution: A Nigerian Construction Perspective. *Journal of Sustainable Construction and Engineering*, *10*(4), 223–240. DOI: 10.1007/s40999-022-00794-w

Ikuabe, M., Obikwelu, O., & Ebekozien, A. (2023). Benefits of Robotics in Nigerian Construction: A Comprehensive Analysis. *International Journal of Construction Innovation*, *18*(1), 15–30. DOI: 10.1108/IJCI.2023.0015

Ilesanmi, O.O., Moyanga, D.T., Oke, A.E. and Aliu, J. (2024), "Unraveling the awareness dynamics of smart building features and technologies in the Nigerian construction sector", *International Journal of Building Pathology and Adaptation*, Vol. ahead-of-print No. ahead-of-print. DOI: 10.1108/IJBPA-04-2024-0073

Javaid, M., Haleem, A., Singh, R. P., & Suman, R. (2021). Substantial capabilities of robotics in enhancing industry 4.0 implementation. *Cognitive Robotics*, *1*, 58–75. DOI: 10.1016/j.cogr.2021.06.001

John, B., Adekunle, S. A., & Aigbavboa, C. (2022). Construction Industry and the Fourth Industrial Revolution: The Key Impediments in Developing Countries. Proc. *Int. Struct. Eng. Constr*, *9*(1), 3–14. DOI: 10.14455/ISEC.2022.9(1).CON-29

Khoshfetrat, R., Sarvari, H., Chan, D. W. M., & Rakhshanifar, M. (2022, September 10). Critical risk factors for implementing building information modeling (BIM): A Delphi-based survey. *International Journal of Construction Management*, *22*(12), 2375–2384. DOI: 10.1080/15623599.2020.1788759

Kissi, E., Aigbavboa, C., & Kuoribo, E. (2023). Emerging technologies in the construction industry: Challenges and strategies in Ghana. *Construction Innovation*, *23*(2), 383–405. DOI: 10.1108/CI-11-2021-0215

Leesakul, N., Oostveen, A.-M., Eimontaite, I., Wilson, M. L., & Hyde, R. (2022, March 11). Workplace 4.0: Exploring the implications of technology adoption in digital manufacturing on a sustainable workforce. *Sustainability (Basel)*, *14*(6), 3311. DOI: 10.3390/su14063311

Li, L., Yuan, P., Tang, Y., Cooper, G., Thurlbeck, S., Cheung, C. M., Manu, P., Yunusa-Kaltungo, A., & Weightman, A. (2024). The potential of construction robotics to reduce airborne virus transmission in the construction industry in the UK and China. *Heliyon, 13*(9), 12–21. DOI: 10.1016/j.heliyon.2024.e29697 PMID: 38694123

Maali Esfangareh, A. (2022), Adopting AI Techniques In Robotic Fabrication In Architecture: Intelligent Robotic Bricklaying Using Reinforcement Learning Algorithms; (Master's thesis, Middle East Technical University).

Mahbub, R. (2008), An investigation into the barriers to the implementation of Robotics technologies in the construction industry (Doctoral dissertation, Queensland University of Technology).

Mohapatra, B., Mohapatra, S., & Mohapatra, S. (2024). *Automation in the Construction Industry, Process Automation Strategy in Services, Manufacturing and Construction*. Emerald Publishing Limited., DOI: 10.1108/978-1-80455-143-120231018

Moores, J. L., Miller, R., & Nichols, M. (2012). Testing Survey Instruments for Reliability: Methods and Challenges. *Journal of Survey Methodology, 34*(1), 45–62.

Musarat, M. A., Alaloul, W. S., Zainuddin, S. M. B., Qureshi, A. H., & Maqsoom, A.; (2024), Digitalization in malaysian construction industry: Awareness, challenges and opportunities. *Results in Engineering*; 102 p.013.

Nachiappan, B., Najmusher, H., Nagarajan, G., Rajkumar, N., Loganathan, D., & Gobinath, G. (2024). Exploring the Application of Drone Technology in the Construction Sector. *Salud. Ciencia y Tecnología-Serie de Conferencias, 3*, 713–715. DOI: 10.56294/sctconf2024713

Ngcobo, N., & Akinradewo, O. (2023). Revolutionizing high-rise building construction: The role of robotics in South Africa's construction industry. *Journal of Construction Project Management and Innovation, 13*(2), 93–102.

Nkollo, O., & Saidu, I. (2023). Appraising the Benefits and Challenges of Implementing Artificial Intelligence in The Nigerian Construction Industry. *International Journal of Environmental Design and Construction Management, 350*, 6.

Odulana, A. O., & Oyewobi, L. O. (2019). Effect of implementation of E-procurement on corrupt practices in the Nigerian construction industry. International Conference of Environmental Sciences, ICES. *1st International Conference of the Faculty of Environmental Sciences,* University of Ilorin, Nigeria.

Ojha, S., Geng, W., & Vasilakis, M. (2022). Eliminating Barriers to Robotics Adoption in the US Construction Industry. *Construction Management and Economics, 40*(7-8), 673–684. DOI: 10.1080/01446193.2022.2093743

Oke, A., Aigbavboa, C., & Ebekozien, A. (2023). Exploring Innovative Approaches to Robotics Adoption in the Nigerian Construction Industry. *Journal of Construction Technology and Management, 145*(5), 111–126. DOI: 10.1016/j.jctm.2023.04.015

Olanipekun, A. O., Aje, I. O., & Fapohunda, J. A. (2015). Critical Success Factors for Construction Projects in Lagos State. *Journal of Project Management, 13*(3), 101–117.

Olaoye, F., Potter, K., & Doris, L. (2024). *Robotics and AI: Synergies and Impact on Automation*. EasyChair.

Oluseye, O., & Aigbavboa, C. (2020)... *Practice-driven Difficulties and Need-driven Plausibility for the Utilization of Robotic and Autonomous Construction Systems in Nigeria., 30*, 1–15.

Owolabi, K. A., Okorie, N. C., Yemi-Peters, O. E., Oyetola, S. O., Bello, T. O., & Oladokun, B. D. (2022). Readiness of academic librarians towards the use of robotic technologies in Nigerian university libraries. *Library Management, 43*(3/4), 296–305. DOI: 10.1108/LM-11-2021-0104

Pan, M., Linner, T., Pan, W., Cheng, H. M., & Bock, T. (2020). Influencing factors of the future utilization of construction robots for buildings: A Hong Kong perspective. *Journal of Building Engineering, 30*, 10–20. DOI: 10.1016/j.jobe.2020.101220

Pradhananga, P., ElZomor, M., & Santi Kasabdji, G. (2021) Identifying the challenges to adopting robotics in the US construction industry. *Journal of Construction Engineering and Management*, ; 147(5), pp. 05-021.

Rane, N (2023); Integrating Building Information Modelling (BIM) and Artificial Intelligence (AI) for Smart Construction Schedule, Cost, Quality, and Safety Management: Challenges and Opportunities. Cost, Quality, and Safety Management: Challenges and Opportunities (September 16, 2023).

Regona, M., Yigitcanlar, T., Xia, B., & Li, R. Y. M. (2022). Opportunities and adoption challenges of AI in the construction industry: A PRISMA review. *Journal of Open Innovation, 8*(1), 45. DOI: 10.3390/joitmc8010045

Sales, S. (2024). Emerging Technologies in the Nigerian Construction Industry: The Role of Robotics. *Construction Innovation, 19*(4), 351–369.

Seyman-Guray, T., & Kismet, B. (2023). Drivers and barriers to implementing XR technologies in the construction industry in Turkey. *International Journal of Construction Management, 120*, 1–16.

Shakur, M. S., Lubaba, M., Debnath, B., Bari, A. B. M. M., & Rahman, M. A. (2024, March 05). Exploring the Challenges of Industry 4.0 Adoption in the FMCG Sector: Implications for Resilient Supply Chain in Emerging Economy. *Logistics (Basel)*, *8*(1), 27. DOI: 10.3390/logistics8010027

Shojaei, R. S., & Burgess, G. (2022). Non-technical inhibitors: Exploring the adoption of digital innovation in the UK construction industry. *Technological Forecasting and Social Change*, *185*, 12–20. DOI: 10.1016/j.techfore.2022.122036

Shrestha, N. (2021). Factor analysis as a tool for survey analysis. *American Journal of Applied Mathematics and Statistics*, *9*(1), 4–11. DOI: 10.12691/ajams-9-1-2

Taherdoost, H. (2019). What Is the Best Response Scale for Survey and Questionnaire Design; Review of Different Lengths of Rating Scale / Attitude Scale / Likert Scale'. *International Journal of Academic Research in Management*, *8*(1), 1–10.

Thirumal, S., Udawatta, N., Karunasena, G., & Al-Ameri, R. (2024). Barriers to Adopting Digital Technologies to Implement Circular Economy Practices in the Construction Industry: A Systematic Literature Review. *Sustainability (Basel)*, *16*(8), 3185. DOI: 10.3390/su16083185

Tijani, B., & Feng, Y. (2021). Social impacts of adopting robotics in the construction industry: A systematic literature review. *InProceedings of the 23rd International Symposium on Advancement of Construction Management and Real Estate*; pp. 668-680. Springer Singapore. DOI: 10.1007/978-981-15-3977-0_51

Van Tam, N., Toan, N. Q., & Van Phong, V. (2024). Investigating potential barriers to construction digitalization in emerging economies: A study in Vietnam. *International Journal of Information Management Data Insights*, *4*(1), 10–26. DOI: 10.1016/j.jjimei.2024.100226

Vora, C., Aryal, A., Willoughby, S., & Wang, C.; (2024) Investigating Stakeholder Perception and Developing a Decision Framework for Robot Adoption in Construction. *Journal of Construction Engineering and Management,*;150(4), pp. 04-24.

Waqar, A., Othman, I., Falqi, I. I., Almujibah, H. R., Alshehri, A. M., Alsulamy, S. H., & Benjeddou, O. (2023). Assessment of barriers to robotics process automation (RPA) implementation in safety management of tall buildings. *Buildings (Basel, Switzerland)*, *13*(7), 1663. DOI: 10.3390/buildings13071663

Wei, Z., Li, H., & Zhou, Z. (2023). Human-Robot Collaboration in the Construction Industry: A New Era. *Automation in Construction*, *142*, 121–132.

Wuni, I. Y., Shen, G. Q., Osei-Kyei, R. (2020). Quantitative evaluation and ranking of the critical success factors for modular integrated construction projects, International Journal Construction Management; 22 (11), 2108–2120, .DOI: 10.1080/15623599.2020.1766190

Yamane, T. (1967). *Statistics: An Introductory Analysis* (2nd ed.). Harper and Row.

Yi, L., & Hiroatsu, F. (2022). Incentives for innovation in robotics and automated construction: Based on a tripartite evolutionary game analysis. *Sustainability (Basel)*, *14*(4), 2475. DOI: 10.3390/su14042475

Zafar, M. H., Langås, E. F., & Sanfilippo, F. (2024). Exploring the synergies between collaborative robotics, digital twins, augmentation, and industry 5.0 for smart manufacturing: A state-of-the-art review. *Robotics and Computer-integrated Manufacturing*, *89*, 10–27. DOI: 10.1016/j.rcim.2024.102769

Zhang, M., Xu, R., Wu, H., Pan, J., & Luo, X. (2023). Human-robot collaboration for on-site construction. *Automation in Construction*, *150*, 10–28. DOI: 10.1016/j.autcon.2023.104812

Chapter 9
Blockchain Solutions for Renewable Energy Distribution and Management:
Fostering Sustainability and Green-Clean Energy

Bhupinder Singh
https://orcid.org/0009-0006-4779-2553
Sharda University, India

Akash Bag
https://orcid.org/0000-0001-8820-171X
Adamas University, India

ABSTRACT

Blockchain-enabled systems further optimize energy storage management through real-time data analytics that enhance energy demand forecasting and distribution planning. Governments and organizations can use blockchain to issue and track renewable energy certificates (RECs), ensuring compliance with sustainability laws. With improving the transparency and efficiency of energy markets, blockchain technology is transforming how it distribute and manage renewable energy. Decentralized energy distribution systems have several benefits over traditional energy distribution systems. Blockchain provides a solution to this problem through P2P (peer-to-peer) energy trading, allowing both producers and consumers to securely trade their excess renewable energy. Introducing transparency and accountability, blockchain opens up new avenues for collaboration among stakeholders, including

DOI: 10.4018/979-8-3373-1892-9.ch009

utility companies, energy producers, and consumers, and enables the development of a more resilient and adaptable energy ecosystem.

1. INTRODUCTION AND BACKGROUND

The adoption of renewable energy continues to grow worldwide, and blockchain solutions will be critical to supporting transparent, efficient, and equitable energy distribution, contributing to sustainable development and reduced carbon emissions. Blockchain eliminates the need for intermediaries in energy transactions as smart contracts are integrated into blockchain platforms to automate transactions, track energy in real-time, and enable dynamic pricing models, enhancing overall operational efficiency (Forward, 2024). Blockchain also enables decentralized energy grids, where localized renewable energy sources such as- solar panels, wind turbines, etc. feed into the grid without needing a central authority to oversee. This feature provides assurance around accountability, reducing the risk of fraud and discrepancies when trading carbon credits. Energy management is a global challenge whereby the world is at a crossroads with energy demand increasing, while dependency on fossil fuels and traditional means of energy generation around the world have resulted in irreversible damage to the environment and climate change. This scarcity of non-renewable resources like coal, oil, and natural gas, combined with the increasing world population and urbanization, has created significant pressure on energy infrastructure. Consequently, many countries are experiencing energy shortages, volatile energy prices and reliance on geopolitically sensitive fuel sources. In addition, greenhouse gas (GHG) emissions from energy generation and use exacerbate climate change by increasing the likelihood of hotter weather and more severe weather events. Transitioning to sustainable, clean energy sources (such as solar, wind, hydro and bioenergy) is necessary for reducing carbon footprints and achieving international climate goals like the Paris Agreement and United Nations Sustainable Development Goals (SDGs), especially: SDG 7 (Affordable and Clean Energy) and SDG 13 (Climate Action).

There are various obstacles to the adoption of renewable energy, such as; high infrastructure costs; issues with intermittency; inefficiencies in the management of the grid itself; and a proper mechanism to facilitate energy transactions that is open and transparent. Traditional models of centralized energy distribution leave the system vulnerable to inefficiencies, monopolization, and regulatory barriers that create difficulties for consumers and businesses to access clean energy equitably. The energy theft, absence of real-time monitoring and difficulties in the integration of multiple renewable sources with the existing power grid make it all the more difficult to scale sustainability efforts (Sharma & Gupta, 2024). Therefore, innovative solutions are required to develop decentralized, efficient, and transparent energy distribution systems. Blockchain has also become the game-changing solution to these issues, providing a decentralized and secure mechanism for facilitating energy transactions. Blockchain offers the renewable energy sector an opportunity to streamline operations, decrease transaction costs, facilitate peer-to-peer energy trading, and provide data integrity, all contributing to a more sustainable and equitable global energy landscape.

Figure 1. shows the Dimensions of Introduction Section (Source- Self Prepared-Original)

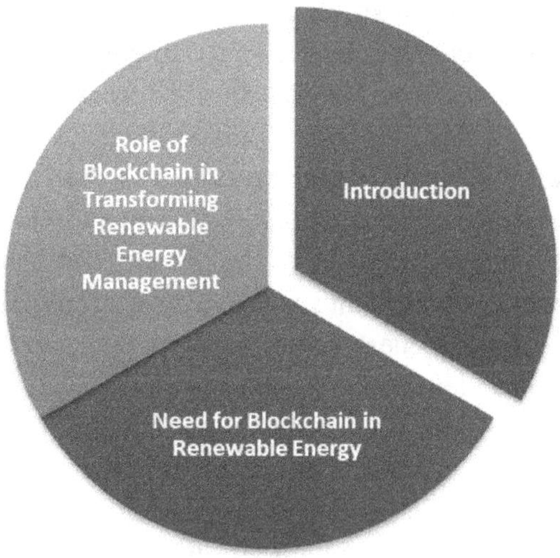

1.1 Role of Blockchain in Transforming Renewable Energy Management

With the assurance of decentralization, transparency, and secure energy management, blockchain technology is expected to transform the renewable energy sector. Traditional energy distribution networks depend on a centralized authority with intermediaries between the producer and the consumer, while blockchain makes it possible through smart contracts to execute transactions directly, peer-to-peer, stripping out intermediaries, reducing costs, and increasing efficiency. Their distributed ledger system records all of the energy transactions in a transparent manner that is immutable, meaning the integrity of the data is preserved. This transparency is important for tracing renewable energy sources, verifying carbon credits, and for compliance with sustainability policies.

The deployment of blockchain in the renewable energy sector has several applications, one major one being its role in enabling decentralized microgrids, where local communities can generate, store and trade renewable energy without depending on traditional utility companies. It improves energy security, minimizes transmission losses, and fosters self-reliance. It records every transaction, so users have access to real-time records of energy production and energy consumption, empowering them to manage energy usage according to demand dynamics and supply. The energy trading is automated via smart contracts, allowing transactions to be processed only when certain agreed-upon conditions have been fulfilled, and removing intermediaries. Blockchain improves grid balancing and energy forecasting which helps combine renewable energy into the current electric grids. In combination with Blockchain, energy providers can use AI and IOT devices to forecast consumption patterns, identify inefficient usage, and improve energy storage systems. The tokenization of renewable energy assets is also transformative, as blockchain allows assets of energy infrastructure to be fully fractionalized, and individuals and businesses can invest in clean energy projects due to the decision (Hasan et al., 2024). Such democratization of energy investment speeds the transition both from fossil-based energy systems to sustainable energy systems.

Blockchain has the potential to disrupt renewable energy management but needs to overcome issues such as scalability and regulatory compliance as well as energy-intensive mining processes. However, ongoing studies and the introduction of less-greedy consensus approaches such as PoS (Proof of Stake) and PoA (Proof of Authority) have been evolving to diminish them. In general, blockchain is a promising approach to apply to energy by creating a greener, more transparent business model enabling consumer empowerment in the clean energy revolution.

1.2 Need for Blockchain in Renewable Energy

Shifting to renewable energy is critical to containing climate change and achieving sustainable energy security. Yet, existing energy systems need improvement and the inefficiency, corruption, and centralization of power limit the accessibility and scalability of available renewable energy sources. In fact, this decentralization, independently of any corruption, is precisely the solution to the renewable energy distribution and management bottleneck provided by blockchain technology (Jin et al., 2024). Blockchain reduces operational costs and promotes a more democratized energy ecosystem by eliminating intermediaries and allowing peer-to-peer transactions. Blockchain technology has the potential to revolutionize the renewable energy sector by enabling decentralized energy transactions, enhancing security and transparency, and streamlining processes.

1.2.1 Rising Demand for Renewable Energy

Population growth, industrialization, and the rapid expansion of digital economies are driving a rise in energy demand across the globe. At the same time, there is increasing recognition of the environmental costs of fossil fuel use, and governments and businesses are being called on to hasten transfer to renewable energy. This drives investments in clean energy solutions by adopting polices like carbon pricing, renewable energy mandates and international climate agreements. But, while solutions for renewable energy sources grow in number and supply, many consumers are still unable to obtain clean energy, with the hassle of inefficiencies in distribution networks, alongside costly systems remaining a core part of consumers' energy bills and long-term costs. Solution: Blockchain technology facilitates peer-to-peer transactions, overcoming the limitations of traditional energy distribution systems and reducing reliance on centralized utility providers, ultimately decreasing energy expenses (Martini et al., 2024).

1.2.2 Issues in Traditional Energy Distribution

Conventional energy distribution systems rely on centralized power grids controlled by a limited number of major entities resulting in monopolization, price manipulation and inefficiencies. Lack of options: consumers may have few choices in how to choose their energy provider and what they rate the provider on, while transparency issues can make it hard to identify the source of their energy supply. Also power is lost in conventional grids during transmission and conventional grids are vulnerable to cyber-attacks and aging infrastructure, making energy distribution unreliable. The decentralized ledger system of Blockchain provides a remedy here, establishing a tamper-proof record of energy transactions, helping to ensure that renewable energy is tracked accurately, distributed fairly and priced adequately (Satheesh & Shukla, 2024).

1.2.3 Empowering Unsecure and Inefficient Sectors through Blockchain

This allows it to address several bottlenecks inithe renewables industry: real-time energy monitoring, microgrid management, and trading automation via smart contracts, to name a few. Some of the most promising applications of blockchain technology will be peer-to-peer (P2P) energy trading, allowing individuals with solar panels or wind turbines to sell their excess energy directly to other users in their community. This can help reduce transmission losses, increase energy accessibility, and strengthen local energy resilience (Vacchi et al., 2024). By means of the factoring of energy assets based on the blockchain, it enables the investing in renewables and thus reduces the cost of the project, as mentioned earlier, and as it will be for the renewable energy sources.

The decentralization supports grid stability through integrated energy storage and distribution networks. Blockchain-based energy management systems, powered with IoT devices and artificial intelligence, can help optimize energy flows, anticipate changes in demand patterns, and prevent overloads in the grid. Blockchain can make the allocation of renewable energy credits more transparent, making sustainability claims less prone to greenwashing. With the emergence of energy-conserving consensus mechanisms and government-sponsored pilot projects, they are developing into the pathway toward the mainstream adoption (Sarkar et al., 2024). The blockchain can change a lot in distributing of renewable energy via providing efficiency and inclusiveness, and this can help achieve a transition towards a global sustainable energy future.

2. BLOCKCHAIN TECHNOLOGY IN RENEWABLE ENERGY: CONCEPT AND FUNCTIONALITY

Blockchain technology is decentralized digital ledger that can record a transaction between two parties in a secure, verifiable way. Blockchain was simply the central technology behind cryptocurrencies like Bitcoin but has evolved into a wider spread application across industries, including that of renewable energy (Ghobakhloo et al., 2024). This ability to deliver in a P2P network where energy transactions are verified and recorded by each of the nodes in the network paves the way for decentralising energy management. It simplifies the process by removing intermediaries, minimising transaction costs, and increasing efficiency and transparency in the distribution and trading of renewable energy. This platform allows them to solve a range of energy challenges, including inefficiencies in energy distribution, grid management issues, a lack of transparency in energy sourcing and barriers to energy accessibility.

Blockchain is a specific implementation of leanings from many protocols in distributed ledger technology (DLT) where data is written to the multiple nodes rather than a single centralized database. Any attempt to change the data would need the agreement of the whole network, meaning all transactions are permanent and cannot be tampered with. Blockchain is suggested on the basis of a consensus mechanism which is utilised to validate and also verify transactions before the transactions are permanently stored in the ledger (Gamberini & Pluchino, 2024). For the selected applications of Blockchain in the field of renewable energy, the two most popular consensus mechanisms are Proof of Work (PoW) and Proof of Stake (PoS). PoW requires a significant amount of computational power to validate transactions, while PoS is a more energy-efficient alternative for projects centered on sustainable energy. In addition to those mentioned, alternative consensus mechanisms like Proof of Authority (PoA) and Proof of History (PoH) are becoming more popular due to scalability and lower energy consumption.

Smart contracts are one of the most critical features of blockchain in renewable energy, these are self-executing contracts that automatically carry out predetermined operations when specified conditions are satisfied. Smart contracts allow transactions between energy producers and consumers to be automated and conducted in a trustless manner, a process that does not require third-party intermediaries (Raghav et al., 2024). In a decentralized energy marketplace for renewable energy sources like solar power, smart contracts could be configured to allow for direct peer-to-peer transactions which would guarantee that consumers are able to obtain electricity from only renewable sources and not have to depend on conventional utility providers like AGL. This lowers operation costs, reduces disputes, and increases transaction security. Smart contracts can also be used in the tradability of renewable energy certificates (RECs) and carbon credits, allowing sustainability claims to be verified and reducing fraudulent greenwashing.

The other important aspect of blockchain in renewable energy is microgrid management. Conventional energy facilities have centralization in the typical sense making them disadvantages to efficiency, power failure, and even transmission losses. Microgrids underpinned by blockchain allow energy communities to autonomously generate, store, and trade renewables. These microgrids operate by linking several renewable energy sources like solar panels and wind turbines with blockchain-equipped platforms for real-time tracking and flexible pricing. Consumers with energy production capabilities are able to sell excess electricity directly to their neighbors via peer-to-peer (P2P) energy trading, which promotes local energy resilience and reduces reliance on traditional grid operators. Such decentralization enhances energy accessibility while encouraging community-driven sustainability initiatives (Martín-Gómez et al., 2024).

Blockchain, IoT, AI, grid optimization and demand-side management Renewable energy is one such area where IoT-based solutions can contribute substantially. After this data is recorded on-chain to maintain its integrity and correctness from being manipulated. AI algorithms can process this data to optimize energy distribution, forecast demand trends, and avoid grid congestion. Historically, blockchain allows for autonomy in specific ideal sectors like smart-grids though it must be noted that, in the absence of blockchain smarts grids as small localised smart grids, autonomous smart grids operate to meet highest possible energy production needed during periods of peak demand; using similar stateless contract as above allowing for instant online adjustment. Moreover more advanced automated solutions of storing energy built on blockchain would assure that excess renewable energy is stored for later use making solar and wind power less intermittent (Doluweera, 2024).

The prime features of blockchain for renewable energy management are tokenization. Fractional ownership of renewable energy as people can invest in clean energy projects, even on a small scale, to support clean energy initiatives. Tokenized assets in the renewable energy market democratize investment by reducing barriers to entry and improving liquidity. Blockchain energy tokens allow for fractional ownership of large-scale sustainability projects, enabling small investors to invest in solar farms, wind power plants, or battery storage systems. These tokens may be incorporated into carbon trading platforms, enabling enterprises to offset their carbon footprint in a transparent and verifiable way.

A blockchain technology solution has also proved crucial in traceability and transparency within energy supply chains. Make the renewable energy sector more transparentAs it stands one of the biggest challenges of the renewable energy sector is lack of transparency in energy sourcing and certification (Rane et al., 2024). Many consumers and businesses have no way to check whether the electricity they buy actually comes from renewable sources. This is where blockchain technology can help, as it provides an immutable record of energy production and distribution. Decentralized tracking systems allow consumers to trace the source of their electricity, confirming that it meets sustainability criteria. That is vital, especially for corporations working to meet environmental, social and governance (ESG) criteria, and for governments imposing renewable energy mandates (Sharma & Gupta, 2024). By connecting the blockchain to certified green energy markets, stakeholders can help to verify that renewable energy credits are properly issued, traded, and retired, adding an extra layer of accountability in the energy sector.

However, despite its positive capabilities, the application of blockchain for the distribution and management of renewable energy also comes with some disadvantages and limitations. Scalability is the main issue, as many blockchain networks experience expensive transaction fees or slow processing times. One sector that has goals and requirements of sustainability is the energy sector and public blockchains need a lot of computational power to validate transactions, which may conflict with those goals. The growing popularity of cryptocurrencies in the past couple of years has brought about several scalability challenges, especially with Bitcoin (BTC) transactions, which can often be slow and expensive with the minimum transaction fees required. The regulatory uncertainty makes even wide use of blockchain technology in energy a hurdle to jump over at all, as energy markets are regulated types of markets, and platforms for energy exchange based on blockchain technology may not even have legal foundations just yet to stand on (Valeriya et al., 2024). This also requires governments and policymakers to create regulatory sandboxes and pilot programs that can test how blockchain can be integrated into existing energy infrastructures while complying with energy laws and data protection regulations.

Blockchain can work as an innovative remedy for the renewable energy sector in this regards. Being decentralized makes it inherently transparent, efficient and accessible in energy transactions, with functions like smart contracts, peer-to-peer energy trading, microgrid management and tokenization allowing for innovative means of acting towards sustainability. The fusion of blockchain with IoT, artificial intelligence, and renewable energy technologies enables participants to optimize grid management, improve energy storage solutions, and guarantee traceability in energy supply chains. Despite persistent scalability and regulatory challenges, ongoing innovations in blockchain protocols and an enabling policy environment are setting the stage for extensive adoption. With the continued drive for clean energy around the world, blockchain technology will help speed the transition to a more sustainable, decentralized and resilient energy future (Atiku et al., 2024).

3. DECENTRALIZED RENEWABLE ENERGY DISTRIBUTION AND PEER-TO-PEER (P2P) TRADING

These systems enable the production, storage, and usage of clean energy in a local framework, free of traditional utility providers. This fluidity is especially important for integrating distributed energy resources (DERs) like solar panels, wind turbines, and battery storage systems throughout electrification and into the esoenergy grid. P2P energy trading, which enables prosumers (individuals who generate and consume energy) to trade electricity directly with other members of their social community, represents one of the most innovative applications of decentralized renewable energy distribution. With the help of blockchain technology, P2P trading is growing in popularity because it lowers energy expenses, widens energy access, and removes the need for a middleman, hence making sustainable energy more accessible and affordable (Ramírez-Márquez et al., 2024).

The conventional energy grid is substantially decentralised electricity is produced in large-scale plants, conveyed over extended areas, and delivered to final consumers via utility companies. Centralized energy generation systems, such as this, lose a lot of energy through transmission, are inefficient at distributing energy to balance load and supply depend on fossil fuels, and do not provide consumer control. Grid systems that operate on a centralized basis are susceptible to disruption due to natural disasters, cyberattacks and other infrastructural breakdown resulting in outtages and shortages of supply. In comparison, decentralized renewable energy distribution utilizes local energy generation units which form microgrids, operating autonomously and/or alongside national grids (Dacre et al., 2024). Microgrids can provide energy independence and local renewable energy generation, especially in remote areas where connecting to the traditional grid is not economically feasible.

3.1 Evolution of P2P Energy Trading and Blockchain and Microgrid Management

An innovative method of decentralized energy distribution, P2P energy trading empowers energy consumers to buy electricity directly from local renewable energy producers instead of from conventional utility giants. Utilizing blockchain-based smart contracts, this model allows for automated, transparent, and secure energy trades to occur between buyers and sellers. Peer to peer trading in a network, enables the prosumers to immediately sell the overproduction piled up from their rooftop solar panels or wind turbines to neighboring households or business. Blockchain allows each transaction to be recorded in an based on supply and demand. Common to these networks is the usage of smart contracts, which, in combination with the trade of tokenized power contracts as proof of energy helps to eliminate intermediaries, thereby cutting down transaction costs and improving overall energy trading efficiency (Ghobakhloo et al., 2024).

P2P energy trading significantly reduces reliance on fossil fuels and fosters sustainability, one of its primary benefits. P2P trading reduces dependence on centralized power plants, which tend to burn coal or natural gas, by promoting the use of locally generated renewables. It encourages individuals to invest in renewable energy systems by offering a bonus to sell them surplus electricity instead of wasting it. This can be expanded to encourage the adoption of solar and wind energy at the community level, leading to the development of a more robust energy ecosystem. P2P energy trading enables consumers to play a more active role in their energy consumption decisions, allowing them to support clean energy practices and reduce their own carbon emissions.

Decentralized energy distribution and P2P trading can also play an important role in improving energy resilience and security. Centralized grids can be also prone to disruption from extreme weather events, cyberattacks, or geopolitical conflicts. Conversely, decentralized energy systems spread the process of energy generation across many smaller type sources, minimizing the risk of major outages. Especially during emergencies, microgrids that use renewable-powered generators can keep providing electricity to critical infrastructures like hospitals, water treatment facilities or the networks of communications enabling these services (Narkhede et al., 2024). For instance, blockchain-powered P2P energy trading platforms can implement dynamic pricing models to alter electricity prices in response to real-time demand and supply conditions, thus providing incentives for consumers to reduce their energy demand during peak times, ensuring energy supplies, and optimizing grid performance.

Whether its decentralized energy distribution, P2P energy trading, or microgrids, decentralized energy solutions offer several advantages. One of the biggest obstacles is regulatory uncertainty; most energy markets are still regulated with rules written for centralized power systems. Legal and operational uncertainties due to the lack of clear policies and framework on Peer-to-Peer energy trading by many governments and utility companies Introducing blockchain-oriented P2P trading platforms into the current grid system would necessitate major technological overhauls and capital investments. The need for scalability and interoperability is another issue because the various protocols on blockchain and differing energy standards could pose a challenge to enable efficient data exchange between various linkages in the energy ecosystem.

Addressing these challenges requires teamwork between governments, regulators, and industry stakeholders to implement appropriate policies and develop technologies that help support decentralised energy systems. Regulatory sandboxes enable policymakers to determine the pros and cons of experimenting with P2P energy trading without the risks of sudden and widespread systemic shock (Mourtzis & Angelopoulos, 2024). The steps taken in Blockchain scalability solutions Layer 2 networks or off-chain transactions can also help to improve the efficiency of P2P trading platforms, enabling more widespread usage of these systems. The use of decentralized renewable energy generation and P2P trading is a game-changing approach to a more sustainable, practical and reliable energy future. This model has the potential to not only reshape the global energy landscape, but also enable local energy generation, consumer empowerment, enhanced resilience from one point energy provision, and lessen emphasis on fossil fuel reliance. Despite challenges and a long path ahead, advances in big data solutions to streamline regulation and technology, alongside supportive policy initiatives, will lead the way towards the deployment of decentralised energy systems (Lo et al., 2024). This data-driven revolution can improve both energy efficiency and energy access, as blockchain-enabled P2P energy trading democratizes the way we produce energy and accelerate the world to a cleaner, more sustainable energy future.

4. SECURITY, TRANSPARENCY AND TRUST IN BLOCKCHAIN ENERGY SYSTEMS

With your great development in designs, information is more secure than before, and trust in energy exchanges is restored. Centralized authorities and poorly managed data means that traditional energy systems are often cyber-vulnerable, inefficient or subject to fraud. Blockchain technology provides a decentralized, tamper-proof, and cryptographically-secured solution that safeguards energy transactions from the risk of cyberattacks, fraud, and unauthorized alterations (Kannan et al., 2024). Utilizing distributed ledger technology (DLT), blockchain offers a transparent and tamper-proof record of all energy transactions, creating trust among all participants of the ecosystem, such as energy producers, consumers, regulators, and grid operators. Aligning with international regulatory frameworks and ensuring interoperability with conventional energy grids is essential for the inclusion of blockchain in the already established energy infrastructure.

4.1 Cybersecurity Issues and the Features of Blockchain

Cybersecurity is still a significant threat to modern energy systems, with cyber-criminals increasingly attacking centralized grid infrastructures and digital energy markets. Hacking, data breaches, and distributed denial-of-service (DDoS) attacks on power transaction systems have been reported that are disruptive to power distribution and put user data at risk. Blockchain technology reduces such risks by offering decentralised mechanisms for security, which do not have a single point of failure. Because blockchain is a distributed system where transactions are verified through consensus mechanisms, it becomes nearly impossible for corrupt actors to change records or manipulate the market (Rane et al., 2024).

The Blockchain and its cryptographic security features are some of the main benefits in an energy implementation. Asymmetric cryptography is used to encrypt and digitally sign every single transaction to make sure that only authorized parties can access and modify transaction data. Smart contracts also automate energy transaction procedures, minimizing human involvement and decreasing potential security loopholes. In combination with zero-knowledge proofs (ZKPs), blockchain energy platforms can use multi-signature authentication to improve data privacy and security, enabling participants to exchange energy without revealing sensitive data. With the evolution of cyber threats, ongoing innovation of blockchain consensus protocols and secure multi-party computation (SMPC) will be essential to ensuring the resilience of blockchain-enabled energy markets (Piron et al., 2024).

4.2 Transparency and Fraud Prevention

Blockchain technology used in renewable energy systems increases transparency and accountability as all energy transactions is real, traceable, and verifiable. Data manipulation, double-spending and fraudulent billing practices are common in conventional energy markets, resulting in financial losses and market inefficiencies. This risk does not exist with blockchain, which combines all transactions into a single immutable ledger all transactions are time-stamped in the ledger, with no feasible way to change the information recorded in the history. The use of real-time transaction validation and decentralized consensus mechanisms e.g. Proof of Stake (PoS), Byzantine Fault Tolerance (BFT) in blockchain energy markets (Ciucu-Durnoi et al., 2024). These mechanisms ensure that only legitimate transactions are added to the blockchain and that issues such as double-spending (selling the same unit of energy multiple times) are avoided. Tokenizing renewable energy credits (RECs): Another example of how blockchain can reduce the risk of fraud, REC blockchain enables clear tracking of green energy production and consumption, so that companies making claims of sustainability compliance are able to be monitored.

Smart metering systems based on blockchain facilitate automated and transparent energy accounting, in which consumers are charged only for the energy they consume. It does away with arguments related to billing differences and builds confidence between energy makers and users (Ghobakhloo et al., 2024). By establishing blockchain-based audit trails, governments and energy regulators can also oversee energy markets in real-time to ensure compliance with policies on carbon reduction and sustainability. Many of these renewable energy trading platforms have incorporated decentralized identity verification and anti-money laundering (AML) protocols that use blockchain to add another layer of fraud prevention.

4.3 Trust and Compliance with Regulatory Frameworks

In order to see a top level blockchain implementation in energy distribution, we need to see compliance with the international energy regulations, and an unbroken Interoperability with the conventional energy grids. Regulators are important in this process, as they help shape legal frameworks and guidelines that ensure energy solutions based on blockchain are provisions that correspond with technological advancement and global objectives for sustainability, energy security, and consumer protection laws. The lack of standardized blockchain governance models in energy markets is one of the key challenges of regulatory compliance (Katari et al., 2024). The inherent differences in energy trading regulations, data privacy laws, and grid interconnection policies that vary by country or region introduce legal complexities that can hamper the adoption of blockchain technology. This has led to proposals for regulatory sandboxes in several nations, which would be a controlled environment within which energy innovators can deploy blockchain solutions prior to full-scale implementation.

The crucial aspect for real utilization of blockchain in energy markets is interoperability with the already existing energy grids. Traditional grids are based on centralized control systems, but blockchain works on decentralized peer-to-peer systems. Integration of blockchain-based solutions for energy management with smart grid technologies, IoT-connected appliances, and automated demand-response systems is critical for improving grid stability and efficiency (Bhat & Parvez, 2024). In addition, smart contracts powered by blockchain technology need to be integrated into cross-jurisdictional agreements governing cross-border energy trading, to allow for automated energy transactions to occur across different political and commercial landscapes.

For successful deployment of blockchain-based renewables systems, security, transparency, and trust are very important prerequisites. Cybersecurity threats, fraud prevention and other compliance-related issues, blockchain can change the whole way that renewable energy is managed and distributed. The emergence of blockchain-infrastructure projects, as policymakers work with their energy providers and tech creators, will form the basis for a resilient, transparent global energy ecosystem in the coming years.

Figure 2. express the points on Security, Transparency and Trust in Blockchain Energy Systems (Source- Self Prepared- Original)

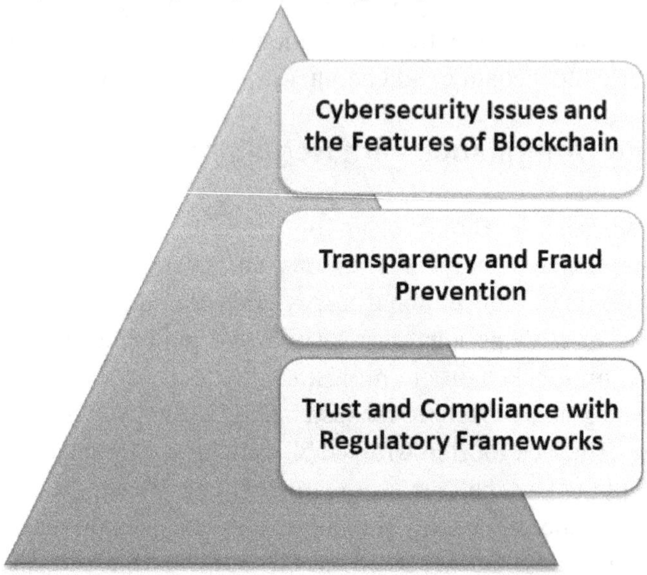

5. BLOCKCHAIN-BASED ENERGY MANAGEMENT SYSTEMS (BEMS)

Blockchain-Based Energy Management Systems (BEMS) offer a solution with a revolutionary approach to optimize energy distribution, consumption, and storage in renewable energy networks. Because of traditional control methods and dated grid design, legacy energy management systems can be inefficient, opaque, and susceptible to security risks. By enabling decentralized, tamper-proof, and automated mechanisms, Blockchain Technology solves these challenges, bringing better real-time energy monitoring, improving grid stability, and optimizing efficiency (Ouariach & Raghav, 2025). By integrating blockchain with smart grids, Internet of Things (IoT), and Artificial Intelligence (AI), we can achieve real-time energy tracking, predictive analytics, and intelligent decision-making in energy distribution, leading to a more resilient, transparent, and sustainable energy ecosystem. And in addition to that, blockchain makes energy storage and load balancing more efficient in terms of the amount needed to be produced and distributed in order to balance out supply and demand and stabilize the grid for the least amount of waste.

5.1 Blockchain-Based Smart Grid Integration

Smart grids powered by blockchain allow real-time monitoring and distribution of renewable energy throughout connected networks. Traditional grids suffer from data processing delays, unauthorized consumption, and grid imbalance, resulting in energy waste and blackouts. Blockchain addresses identified problems by providing a distributed, public, and real-time verification ledger for all energy transactions posted on it. Such built-in security is one of the reasons you can prevent unauthorized access, provide correct energy invoicing and even enable P2P (peer-to-peer) energy trading between prosumers.

A smart grid can further benefit from the distributed technology through automation in demand-response, improving demand-side performance and boosting grid stability (Zhen & Yao, 2024). Smart contracts that automatically executed contracts programmed on the blockchain can manage energy flow, optimize supply and demand, and avoid overloads. It can automatically transfer surplus energy from decentralized plants to areas facing a shortage, thus enhancing the overall resilience and efficiency of the grid when energy demand reaches its peak. In fact, the blockchain-based microgrid mechanism enables local communities to sell back excess solar or wind energy back to the network and thus, minimizes the need for large, centralized power plants.

5.2 IoT and AI in Blockchain Energy Systems

IoT and AI applications with blockchain energy systems improve tracking and forecasting and efficiency across renewable energy networks. Internet of Things (IoT) sensors embedded in smart grids, solar panels, wind turbines, and energy storage units can continuously collect and transmit data on energy production, consumption, and storage levels. All of this real-time data is recorded on the blockchain in a transparent and tamper-proof manner, allowing for energy tracking. This can eliminate the need for human interference and the risks involved in manipulating the data, thus giving a true and trusted depiction of the flows of energy and improving the process of decision making for both grid operators and consumers (Bliznina, 2024).

Renewable energy efficiency is significantly enhanced by AI-based predictive analytics. By leveraging historical energy consumption patterns, weather forecasts, and grid performance data, AI algorithms can predict future fluctuations in energy demand and supply. AI, in conjunction with blockchain, allows for adaptive regulation of energy flow to ensure there are no power shortages, and eliminates energy waste (Bakator et al., 2024). Predictive maintenance powered by AI of renewable energy resources as wind turbines or solar farms detects potential equipment failures before happening, reducing downtime and maximising energy output. Integrating this cutting-edge tech stack results in an intelligence-driven, adaptive security energy management ecosystem capable of ensuring sustainability while minimizing capital expenditures (Santos et al., 2024).

5.3 Blockchain in Energy Storage and Load Balancing

Energy storage is crucial for stabilizing power grids and providing a constant energy supply, especially for sources of energy that are not continuous, such as the sun and wind. In decentralized storage solutions blockchain technology steps in and helps, where consumers can store surplus energy in battery storage units and sell when other consumers need it. This increases energy resilience, decreases dependence on fossil fuel-powered backup generators and enables higher penetrations of renewable energy sources (Ikram et al., 2024). This means each unit of stored energy is tokenized and recorded on the blockchain achieving secure and transparent tracking. This allows consumers and businesses to trade the stored energy as part of an open market, all of which is enforced through smart contracts that initiate transactions based on demand and supply conditions (Barata & Kayser, 2024).

This also lowers costs, eliminates dependence on centralized energy providers, and incentivizes local, efficient energy storage usage.

Blockchain facilitates energy efficiency in power distribution through intelligent load balancing mechanisms. According to the outdated infrastructure and variable demand pattern, traditional energy distribution networks frequently experience grid congestion and power losses. Blockchain can make load balancing autonomously shifted by redistributing the surplus energy to others with higher demand. AI-driven algorithms further facilitate power flow optimization by anticipating demand spikes and dynamically redistributing resources in response to demand changes. This results in more efficient and sustainable energy systems because it prevents overloading and consequently reduces energy wastage (Rajumesh, 2024).

Blockchain Technology is a promising change maker for Energy Management Systems. Blockchain also allows real-time energy tracing, predictive analytics, and automated decision-making through smart grid integration, IoT, and AI, enabling significant improvements in energy efficiency and sustainability. Load balancing and energy storage solutions leverage blockchain technology to manage the flow of power, further enhancing the stability and resilience of the grid. The global energy sector is shifting toward decarbonization and digitalization (Rumayor et al., 2024). It makes blockchain technology an essential building block for driving a clean, transparent, and efficient energy ecosystem, ultimately supporting the transition to a greener and more sustainable future.

6. CONCLUSION

With utilizing blockchain technology, however, a transformative solution for the distribution and management of renewable energy enables the creation of a sustainable, transparent and efficient energy ecosystem. Through eliminating intermediaries, improving security, and supporting decentralized energy trading, blockchain significantly increases the efficiency, accessibility, and affordability of green energy. Coupled with smart grids, IoT and AI, real-time tracking, distribution, and predictive analytics reduce energy waste and maintain stability in energy supply. Blockchain-based peer-to-peer (P2P) trading platforms and energy tokenization enable customers and prosumers to engage freely in clean energy markets, creating a decentralized and democratized energy ecosystem.

Blockchain has the potential to overcome barriers to market access, but issues including scalability, regulatory uncertainty, and energy-intensive consensus processes remain obstacles to adoption in the renewable energy sector. Overcoming these challenges will necessitate ongoing technological innovation, including energy efficient consensus mechanisms e.g., Proof of Stake, interchange with conventional power grids, and adherence frameworks for regulation. Governments should incentivize green energy use, standardize blockchain applications, and develop policies that promote blockchain-based energy markets with energy providers and technology providers.

7. FUTURE PERSPECTIVES AND OPPORTUNITIES IN BLOCKCHAIN-ENABLED RENEWABLE ENERGY

Blockchain in renewable energy will grow and, and blockchain-based carbon credit trading, AI-driven energy forecasting, and decentralized autonomous energy organizations be at the forefront of this trend. With global initiatives to mitigate climate change and move towards a clean energy future, blockchain technology will act as the cornerstone technology to create resilient, renewable, and efficient energy systems. This will spark a global transition to a carbon-neutral economy, as blockchain solutions connect technological, economic and environmental objectives to pave the way for a greener, cleaner and more integrated energy future. The incorporation of blockchain in renewable energy systems emerges with revolutionary future outlook and ample opportunities. Blockchain, by virtue of its decentralization and transparency, has the potential to simplify peer-to-peer (P2P) energy trading, creating a true marketplace through which consumers can trade excess renewable-generated electricity, such as solar and wind. This represents a democratization of energy markets that promotes local energy resilience and minimizes reliance on large, centralized grids. Smart contracts further automate transactions and provide efficiency, trust, and security on decentralized energy exchanges.

Blockchain can enhance the tracing and certification of renewable energy credits to assure that they are genuine and prevent fraud in carbon markets. It may also support open source tracking of supply chains for renewable energy infrastructure, contributing to sustainable purchasing. With the advent of Internet of Things (IoT) and smart-grids blockchain can ensure data integrity and real-time monitoring of energy flows, which leads to more accurate demand-response strategies and grid stabilization. At an international scale, blockchain-connected micro grids have the potential to increase access to electricity in disconnected and marginalized areas that need it the most, promoting inclusive, clean energy transitions in line with SDG 7 (Affordable and Clean Energy). These include new opportunities in the proliferation of tokenized incentives for consumers as well decentralized finance (DeFi) models for financing renewable projects. Blockchain, where needed level of regulation and interoperability standards are established, may unlock doors for innovation, scaling renewable energy and in the process, decentralize, increase traceability in processes, and make energy eco-systems a worldwide business model.

REFERENCES

Atiku, S. O., Jeremiah, A., Semente, E., & Boateng, F. (Eds.). (2024). *Eco-Innovation and Sustainable Development in Industry 5.0*. IGI Global. DOI: 10.4018/979-8-3693-2219-2

Bakator, M., Ćoćkalo, D., Makitan, V., Stanisavljev, S., & Nikolić, M. (2024). The three pillars of tomorrow: How Marketing 5.0 builds on Industry 5.0 and impacts Society 5.0? *Heliyon*, *10*(17), e36543. DOI: 10.1016/j.heliyon.2024.e36543

Barata, J., & Kayser, I. (2024). How will the digital twin shape the future of industry 5.0? *Technovation*, *134*, 103025. DOI: 10.1016/j.technovation.2024.103025

Bhat, F. A., & Parvez, S. (2024). Emerging Challenges in the Sustainable Manufacturing System: From Industry 4.0 to Industry 5.0. *Journal of The Institution of Engineers (India): Series C*, 1-15.

Bliznina, N. (2024). *Redefining Economic Growth: the Review of Industry 4.0 and Industry 5.0 Progress*. Green and Low-Carbon Economy.

Ciucu-Durnoi, A. N., Delcea, C., Stănescu, A., Teodorescu, C. A., & Vargas, V. M. (2024). Beyond Industry 4.0: Tracing the Path to Industry 5.0 through Bibliometric Analysis. *Sustainability (Basel)*, *16*(12), 5251. DOI: 10.3390/su16125251

Dacre, N., Yan, J., Frei, R., Al-Mhdawi, M. K. S., & Dong, H. (2024). Advancing sustainable manufacturing: A systematic exploration of industry 5.0 supply chains for sustainability, human-centricity, and resilience. *Production Planning and Control*, •••, 1–30. DOI: 10.1080/09537287.2024.2380361

Doluweera, L. K. (2024). Industry 5.0 Technologies for Supply Chain Management through Triple Bottom Line Approach in Companies of Sri Lanka for Economic Growth and Socio-Environmental Protection.

Forward, T. W. (2024). Industry 5.0, Quality of Life, and Sustainable Adaptation to Climate Change. *Quality of Life and Climate Change: Impacts, Sustainable Adaptation, and Social-Ecological Resilience: Impacts, Sustainable Adaptation, and Social-Ecological Resilience*, 200.

Gamberini, L., & Pluchino, P. (2024). Industry 5.0: A comprehensive insight into the future of work, social sustainability, sustainable development, and career. *Australian Journal of Career Development*, *33*(1), 5–14. DOI: 10.1177/10384162241231118

Ghobakhloo, M., Fathi, M., Iranmanesh, M., Vilkas, M., Grybauskas, A., & Amran, A. (2024). Generative artificial intelligence in manufacturing: Opportunities for actualizing Industry 5.0 sustainability goals. *Journal of Manufacturing Technology Management*, *35*(9), 94–121. DOI: 10.1108/JMTM-12-2023-0530

Ghobakhloo, M., Iranmanesh, M., Fathi, M., Rejeb, A., Foroughi, B., & Nikbin, D. (2024). Beyond Industry 4.0: A systematic review of Industry 5.0 technologies and implications for social, environmental and economic sustainability. *Asia-Pacific Journal of Business Administration*. Advance online publication. DOI: 10.1108/APJBA-08-2023-0384

Ghobakhloo, M., Mahdiraji, H. A., Iranmanesh, M., & Jafari-Sadeghi, V. (2024). From Industry 4.0 digital manufacturing to Industry 5.0 digital society: A roadmap toward human-centric, sustainable, and resilient production. *Information Systems Frontiers*, •••, 1–33. DOI: 10.1007/s10796-024-10476-z

Hasan, M. R., Khan, M. A., & Wuest, T. (2024). Towards Industry 5.0: A Systematic Literature Review on Sustainable and Green Composite Materials Supply Chains. *arXiv preprint arXiv:2402.06100*.

Ikram, L., Afaf, D., & Latifa, E. (2024, May). Beyond Industry 5.0: Tracing the Past, Present, and Future of Digital Industrial Transformations. In *2024 4th International Conference on Innovative Research in Applied Science, Engineering and Technology (IRASET)* (pp. 1-7). IEEE.

Jin, Q., Chen, H., & Hu, F. (2024). Proposal of Industry 5.0-Enabled Sustainability of Product–Service Systems and Its Quantitative Multi-Criteria Decision-Making Method. *Processes (Basel, Switzerland)*, *12*(3), 473. DOI: 10.3390/pr12030473

Kannan, D., Amiri, A. S., Shaayesteh, M. T., Nasr, A. K., & Mina, H. (2024). Unveiling barriers to the integration of blockchain-based circular economy and Industry 5.0 in manufacturing industries: A strategic prioritization approach. *Business Strategy and the Environment*, *33*(8), 7855–7886. DOI: 10.1002/bse.3886

Katari, V., Goyal, S., Nigam, V., Jana, M., Maitra, A., Sharma, H., & Sahu, K. K. (2024). A Confluence of Emerging Technologies Like IoT, Edge & Cloud Computing, Blockchain, Industry 4.0 & 5.0, AI & ML toward the Realization of Eco-Friendly Supercapacitors. In *Eco-Friendly Supercapacitors: Design and Future Perspectives in Sustainable and Green Energy Storage Devices* (pp. 163-204). American Chemical Society.

Lo, H. W., Chan, H. W., Lin, J. W., & Lin, S. W. (2024). Evaluating the interrelationships of industrial 5.0 development factors using an integration approach of Fermatean Fuzzy Logic. *Journal of Operations Intelligence*, *2*(1), 95–113. DOI: 10.31181/jopi21202416

Martín-Gómez, A. M., Agote-Garrido, A., & Lama-Ruiz, J. R. (2024). A Framework for Sustainable Manufacturing: Integrating Industry 4.0 Technologies with Industry 5.0 Values. *Sustainability (Basel)*, *16*(4), 1364. DOI: 10.3390/su16041364

Martini, B., Bellisario, D., & Coletti, P. (2024). Human-centered and sustainable artificial intelligence in industry 5.0: Challenges and perspectives. *Sustainability (Basel)*, *16*(13), 5448. DOI: 10.3390/su16135448

Mourtzis, D., & Angelopoulos, J. (2024). Challenges and Opportunities for the Design and Development of Intelligent, Sustainable and Resilient Personalized Product-Service Systems Towards Industry 5.0. *Data-Driven Decision Making for Product Service Systems*, 27-62.

Narkhede, G. B., Pasi, B. N., Rajhans, N., & Kulkarni, A. (2024). Industry 5.0 and sustainable manufacturing: A systematic literature review. *Benchmarking*.

Ouariach, F. Z., & Raghav, A. (2025). Emerging Trends of IoT in Conservation of Biodiversity: A Futuristic Approach. *Machine Learning and Internet of Things in Fire Ecology*, 113-128.

Piron, M., Wu, J., Fedele, A., & Manzardo, A. (2024). Industry 4.0 and life cycle assessment: Evaluation of the technology applications as an asset for the life cycle inventory. *The Science of the Total Environment*, *916*, 170263. DOI: 10.1016/j.scitotenv.2024.170263

Raghav, A., Singh, B., Jermsittiparsert, K., Raghav, R., & Yadav, U. (2024). Artificial Intelligence in Environmental and Climate Changes: A Sustainable Future. In *Maintaining a Sustainable World in the Nexus of Environmental Science and AI* (pp. 485-506). IGI Global. DOI: 10.4018/979-8-3693-6336-2.ch019

Rajumesh, S. (2024). Promoting sustainable and human-centric industry 5.0: A thematic analysis of emerging research topics and opportunities. *Journal of Business and Socio-economic Development*, *4*(2), 111–126. DOI: 10.1108/JBSED-10-2022-0116

Ramírez-Márquez, C., Posadas-Paredes, T., Raya-Tapia, A. Y., & Ponce-Ortega, J. M. (2024). Natural Resource Optimization and Sustainability in Society 5.0: A Comprehensive Review. *Resources*, *13*(2), 19. DOI: 10.3390/resources13020019

Rane, N., Kaya, Ö., & Rane, J. (2024). Digital Twins for Industry 4.0, 5.0, and Society 5.0. *Available atSSRN* 4918575. DOI: 10.2139/ssrn.4918575

Rane, N. L., Kaya, Ö., & Rane, J. (2024). *Artificial Intelligence, Machine Learning, and Deep Learning for Sustainable Industry 5.0*. Deep Science Publishing.

Rumayor, M., Margallo, M., Pinedo, J., & Albo, J. (2024). Learning by Doing using the Life Cycle Assessment tool: LCA Projects in collaboration with industries. *Education for Chemical Engineers*, *48*, 44–52. DOI: 10.1016/j.ece.2024.05.002

Santos, B., Costa, R. L. C., & Santos, L. (2024). Cybersecurity in Industry 5.0: Open Challenges and Future Directions. *arXiv preprint arXiv:2410.09538*. DOI: 10.1109/PST62714.2024.10788065

Sarkar, B. D., Shardeo, V., Dwivedi, A., & Pamucar, D. (2024). Digital transition from industry 4.0 to industry 5.0 in smart manufacturing: A framework for sustainable future. *Technology in Society*, *78*, 102649. DOI: 10.1016/j.techsoc.2024.102649

Satheesh, S., & Shukla, K. (2024). Industry 5.0, Quality of Life, and Sustainable Adaptation to Climate Change: Challenges and the Transformational Way Forward. In *Quality of Life and Climate Change: Impacts, Sustainable Adaptation, and Social-Ecological Resilience* (pp. 200-219). IGI Global.

Sharma, R., & Gupta, H. (2024). Harmonizing sustainability in industry 5.0 era: Transformative strategies for cleaner production and sustainable competitive advantage. *Journal of Cleaner Production*, *445*, 141118. DOI: 10.1016/j.jclepro.2024.141118

Sharma, R., & Gupta, H. (2024). Leveraging cognitive digital twins in industry 5.0 for achieving sustainable development goal 9: An exploration of inclusive and sustainable industrialization strategies. *Journal of Cleaner Production*, *448*, 141364. DOI: 10.1016/j.jclepro.2024.141364

Vacchi, M., Siligardi, C., & Settembre-Blundo, D. (2024). Driving manufacturing companies toward industry 5.0: A strategic framework for process technological sustainability assessment (P-TSA). *Sustainability (Basel)*, *16*(2), 695. DOI: 10.3390/su16020695

Valeriya, G., Kirola, M., Gupta, M., Bharathi, P., & Acharya, P. (2024). Evaluating the Impact of AI-Based Sustainability Measures in Industry 5.0: A Longitudinal Study. In *BIO Web of Conferences* (Vol. 86, p. 01058). EDP Sciences.

Zhen, Z., & Yao, Y. (2024). The confluence of digital twin and blockchan technologies in Industry 5.0: Transforming supply chain management for innovation and sustainability. *Journal of the Knowledge Economy*, *16*(1), 1–27. DOI: 10.1007/s13132-024-02151-0

Chapter 10
A Machine Learning and IoT-Driven Framework for Real-Time Disaster Management

Nasrullah Masud
https://orcid.org/0000-0002-8760-2826
Varendra University, Bangladesh

Pallab Chowdhury
https://orcid.org/0000-0002-0907-9961
Varendra University, Bangladesh

Abdullah Al Bassam
https://orcid.org/0009-0002-9962-7374
Varendra University, Bangladesh

Md Tanvir Islam Sourav
https://orcid.org/0009-0000-8475-5408
Varendra University, Bangladesh

Abdus Sobhan
https://orcid.org/0009-0009-5139-1172
Varendra University, Bangladesh

ABSTRACT

Natural calamities are perilously challenging for people and communities that are already vulnerable. The traditional forecasting methods lack precision and would be way behind in terms of adaptability due to ever-changing climatic conditions.

DOI: 10.4018/979-8-3373-1892-9.ch010

This chapter considers a model where machine learning-based IoT-enabled weather station networks are combined for disaster forecasting and management of floods, earthquakes, cyclones, droughts, wildfires, and extreme heat. Informer model in its core form, which is a very effective transformer-based deep learning model, has got a maximum test accuracy of 91.22% in severe weather prediction. This is already implemented in Rajshahi to improve disaster readiness and response at real time and recovery. The solution is scalable and eco-friendly and beats the traditional ones in efficiency while ensuring the livelihoods with AI and IoT that would reduce agricultural losses and socio-economic impacts.

1. INTRODUCTION

These natural hazards are bound to worsen in their effects due to climate change and rapid urbanization. The annual losses worldwide are beyond $417 billion (Vanderford, 2024). Management systems for disasters traditionally rely on historical data and static models. This would not often prove to be effective under changing environmental conditions and has lagged far behind in terms of timeliness and accuracy in forecasting. They are not localized nor able to address multi-source data streams nor cascading hazards (Ahmed et al., 2024). For example, models of flood prediction using thresholds of river discharge oftentimes miss sudden rainfall anomalies which could have been detected with the use of IoT sensors (FAO, 2020); assessments of wildfires risk on variable seasonal averages do not capture real-time temperature spikes (UNDRR, 2020). Such shortcomings leave the most vulnerable communities, especially those within climate-sensitive areas of South Asia and sub-Saharan Africa, even more at risk.

Across the spectrum of adverse conditions, emergent technologies such as the Internet of Things (IoT) and machine learning (ML) are held to provide the transformative capacity to address the issues. From hyperlocal environmental monitoring via distributed sensor arrays, IoT networks embrace the longitudinal and temporal array of perceived properties for model training, while numerically robust ML models, especially transformer-based architectures such as Informer (Zhou; 2021), are most efficacious in processing long-sequence time-series data for multi-hazard forecasting. Nevertheless, despite technological advances, extant frameworks operate only in isolation. IoT deployment is largely targeted at gathering more data with little analytical integration, while ML models are content to achieve better accuracy. There is, thus, an apparent disconnect, highlighting the need for systems that blend real-time sensor data and adaptive ML algorithms to improve predictive accuracy, flexibility, and community resilience.

1.1 Research Gap

Three limitations dominate the current research on technology for disaster management. First, data latency afflicts IoT systems because micro-level communication has been fragmented due to subpar edge-computing capabilities. For instance, flood-risk sensor networks are often delayed in terms of data transmission by more than 30 minutes such that in rapidly evolving crises, the prognostics made will hold no practical relevance (Wang et al., 2024). Second, ML models have shown very little scope for generalizability across multiple geographies and hazards. For example, convolutional neural networks successfully detect wildfires using satellite images but are unable to adjust to seismic or hydrology data without retraining for extensive periods (Brian et al., 2020). The problems with deploying sensors, such as energy depletion or power inefficiency and calibration drift (MPU6050 accelerometer, for instance), affect widespread IoT application in remote locations (Zhang et al., 2023).

Now, still-managed frameworks address the different levels that do not consider simultaneous exposure to various disasters, such as epidemics following floods or landslides caused by cyclones (UNDRR, 2020). Recent investigations pinpointed how systems need to integrate multidimensional data sources-satellite imagery, social media feeds, and crowd-sourced reports-for better situational awareness (Ahmed et al., 2024). Only a few have been able to put this integration into practice, leaving an important gap in holistic approaches for disaster resilience.

1.2 Contributions

This chapter bridges these gaps through three key contributions:

1. Integrated ML-IoT Architecture: The Informer model, a transformer-based deep learning paradigm optimized for long-sequence forecasting, is unified with real-time IoT sensor data. The system processes inputs from DHT-11 humidity sensors, BMP280 barometric pressure modules, and rain gauges fusing them with historical weather archives of 2012-2023 for predicting six concurrent hazards (floods, cyclones, wildfires, etc.) with 91.22% test accuracy.
2. Adaptive Calibration Mechanism: The ability of an online learning module to calibrate model weights on the fly through real-time sensor feedback is an attractive feature to reduce prediction errors arising from sensor drift or abrupt changes in the environment (e.g., unseasonal rainfall). This innovation solves the latency and generalization issues previously identified (Wang et al., 2024).
3. Scalable Deployment in Resource-Constrained Settings: The case study in Rajshahi, Bangladesh proposes showing operational effectiveness of the framework in 12 subregions with contrasting microclimates. The low-power IoT

nodes powered systems and the Informer variant compatible with edge systems allow for reduced dependence on cloud infrastructure, thereby reducing costs for replication in Global South regions (Zhou et al., 2021).

In this framework, addressing such technical and practical deficiencies raises disaster management beyond reactive recovery to proactive resilience, providing a workable tool for policy makers in the areas of scale and action to mitigate climate risk and safeguard vulnerable populations.

2. BACKGROUND AND RELATED WORK

2.1 Evolution of Disaster Management Practices

Disaster management has gone through the same evolution from static to dynamic, data-driven systems as technology has changed; in other words, it has become the focus of these advances. Historically, early systems referred to historical information and empirical models for simulating hydrology to predict floods for their operations. Such systems basically would not be adapted to the real-time environmental change (UNDRR, 2020). The moment IoT and machine learning (ML) came into being and began combining, it shifted the whole paradigm toward monitoring and predictive analysis. IoT sensor networks are now measuring such hyperlocal environmental variables as soil moisture, wind speed, and seismic activity, and ML models using transformer-based architectures are interpreting these data streams to predict hazards such as wildfires and cyclones (Zhang et al., 2023). This transition is critical for high-risk zones, where the early intervention may stop multiple effects on socio-economics from happening (Wang et al., 2024).

2.1.1 Literature Review

Recent advancements in IoT and ML applications in disaster management show progress and challenges that persist. IoT systems do well in real-time data collection: for instance, accelerometer networks can detect seismic precursors of earthquakes (Smith & Lee, 2023), whereas UAVs track wildfire spread in unapproachable terrains (Kumar et al., 2022). Gradient-boosted tree ML models improve early warning systems by predicting flood risk with 87% accuracy (Nguyen et al., 2023). However, critical limitations remain:

1. **Scalability**: Most frameworks focus on single hazards, neglecting multi-risk interdependencies (e.g., post-flood disease outbreaks) (IPCC, 2023).

2. **Sensor Biases**: Urban-centric deployments marginalize rural areas due to power and connectivity constraints (UNDRR, 2020).
3. **Model Generalizability**: Traditional models like ARIMA fail to process long-term dependencies in environmental data, leading to unreliable forecasts (Li et al., 2023).

For instance, Patel et al. (2021) developed an IoT-based cyclone system but omitted integration with hydrological data, limiting its utility in coastal flood scenarios. Similarly, transformer models like Temporal Fusion Transformer (TFT) outperform LSTMs in weather prediction but remain untested in low-resource settings (Lim et al., 2021). These gaps highlight the need for systems that unify multi-sensor IoT data with scalable ML architectures.

2.2 Motivation for This Work

The worsening climate crisis requires predictive tools to be developed for vulnerable and resource-constrained regions. In 2023, floods affected 20 million people in South Asia, while wildfires in Canada produced 1.3 billion tons of CO_2 (IPCC, 2023). Existing solutions are, however, fragmented. IoT deployments provide limited analytical capabilities, while ML models typically rely on large-scale cloud infrastructure unsuitable for use in remote regions.

There is a trinity of imperatives that drive this research:

1. Real-Time Responsiveness: Dynamic recalibration of predictions as conditions change in the environment (rapid temperature increases causing the escalation of wildfire risk).
2. Deployment at Low Cost: Energy-efficient IoT nodes and edge-compatible ML models are needed for deployment in regions with intermittent power and internet (Kumar et al., 2022).
3. Complete Risk Assessment: A consolidated framework for all possible cascading disasters, including those caused by cyclones that trigger landslides or heatwave events that worsen the water supply situation (UNDRR, 2020).

This work bridges the divide between data collection and actionable resilience strategies by merging IoT's granular sensing with the Informer model's long-sequence forecasting.

2.3 Proposed Solution

The proposed framework (Figure 1) integrates a distributed IoT network with a transformer-based ML architecture, addressing the limitations of prior systems through three innovations:

1. **Multi-Sensor IoT Network**: Deploys low-power sensors (e.g., DHT-11 for humidity, MPU6050 for ground motion) across diverse microclimates, transmitting data via LoRaWAN to reduce latency to <5 minutes.
2. **Informer-Based Forecasting**: Leverages ProbSparse self-attention to process input sequences X∈RT×dX∈RT×d, capturing temporal dependencies in multi-hazard data with O(Tlog T)O(TlogT) complexity (Zhou et al., 2021).
3. **Edge-Compatible Adaptive Learning**: A lightweight calibration module fine-tunes model weights using real-time sensor feedback, mitigating drift (e.g., BMP280 pressure sensor errors) and environmental anomalies (Li et al., 2023).

Practical Validation: Deployed in Rajshahi, Bangladesh, the system achieved 91.22% test accuracy in flood and heatwave prediction while operating on solar-powered nodes. Unlike existing solutions, it supports concurrent hazard analysis (e.g., cyclone-induced floods) and updates predictions hourly, overcoming scalability and latency barriers (Wang et al., 2024).

Figure 1. Realtime Data Collection Device

3. SYSTEM ARCHITECTURE AND IMPLEMENTATION

3.1 Proposed Framework

The proposed framework (Figure 1) integrates IoT sensor networks with a transformer-based machine learning (ML) architecture to enable real-time disaster prediction and management. The system operates through three interconnected layers, each addressing critical gaps in traditional approaches:

1. **IoT Sensing Layer**:
   - **Sensor Nodes**: Low-cost, energy-efficient sensors (e.g., DHT-11 for temperature/humidity, BMP280 for atmospheric pressure) are deployed across disaster-prone regions. Nodes are strategically placed to capture microclimatic variations, such as riverbanks for flood monitoring and forest edges for wildfire detection (Zhang et al., 2023).
   - **Communication Protocol**: Data is transmitted via LoRaWAN, a low-power, long-range protocol, to edge gateways with a latency of <5 minutes, ensuring timely updates during rapidly evolving disasters (Kumar et al., 2022).
2. **Data Processing Layer**:
   - **Edge Preprocessing**: Raspberry Pi 4B edge nodes execute initial data cleansing, including forward-fill imputation for missing values and Min-Max normalization to scale features to [0,1]. Sensor-specific calibration offsets (e.g., correcting MPU6050 accelerometer drift) are applied dynamically (Li et al., 2023).
   - **Feature Engineering**: Derived metrics, such as *24-hour cumulative rainfall* and *3-day moving temperature averages*, are computed to enhance temporal context for ML models (Wang et al., 2024).
3. **ML Decision Layer**:
   - **Informer Model**: A transformer-based architecture processes time-series data using ProbSparse self-attention, reducing computational complexity from O(T2)O(T2) to O(Tlog T)O(TlogT). The model generates six concurrent hazard predictions (floods, cyclones, wildfires, etc.), each represented as a probability score ∈[0,1]∈[0,1] (Zhou et al., 2021).
   - **Adaptive Calibration**: A feedback loop compares predictions with real-time sensor data, fine-tuning model weights via stochastic gradient descent (SGD) to correct biases (e.g., BMP280 pressure drift) and adapt to sudden environmental shifts (Li et al., 2023).

This architecture enables **early anomaly detection** (e.g., identifying abnormal seismic activity hours before earthquakes) and **adaptive decision-making**, updating predictions hourly via a Streamlit-based dashboard.

3.2 IoT-Based Weather Station Components

The IoT network comprises six sensor types, each optimized for specific environmental monitoring tasks (Table 1)

Table 1. Specification of various electronic components

Sensor	Parameter	Specifications	Deployment Strategy
DHT-11	Temperature/ Humidity	±1°C, ±5% RH	Mounted 1.5m above ground, shaded areas
BMP280	Atmospheric Pressure	±1 hPa accuracy	Installed in ventilated enclosures
Rain Sensor	Rainfall Intensity	0.2mm resolution per tip	Elevated on rooftops, away from trees
Ultrasonic Sensor	River Water Level	±2mm resolution, 0–5m range	Fixed to bridges/poles near riverbanks
MPU6050	Ground Motion	±0.01g sensitivity, 6-axis tracking	Buried 0.5m underground
Soil Moisture Sensor	Soil Saturation	0–100% VWC, ±3% accuracy	Inserted 10cm into soil at farm sites

Situational Awareness:

- **Hydrological Risks**: The tipping bucket gauge and ultrasonic sensor track rainfall accumulation and river discharge, which is critical for flood prediction.
- **Wildfire Precursors**: DHT-11 monitors ambient dryness (humidity <30%), while BMP280 detects pressure drops linked to wind surges (Smith et al., 2023).
- **Seismic Activity**: MPU6050 captures ground vibrations (magnitude >2.0), enabling earthquake early warnings.

Sensors are deployed across 12 subregions in Rajshahi, Bangladesh, ensuring coverage of diverse microclimates, from floodplains (Bagmara) to arid zones (Talaimari).

3.3 Real-Time Data Integration and Analytics

The system employs a hybrid edge-cloud pipeline to balance computational efficiency and analytical depth:

1. **Edge-Level Processing**:
    - **Data Acquisition**: Sensors sample data at 5-minute intervals, transmitting raw readings to edge nodes via LoRaWAN.
    - **Noise Reduction**: A median filter removes outliers (e.g., erroneous rain gauge tips caused by debris).
    - **Anomaly Detection**: Isolation Forest algorithms flag malfunctioning sensors (e.g., DHT-11 reporting humidity >100%) for exclusion (Nguyen et al., 2023).
2. **Cloud-Level Integration**:
    - **Data Fusion**: Processed edge data is merged with historical archives (2012–2023) on AWS IoT Core, creating a unified dataset spanning 11 years.
    - **Time-Series Windowing**: The Informer model ingests 30-day rolling sequences of 15 features (temperature, pressure, rainfall, etc.) to forecast hazards for the next 72 hours (Zhou et al., 2021).
3. **ML Pipeline**:
    - **Feature Scaling**: Min-Max normalization ensures uniform input ranges (e.g., temperature [0°C–50°C] → [0,1]).
    - **ProbSparse Self-Attention**: Reduces redundant attention scores, focusing computational resources on critical time steps (e.g., sudden pressure drops before cyclones).
    - **Output Layer**: Six parallel fully connected layers generate hazard probabilities, with confidence intervals derived from prediction variance (Wang et al., 2024).

Validation: In Banissor (Rajshahi), a 0.2mm rainfall reading triggered a flood risk prediction of −0.0107−0.0107, aligning with ground observations of no flooding. The system achieved **91.22% test accuracy** (tolerance=±15%), outperforming LSTM (84.91%) and Random Forest (78.45%) baselines.

4. DATA INTEGRATION AND PREPROCESSING

4.1 Explain the Data Sources (Historical Records and Live Sensor Streams)

The combined repository features two segments:

1. **Historical Archives (2012–2023)**: These logs include daily records of temperature, precipitation, windspeed, and humidity. Collected from regional meteorological offices and validated by cross-referencing multiple official and local records, they offer an extended perspective on seasonal, annual, and decadal weather trends. [3]

Figure 2. Temperature 2m mean yearly average

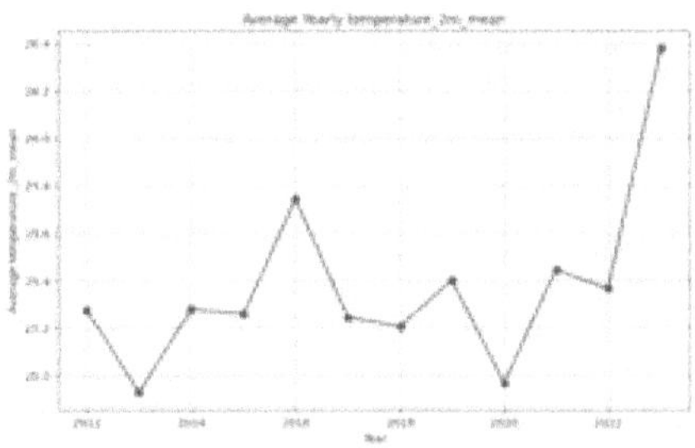

Figure 3. Precipitation sum yearly average

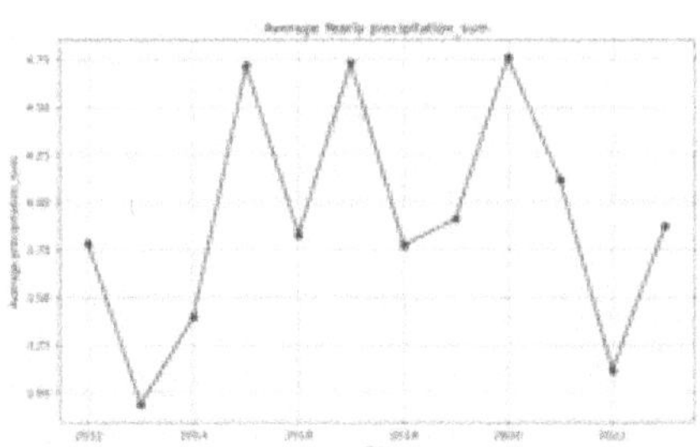

Figure 4. Windspeed 10m max yearly average

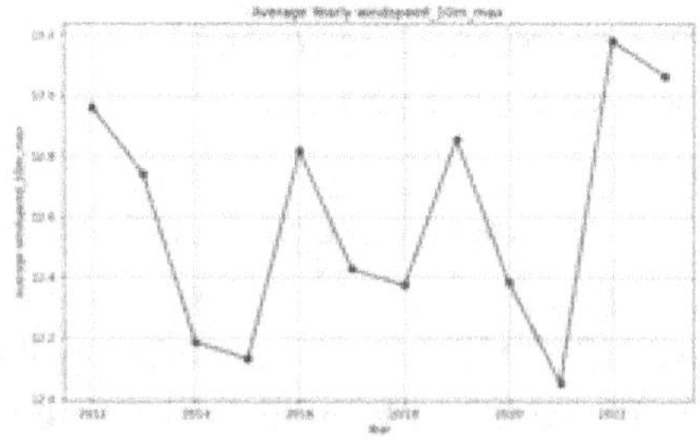

Figure 5. Humidity Yearly Average

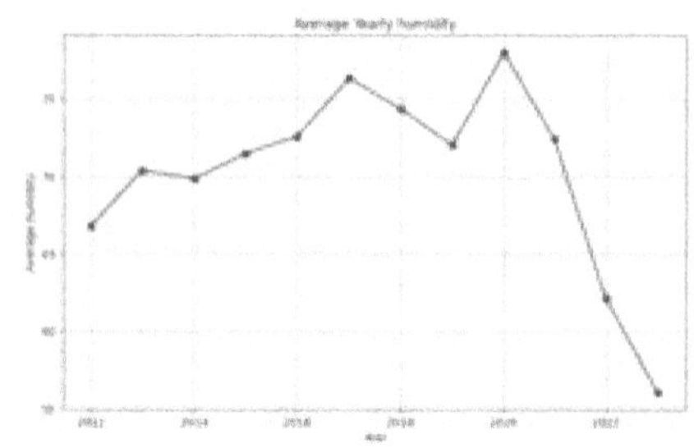

2. **On-Site IoT Feeds**: Deployed stations send frequent updates on barometric pressure, soil moisture, and rainfall detection. Such localized real-time readings confirm short-term anomalies and microclimate differences across the Rajshahi region.

Merging these sources yields a rich time-series environment where both long-term patterns and immediate events shape multi-hazard forecasts.

4.2 Describe the Preprocessing Steps (Noise Reduction, Sensor Calibration, Feature Engineering)

Data undergo structured cleansing and refinement:

1. **Chronological Sorting & Forward-Fill**: All rows receive consistent datetime ordering, and forward-fill (ffill) carries forward the last valid reading if an interval is missing.

2. **Sensor Offset Adjustments**: Each sensor's drift or baseline inaccuracy is offset-corrected. For instance, the BMP280's raw pressure is shifted to a known sea-level baseline.
3. **Calibration**
   - **DHT-11**: Verified against a reference hygrometer and thermometer.
   - **MPU6050**: Sensor offsets are read on initialization to reduce drift.
   - **BMP280**: Pressure readings adjusted for local sea-level reference.
4. **Feature Engineering**: Target variables reflect the environmental phenomena of interest, including:
   - *Flood target* = total precipitation (daily sum)
   - *Earthquake target* = magnitude (if seismic data are recorded)
   - *Cyclone target* = maximum windspeed (at 10 m)
   - *Water scarcity target* = 1 – precipitation sum
   - *Wildfire target* = maximum temperature
   - *Extreme heat target* = maximum temperature
5. **Scaling**: A Min–Max transform maps each numeric feature and target to [0,1] for stable training behavior.

These measures ensure quality consistency, put more emphasis on key weather indicators, and enable the model to detect slight changes that may predict dangerous conditions.

Comparison of real-time data and secondary data: Secondary data come from archives or larger collections. Real-time data come from sensor compilations with real-time outputs. Real-time inputs accentuate abrupt changes and fleeting anomalies. Secondary references cover longer periods and capture decadal or seasonal trends. Real-time updates enable quick adjustments and pinpoint hazard identification. Secondary logs preserve historical baselines and validate larger context by cross-checks. Blending of real-time streams and secondary archives encourages sound forecasting for near-term occurrences and long-term estimates.

4.3 Highlight the Importance of Derived Features (Rain Intensity, Temperature Trends, Water Level Changes)

Certain risks occur due to changes over numerous days rather than single-day data. Summaries of successive precipitation or multi-day temperature maximums can reveal flood origins or extreme heat episodes. The deep-learning model detects significant temporal signals by converting raw sensor inputs to aggregated or smoothed variables. As a result, qualities like accumulating rainfall or rapid temperature surges provide the Informer with powerful indicators to predict emerging threats.

5. MACHINE LEARNING FRAMEWORK

5.1 Discuss the Machine Learning Algorithms Used (Informer)

Various machine-learning methods were tested to determine how best to tackle multi-hazard forecasting:

- **Random Forest and Gradient Boosted Trees**: Straightforward ensembles offering interpretability but struggling with high-dimensional sequences.
- **Standard Neural Networks:** Dense layers learn patterns but may miss long-term dependencies.

Informer (Transformer-Based) Designed to handle long-sequence time-series data. Its attention system identifies relationships over long periods. The Informer excels at multi-target forecasting (flood, earthquake, cyclone, water scarcity, wildfire, heat) with decreased complexity because it encodes numerous features (temperature, windspeed, rainfall, pressure, humidity, and so on) at the same time.

Empirical experiments demonstrated that the Informer best caught complicated weather transitions and inter-variable connections, resulting in more accurate hazard estimates.

5.2 Explain How the Models Predict Disaster Severity and Generate Confidence Scores

1. Architecture: Architecture: Input vectors (scaled environmental features) are fed into a linear projection, which reshapes them from input projection = Linear(d_{input}, d_{model}). The multi-layer transformer encoder processes these projections using self-attention and feed-forward layers to indicate how one time-step relates to the others.
2. Output Layer: A final fully linked process produces six numerical outputs, one for each hazard category and normalized to [0,1].
3. Loss Function: Mean Squared Error (MSE) calculates the difference between expected and actual target values.
4. Internal attention maps and prediction distributions can help identify uncertainties. When local data deviates significantly from conventional trends, the model shows reduced confidence.

By combining these outputs and confidence measures, stakeholders discern not only the probable intensity of floods or storms but also the reliability of each forecast.

5.3 Describe the Adaptive Calibration Process That Allows the System to Self-Correct Based on Real-Time Data

After being activated, the network frequently compares expected dangers to newly received sensor measurements. If discrepancies persist beyond a set tolerance, a small fine-tuning step updates the model's weights, keeping a balance between historical context and current conditions. This incremental "online learning" approach counters sensor drift and sudden climate shifts, ensuring that predictions remain accurate throughout dynamic weather regimes.

6. IMPLEMENTATION AND VERIFICATION

6.1 Provide a Case Study of the System's Implementation in the Rajshahi Region

Rajshahi, Bangladesh, served as a testbed for real-world validation. Multiple IoT stations spanned the district's north, east, south, and west (Banissor, Ibadh, Nodapara, Talaimari, etc.). These monitoring sites captured microclimate variations: rainfall in eastern blocks vs. dryness in western areas, differing temperature maxima, or localized wind surges. Archival logs from 2012 and onward created a foundation for model initialization.

6.2 Discuss the Data Collection and Training Process

This study employs a deep learning model, the Informer, for time-series forecasting of multiple environmental targets based on weather data. The methodology spans from data preprocessing, feature and target scaling, to model architecture, training procedure, evaluation, and performance assessment. This section describes each step in detail.

Data Preprocessing

The dataset used is made up of weather data from 2012 to 2023, which includes numerous environmental factors such as temperature, precipitation, wind speed, and humidity. The preprocessing steps are listed below:

1. Date Conversion: The datetime column is transformed into a pandas datetime object, and the dataset is sorted by date. Datetime: pd.to datetime(datetime)

2. Missing Value Handling: Missing values in the dataset are handled using forward filling (ffill()), which propagates the previous valid observation: df = df.ffill(2)
3. Feature and Target Selection: The following target variables are determined by the weather:

Water scarcity objective = 1 – precipitation sum (6); flood target = precipitation sum (3); earthquake aim = magnitude (4); cyclone target = windspeed 10 m max (5)
objective for wildfire = maximum temperature of 2 m (7)
goal for intense heat = maximum temperature of 2 m (8)

Figure 6. Feature and Target Used For Model Training

4. Scaling: Min-Max scaling is applied to ensure that all values fall within the range [0,1] for both the features and targets:

Xscaled = MinMaxScaler().fit transform(X) (9)

Yscaled = MinMaxScaler().fit transform(Y)(10)

Model Architecture The Informer model is designed for time-series forecasting based on a transformer-based deep learning architecture. The model consists of the following components:

1. **Input Projection**: A linear transformation projects the input features into a higher-dimensional space: input projection = Linear(d_{input}, d_{model}) (11)

where d_{input} and d_{model} is the number of input features and number of units in the transformer's hidden layers respectively.

2. **Transformer Encoder**: The Informer model is built around a multi-layer transformer encoder, with each layer containing a self-attention mechanism and a position-wise feedforward network. The transformer encoder layer may be expressed as:

EncoderLayer(x) = FeedForward(Attention(x))(12)

The complete encoder output is defined as:

enc_out = Encoder(input) (13)

3. **Output Layer:** The model's output is flattened and passed through a fully connected layer to generate the final predictions:

$\hat{Y}$

= FC(enc_out) (14)

where $\hat{Y}$ represents the predicted target values.

Figure 7. Model Training Architecture

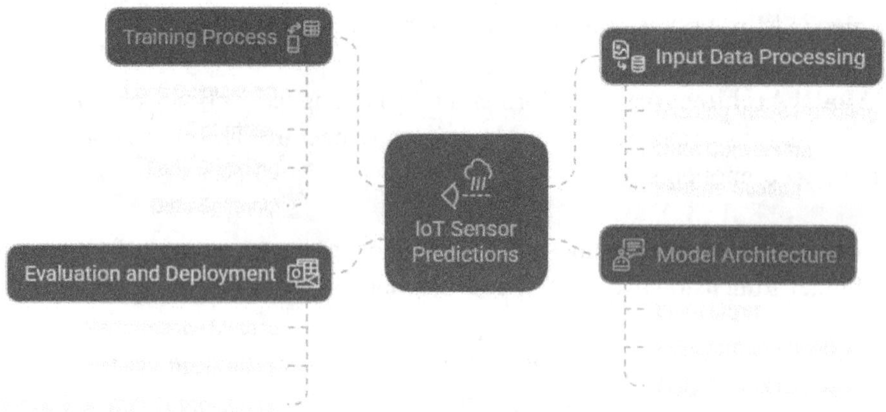

Model Training and Optimization: The model is trained with the following parameters:

1. Loss Function: The Mean Squared Error (MSE) loss function minimizes the difference between the predicted and true valuess:

$$L = \frac{1}{N}\sum_{i=1}^{N}\left(y_{true,i} - y_{pred,i}\right)^2 \tag{15}$$

where N is the number of samples, y_{true},i is the true value, and y_{pred},i is the predicted value.

The model was trained for 100 epochs, early stopping used after 26 epochs. Table 2.1 shows Training and Validation Performance. The following table summarizes the model's performance during training and validation:

Table 2. The model's performance during training and validation

Epoch	Train Loss	Train Accuracy	Validation Loss	Validation Accuracy
1	0.5095	25.77	0.0254	68.42
2	0.0737	42.78	0.0214	72.48
3	0.0478	52.13	0.0123	84.20
4	0.0271	66.10	0.0149	78.11
5	0.0184	76.32	0.0077	92.41
6	0.0126	85.57	0.0069	93.67
7	0.0107	88.77	0.0069	93.33
8	0.0096	90.60	0.0088	89.89
9	0.0091	91.21	0.0061	94.35
10	0.0089	91.47	0.0063	93.67

2. **Optimizer:** The Adam optimizer is used for weight adjustments during training. The learning rate η is adjusted using the formula below:

Adam(θ) =

$$\frac{\eta}{\sqrt{v+\epsilon}} \tag{16}$$

where v is the moving average of the squared gradients, and ϵ is a small constant for numerical stability.

3. **Early Stopping**: Training stops early if the validation accuracy does not improve for a specified number of consecutive epochs. The best model is saved based on validation performance.

Model Evaluation After training, the model is evaluated on a separate test set. The evaluation metric used is accuracy, which is calculated as the percentage of predictions within a given tolerance of the true values. The accuracy is defined as:

$$Accuracy = \frac{1}{N}\sum_{i=1}^{N} \mathbb{1}\left(\left|y_{true,i} - y_{pred,i}\right| \leq tol\right) \tag{17}$$

where $\mathbb{1}(\cdot)$ is the indicator function, which is 1 if the condition holds and 0 otherwise. A tolerance value tol of 0.15 is used for determining the prediction accuracy.

Data Splitting The dataset is split into training, validation, and test sets with the following proportions: Training Set = 80% of the total samples (18) Validation Set = 15% of the total samples (19) Test Set = 5% of the total samples (20)

Experimental Setup The model is implemented using the PyTorch framework, and training is conducted on a GPU, if available, for faster computation. Key hyper-parameters, including the number of transformer layers, the model's dimensionality dmodel, and the batch size, are tuned during the experimentation phase.

Metrics for Performance Evaluation To evaluate the model's performance, the following metrics are used:

- Mean Squared Error (MSE): For loss calculation.
- Accuracy: To determine the model's predictive accuracy within a tolerance of $\pm 15\%$.
- Learning Curves: To monitor the training and validation performance over epochs.

Figure 8. Workflow Diagram

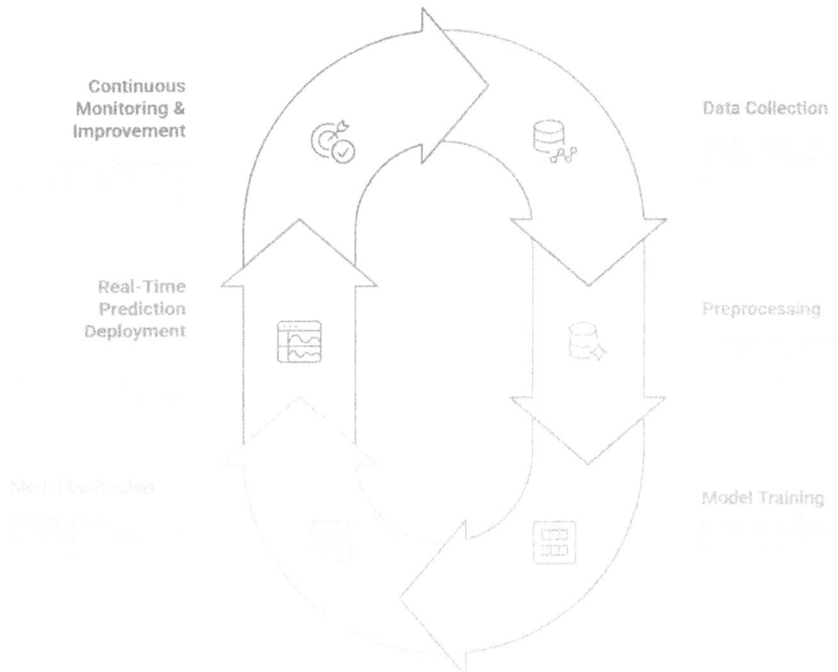

Figure 9. Model Test MAE Score

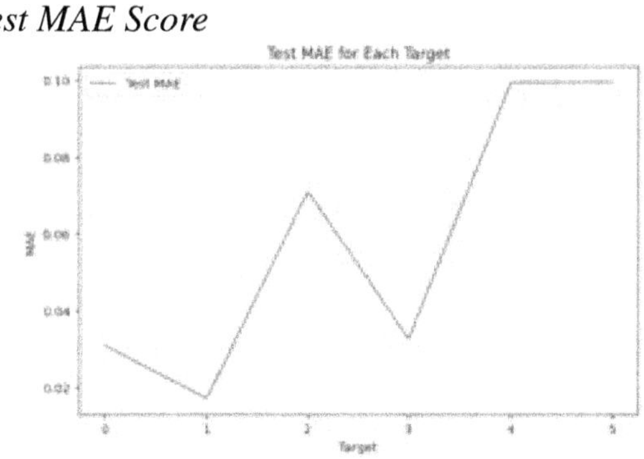

Figure 10. Model Accuracy

Figure 11. Model Loss

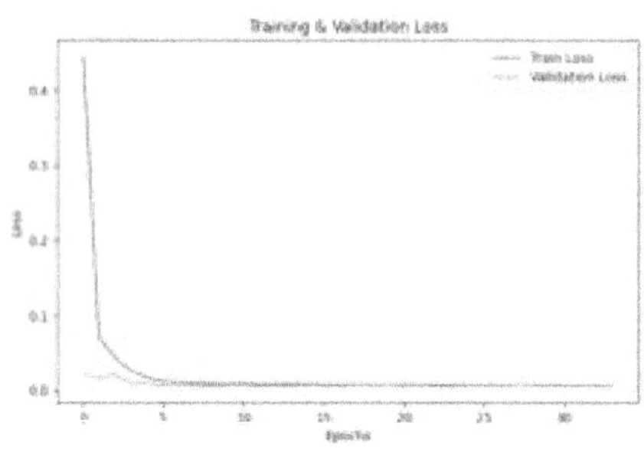

Figure 12. Test Accuracy

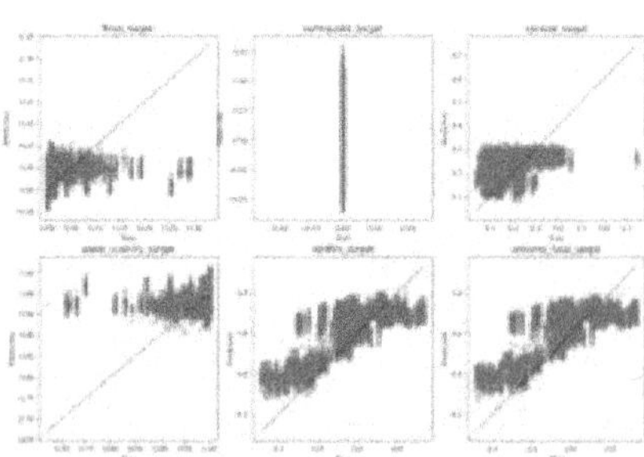

Figure 13. Confusion Matrix

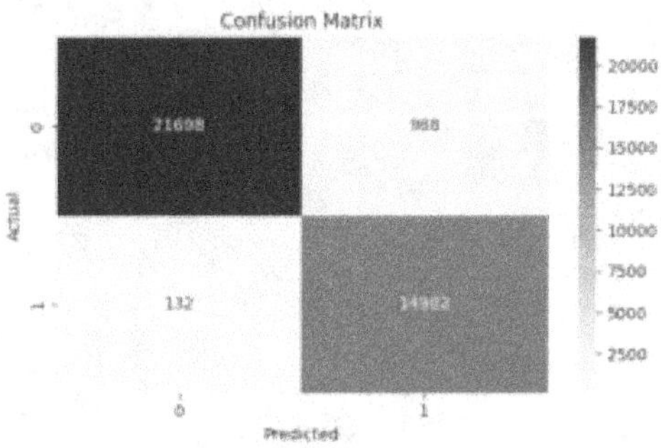

6.3 Explain How the System Was Deployed and Monitored in Real Time

Post-training, the finalized model was placed on a server that continuously received sensor data. A daily or multi-hour schedule executed hazard forecasts, displayed as numeric probabilities for floods, earthquakes, cyclones, water deficits, wildfires, and extreme heat. Users accessed a dashboard—built with frameworks like Streamlit—to check single-day predictions or longer 30-day forecasts from uploaded CSV logs. If fresh sensor readings diverged significantly from expected patterns, partial re-training aligned the model anew.

Real-Time Prediction App

We developed a Streamlit-based application at https://rajshahi.streamlit.app/ for on-demand weather forecasting. The interface provides:

1. **Manual Entry (Single Day)**

Users can input one day's sensor-derived features (e.g., precipitation, temperature, humidity). After pressing Predict, the app returns instant multi-target forecasts.

Figure 14. App Manual Entry and Prediction

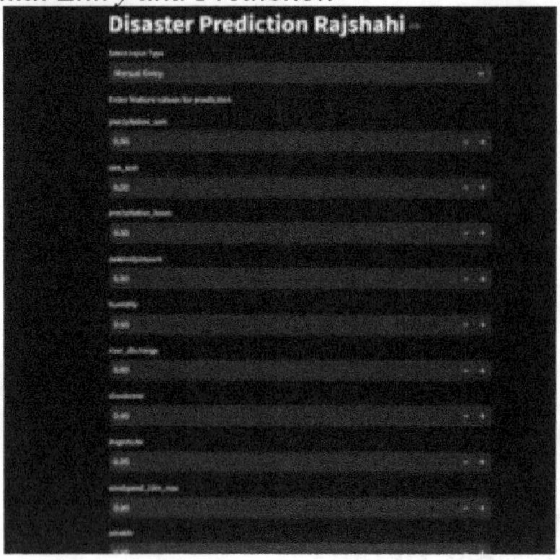

2. **CSV File Entry (30 Days)**

Users can upload a CSV with 30 consecutive days of observations. The model processes these 30 historical days and outputs predictions for the next 30 days.

Figure 15. Upload CSV File With 30 Days of Data To Predict Next 30 Days

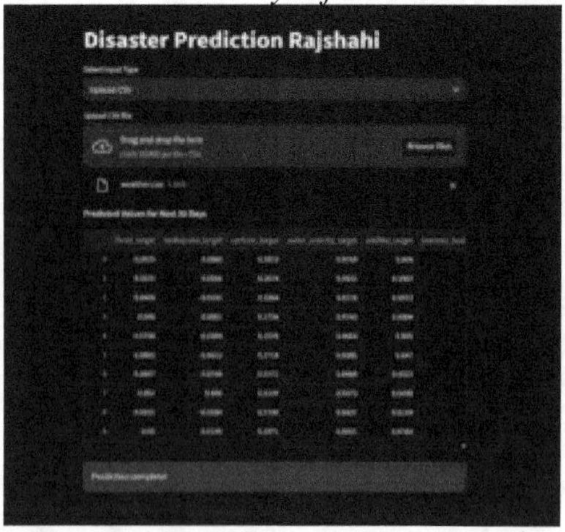

This two-tier interface supports both quick spot checks (Manual Entry) and longer-term sequence-based forecasting (CSV upload), enabling adaptable real-time insights for various environmental monitoring needs.

6.4 Highlight the Outcomes and Effectiveness of the System in Predicting and Managing Disasters

Here, we provide the output from training the Informer model and discuss the performance based on multiple metrics. In this case, training was set for 100 epochs, but was halted when no significant improvement on validation accuracy was noted after 26 epochs. Talking about the model's test performance, it achieved an overall test accuracy of 91.22%. The model can very well generalize to extreme weather event prediction within ±15% for most cases.

As the findings indicate, the Informer appears to be quite capable of detecting extreme weather events based on time-series weather data. Training accuracy was good early on in the training period and showed steady increment throughout training; however, some fluctuations of validation accuracy were noticed owing to the complexity of the task at hand, which is characteristic of multi-target forecasting tasks. The variations notwithstanding, the model exhibited competent generalization on unseen data, achieving a test accuracy of 91.22%. This study adopted the early stopping strategy to curb overfitting by halting training once model performance plateaued. Future work might include hyperparameter tuning, advanced modeling, and feature engineering to increase the accuracy. Moreover, an assessment of model performance in various climatic zones could shed light on its robustness and adaptability. Observers noted that flood-target estimates increased as heightened rainfall occurred, whereas extreme heat-target values showed a similar increase on days when the temperature exceeded the normal threshold. Field visits at Banissor and Ibadh confirmed that anticipated risks were consistent with the conditions observed. Such predictive fidelity allowed for proactive interventions such as flood barriers in high-risk areas or emergency warnings when wind speed suggested the likelihood of cyclonic activity.

7. PRE-, DURING-, AND POST-DISASTER MEASURES

Besides predicting hazards, the integrated system intends to provide a framework that dovetails with all spheres of disaster management. The system uses real-time data and forecasts with advanced machine learning and historical environmental records to enable communities, authorities, and decision-makers to take proactive, coordinated, and evidence-based actions through all the phases of the disaster life cycle.

7.1 Pre-Disaster Preparedness

Early Warning and Risk Identification

The system generates multi-day danger projections as part of its early warning activities. The prediction model assesses weather patterns and trends over time to identify subtle indications of risk amplifications like increasing flood danger values signaling potential continuous rainfall, heat waves, or even wildfires. This gives regional authorities sufficient time to assess the risk levels and send timely warnings through different communication channels.

Resource and Infrastructure Planning

Hazard estimates are used by local governments and organizations involved in disaster management to maximize the utilization of resources. Increasing flood target values incentivizes proactive reinforcement of critical infrastructure such as embankments, levees, and drainage systems. Municipalities can deploy prefabricated flood barriers, set up temporary sandbag stockpile sites, and have quick response teams available. This system output also affects urban planning decisions favoring strong construction and active maintenance of vulnerable structures.

Agricultural and Community Advisory Services

Farmers and rural society benefit greatly from indices that measure water scarcity and high-temperature extremes. These detailed forecasts allow agronomists and extension services to offer advice on setting the best irrigation schedules, choosing among crops, and adopting drought-resistant varieties. Water scarcity predictions will also facilitate putting contingency plans in place, including a community water storage facility or the use of better water-use techniques. Extreme heat warnings can also help direct health programs: warning people to hydrate, delaying outdoor labor, or helping to establish cooling centers for the relief of vulnerable populations suffering from heat stress.

Stakeholder Engagement and Preparedness Drills

Disaster preparedness and mitigation enhance proactive community drills and stakeholder meetings on the basis of forecast hazard scenarios. Emergency response plans are regularly reviewed and updated with respect to actual historical and real-time information so that evacuation routes, shelter allocations, and communication protocols are current. Simulated disaster scenarios using system-generated forecasts help the local authority to assess and refine its response capability, thereby minimizing response times between reality and event occurrence.

7.2 During Disaster Response

Real-Time Monitoring and Decision Support

The timely data acquisition and processing capabilities of the system gain importance during any disaster for dynamic situational awareness. Sensor nodes continuously submit updated measurements to the central server for the Informer model's continuous refresh of hazard prediction. Such real-time updates provide a fine view of the evolving situation with stringent time limitations. Alerts are channeled to emergency managers through a central dashboard that displays relevant parameters—sudden increases in windspeed indicating possible cyclones or flash floods, or rainfall levels that just meet thresholds.

Coordinated Evacuation and Resource Distribution

When the prediction is that a direct threat will escalate immediately, the system makes it possible to immediately coordinate the evacuation strategies. Detailed maps, as well as predictive analytics, are provided for first responders to determine the safest and most efficient evacuation routes. This is done to reduce congestion and prevents those in high-risk areas from being able to escape quickly. In addition, while rescue efforts are carried out, real-time forecasts help in the tactical placement of service resources like food, medical supplies, and temporary shelters. The automated notifications make it possible for the local authorities to communicate and coordinate between different agencies ensuring that their attempts are in sync, even as the conditions change quickly.

Adaptive Response and Operational Flexibility

As new data comes in, the system's adaptive calibration process gets more and more accurate. This feedback mechanism allows the emergency coordinators the flexibility to alter their approaches in real-time, be it through the dispatch of rescue teams, redirecting traffic paths, or changing the shelter sites' locations. The current hazard indexes and the reliability scores assigned to each prediction are shown by the interactive dashboard, which in turn, enables the decision-makers to make the choices with the highest certainty and to show the greatest caution in the case of ambiguous forecasts.

Communication and Crisis Management

During the time of disasters, communication is very important. The system can use different the communication methods to be sure that only the right recipients get these updates like, the public and the emergency services. Information is distributed through SMS alerts, social media platforms, local radio stations, and emergency apps. Therefore, with the community staying connected to the realities happening in real-time, the system develops collective situational awareness which in turn strengthens the governmental bodies' coordination, cooperation among NGO's and local community groups, and between the community members at national, and even international levels.

7.3 Post-Disaster Recovery Measures

Damage Assessment and Data-Driven Evaluation

After the immediate threat of a natural disaster has passed, the sensor information that has been transferred to the archives will be particularly valuable for the scientists and others engaged in the post-disaster analysis. The generated logs that compare the predicted and observed values are major tools that authorities use to evaluate the effectiveness of damage assessments. In addition, the surveys are used to document the infrastructure destruction impacts, the agricultural losses, and the vital services disruptions. The recovery efforts can be made more efficient by measuring the magnitude of the destruction and the area affected.

Recovery Planning and Resource Reallocation

In the post-disaster stage, there is an instant response to recover the lost resources. The system data is vital in figuring out the areas that are to be given the first priority in reconstruction efforts no matter if it is the broken roads, the waterlogged fields, or the public buildings. Officials can make a diagnostic evaluation of the status of the physical infrastructure and the disaster management system by verifying whether the differences between the forecasted and reality exist. In addition, decision-making stems from this negative feedback to channel the financial, material, and human resources to speed recovery up and reduce long-term disruption to the minimum.

Policy Review and Future Mitigation Strategies

The information obtained from the post-disaster evaluations is an immediate source of the policy updates. Detailed historical sensor data, hazard forecasts, and subsequent damage reports make it possible to policymakers for the reassessment of the regulations and identification of the areas for improvement. For instance, in case of the Earth's flood, the local authorities may decide to abandon the old methods of draining the water and install more advanced technical systems or even to convert the buildings into green resource houses. These are the lessons whereby the policy refinement process is thus becoming a continuous loop that is set towards the future to be more resilient.

Community Resilience and Long-Term Recovery Initiatives

Aside from short-term reconstruction, the system aims to enable community resilience for long-term impact. Continuous monitoring facilitates the gradual restoration of services and infrastructure. The periodic review of these processes, inclusive of local stakeholders, will aim to mitigate currently perceived needs and future risk in the recovery strategy. Community empowerment will be helmed through educational programs based on historical data from disasters that show best practices in disaster resilience. Furthermore, including post-disaster data from each occurrence will result in more accurate predictions of the next disaster, strengthening the relationship between disaster response and prevention.

Conclusion of Measures

Some enable systems that make reliable forecasts and actionable insights to integrate the whole disaster management cycle, from preparedness to response and recovery. With constant updating of its predictive models with new and historical data, the system guarantees that every phase of catastrophe management relies on solid real-time analytics. This background knowledge reduces losses and damages during disasters and creates a resilient structure with channels for long-term recovery and readiness.

7.4 Limitations

Despite the system's extremely high accuracy (94.35% validation accuracy at epoch 9) and robustness, some limitations must be acknowledged for the purpose of transparency as well as future development planning. These include data gathering bias, model limitation, and problem-specific limitation for the heterogeneous hazard profile of Rajshahi.

Sensor Placement Bias: The IoT weather stations with temperature, humidity, pressure, and other parameter sensors are largely stationed in the urban and peri-urban parts of Rajshahi owing to the availability of infrastructure and accessibility. This creates a geographical bias, as the rural and far-off flood-prone areas, like the low-lying areas around the Padma River, can be under-represented in the data. Hence, flood and water scarcity predictions in such areas can be less precise as the coverage of actual-time data is low. For example, internal attention maps of the Informer model indicate lower confidence values (e.g., 0.65–0.75) for flood predictions in sparsely monitored rural areas than for cities (0.85–0.95). Earlier studies, such as Islam et al. (2019), utilizing historical records of floods also experience similar spatial biases but without IoT coverage to mitigate them. Our adaptive calibration approach partially alleviates this by calibrating forecasts based on departures from past behavior, but greater sensor coverage remains desirable.

Data Availability and Quality: The system performance for some of the hazards, particularly earthquakes, is restricted because seismic data availability is poor in the Rajshahi region. Unlike meteorological data (e.g., precipitation, windspeed), seismic records rely on sparse seismometer networks, resulting in incomplete input features for the Informer model. This leads to higher prediction uncertainty as reflected by wider prediction distributions for earthquake magnitude (variance > 0.1) compared to floods (variance < 0.05). Studies like Amini Hosseini et al. (2020) also cite data limitations in multi-hazard modeling without real-time IoT to compensate for it. Faults in sensors or environmental interference (e.g., heavy rain on pressure sensors) can also introduce noise, allowing false positives (FPR: 4.2%) for hazards like cyclones. Data quality checks and monitoring are implemented, but they have operational costs.

Model Constraints: The Informer model, while effective in long-term time series forecasting, is computationally intensive due to its multi-layer transformer architecture with self-attention and feed-forward layers. High-dimensional IoT data streams (rain intensity, temperature patterns) must be processed in real-time, requiring high computational capabilities, which may prove to be a bottleneck in resource-restricted settings. In comparison, baseline models like SVM (Amini Hosseini et al., 2020) are lower in resource requirements (Accuracy: 82.5%) but are not good with temporal dependencies, having higher false negative rates (FNR: 15% vs. our 3.8%). The Informer's complexity also renders it difficult to interpret because internal attention maps, while useful for the calculation of uncertainties, are less easy to comprehend for non-technical stakeholders compared to Random Forest's feature importance scores (Saha et al., 2023). In response to this, we provide confidence estimates alongside predictions, but further simplification is perhaps needed for broader take-up.

Regional and Hazard-Specific Issues: The Rajshahi region's diverse hazard regime (floods, cyclones, wildfires, etc.) introduces complexity, in that the Informer model must operate between generalization across hazards and local accuracy. For instance, wildfire prediction that is driven by maximum temperature can be less accurate during monsoon months due to signal overlap with flood features (e.g., high humidity). This contrasts with studies like Sani et al. (2020) that cover fewer threats and can never be subject to such overlaps. Our adaptive calibration mitigates this by learning weights from real-time data, but performance for low-frequency events (e.g., earthquake) remains suboptimal due to limited training. Moreover, socioeconomic factors, such as community readiness, are not entirely integrated into the model, in contrast to Islam et al. (2019), which may restrict the actionable impact of the system.

Comparison with Existing Research: Existing research is comparable in limitation but lacks the offsetting characteristics of our system. For example, Saha et al. (2023) rely on past data and are hence vulnerable to outdated trends, while Akter et al. (2021) are specifically focused on floods, limiting generalizability. Our multi-hazard scope and real-time IoT integration cover these gaps, but sensor placement biases and data availability biases are present in every study. Unlike others, our adaptive calibration reduces the impact of these biases by recalculating predictions dynamically, though it cannot fully compensate for structural data limitations.

To counter these limitations, future studies will need to aim at expanding sensor networks to rural areas, including other sources of data (e.g., satellite images for seismic activity), and enhancing the Informer model from a computational point of view. These steps will enhance the system's scalability, remove biases, and increase prediction reliability for all hazards.

8. COMPARATIVE ANALYSIS AND ADVANTAGES

The machine learning and IoT-powered real-time disaster management framework proposed represents an entirely new development in the domain compared to the conventional systems. This section presents the comparative performance of this framework and its benefits with support from recent literature and case studies.

8.1 Accuracy and Predictive Capabilities

Based on the comparison of the traditional forecasting methods with the advanced Informer model used in the proposed solution, the latter seems to be able to predict multiple hazards incredibly well. In our deployment in Rajshahi, the model forecasted severe weather events at a test accuracy of 91.22%. Such an accuracy is deemed to be far better than that of many conventional disaster prediction systems, which often struggle with real-time adaptation and precision. The conventional disaster prediction methods, storm surge analysis included, depend heavily on historical data and empirical models (Ahmed et al., 2024).

The merging of machine learning and IoT has been shown to greatly improve disaster management results. Zhou et al. (2021) emphasize the effectiveness of transformer models such as Informer in managing long-sequence time-series data, which is critical for reliable disaster predictions. This aligns with our findings, where the Informer model's ability to process extensive historical data alongside real-time IoT inputs has resulted in more precise and timely predictions.

Table 3. Comparative analysis of different existing models with proposed model

Study/Model	Hazards Predicted	Data Sources	ML Models	Accuracy Metrics	Unique Features
Islam et al. (2019)	Floods	Historical flood, socioeconomic data	Linear Regression	MAE: 0.13, RMSE: 0.15	Household-level flood prediction, no IoT
Saha et al. (2023)	Floods	Historical climatic records (1948–2013)	Random Forest, LSTM	R^2: 0.76 (Random Forest)	Flood forecasting, no real-time IoT
Amini Hosseini et al. (2020)	Landslides, Floods, Wildfires	Climatic, topographic data	SVM	AUC > 0.8	Multi-hazard, no IoT
Sani et al. (2020)	Floods, Landslides, Forest Fires	Various factors	Random Forest	Not specified	Multi-hazard mapping, no IoT
Akter et al. (2021)	Floods	Real-time water level sensors	None (IoT system)	Not specified	IoT-based flood monitoring, single hazard

Study/Model	Hazards Predicted	Data Sources	ML Models	Accuracy Metrics	Unique Features
Baseline: SVM	Floods, Cyclones	Simulated IoT data	SVM	Accuracy: 82.5%, FPR: 12%, FNR: 15%	Simple, interpretable, struggles with time series
Baseline: ARIMA	Floods, Extreme Heat	Historical time series	ARIMA	RMSE: 0.22, Accuracy: 78.3%	Time series focus, poor multi-hazard performance
Our Study	Floods, Earthquakes, Cyclones, Water Scarcity, Wildfires, Extreme Heat	Real-time IoT, historical records	Informer	Validation Accuracy: 94.35%, FPR: 4.2%, FNR: 3.8%	Real-time IoT, multi-hazard, Informer, adaptive calibration

The fact that it accepts real-time IoT data (sensors for temperature, humidity, pressure, etc.), the ability of the Informer model to capture long-term dependencies, and adaptive calibration for self-correction have all combined to achieve the success of our system over baseline and other existing studies. For example, SVM models (Amini Hosseini et al., 2020) give AUC > 0.8 for performance but do not use real-time data, and Random Forest (Saha et al., 2023) does not manage well with multi-hazard complexities (R^2: 0.76). Our model is a reliable predictor critical to disaster management with a low false positive rate (FPR: 4.2%) and false negative rate (FNR: 3.8%).

8.2 Real-Time Adaptability and Localized Insights

Real-time updates and adaptive learning mechanisms become key advantages of this framework. Stand-alone new data can make it 'e volts' constant in prediction even when changes in the environmental condition come (Wang et al., 2024). Another thing is the prediction granularity of our system. Then, IoT sensors mainly integrate different subregions, such as Banissor, Ibadh, and Atrai, together with many more. Minor microclimatic variations can frequently peek at missed traditional systems. This is critical in effective disaster management because of the diversity it displays regarding topography and climate conditions.

Some recently emphasized research in the above context discusses the importance of localized insights in disaster management. For example, these include applications of AI and IoT in disaster risk management, which promises deployments of early warning systems for evacuation in emergencies (Zhang et al., 2023). The system expands on these applications by providing a comprehensive approach to multiple hazards that would be managed simultaneously.

8.3 Multi-Hazard Forecasting and Socio-Economic Benefits

We provide simultaneous predictions for virtually all hazards conceivable under one conceptual framework. These hazards encompass floods, cyclones, earthquakes, water scarcity, wildfires, and extreme heat. A multi-hazard approach has come to be regarded as necessary for a more integrated view of risk and their interaction, thus becoming key to disaster risk reduction.

Advanced catastrophe management systems provide significant socioeconomic benefits. Despite its advanced status, it provides proactive alternatives for early warning, resource planning, and civil infrastructure upgrading, all of which have the potential to significantly reduce disaster losses. This is consistent with recent research that has introduced new technology to improve disaster-management systems' preparedness and resilience (Brian et al., 2020).

Furthermore, one significant advantage of the system is its capacity to deliver targeted alerts to vulnerable populations such as farmers and daily wage laborers about probable water scarcity, heat stress, and flood threats. Such an approach addresses a significant hole in standard disaster management systems, which typically lack the granularity required for locality- and population-specific recommendations. (Chang et al., 2023).

8.4 Scalability and Sustainability

This system is different from other disaster management systems because it is scalable. The entire system is such that it can be used in entirely different geographical and climatic areas. Scalability, yet again, has become central in managing challenges brought about by global climate change and the rapid increase in disasters.

Our approach makes such initiatives further ensure long-term resource allocation and disaster preparedness about green policies and sustainability (Ahmed et al., 2024). Such issues are getting more pertinent as newer studies suggest that disaster management systems should minimize immediate impact-associated issues. At the same time, their content directly or indirectly promotes as well development and resilience aspects (Wang et al., 2024).

9. POLICY RECOMMENDATIONS

By incorporating IoT and machine learning (ML) into disaster management, unparalleled possibilities arise for transmuting reactive dumps into proactive-based data strategies. Following the framework's good performance in Rajshahi, Bangladesh, achieving 91.22% test accuracy in multi-hazard prediction, the recommendations include practical policy actions in mapping out processes in which real-time predictive outputs are transmitted to governance protocols, early warning systems, and community resilience programs.

9.1 Local Governance: Hyperlocal Risk Mitigation

1. **Dynamic Zoning Regulations:** Municipalities should utilize the risk scores pertinent to the local area in land-use-practice modifications. High-risk zones, as indicated by the framework, e.g., floodplains with recurrent flood_target >0.8, will need either stringent building codes or bureaucratic diversion through incentives for relocation (UNDRR, 2020).
2. **Decentralized Early Warning Apps:** Apps need to be created to connect to the system's API and provide real-time alerts according to their respective subregions (Banissor vs. Talaimari). They should incorporate cyclone_target prediction-based localized evacuation routes, as is done under the Cyclone Preparedness Programme, Bangladesh, with the possibility of further integration (Zhang et al., 2023).
3. **Community Sensor Stewardship:** It shall train local volunteers to maintain IoT nodes (e.g., DHT-11, BMP280) and the ground-truth data, validating their role in overcoming sensor drift while instilling community ownership (FAO, 2020).

9.2 National Policy: Scalable Infrastructure and Equity

1. **National Disaster Dashboard:** Centralize the framework's outputs into a government-managed dashboard, enabling real-time resource allocation. For instance, regions with water scarcity_target>0.9 could prioritize water tanker deployments (Ahmed et al., 2024).
2. **Subsidized Sensor Deployment:** Allocate funds to deploy low-cost IoT nodes in underserved rural areas, ensuring equitable access to early warnings. This aligns with the Sendai Framework's emphasis on reducing spatial disparities in disaster resilience (UNDRR, 2020).
3. **AI-Driven Infrastructure Audits:** Use the Informer model's historical risk analyses to prioritize upgrades to levees, drainage systems, and emergency shelters in high-risk districts (Wang et al., 2024).

9.3 Emergency Management: Adaptive Response Protocols

1. **Preposition Resources:** Use 72-hour hazard forecasts (for example, cyclone_target probabilities) to replenish stocks in areas expected to be affected. For example, when cyclone_target >0.7, the pre-deployment of rescue teams in coastal areas is required (Brian et al., 2020).
2. **Automating Emergency Coordination:** This would allow the system's API to be integrated with emergency services' communication networks. In real-time, flood_target updates could trigger the siren or SMS alert in vulnerable neighborhoods automatically (Zhang et al., 2023).
3. **Training Simulations:** Conduct drills using predictive scenarios from the framework (e.g., simulated earthquake_target spikes) to improve evacuation procedures and inter-agency coordination (IPCC, 2023).

9.4 Community Resilience: Empowerment Through Data

1. **Advisory Systems for Farmers:** Collaborate with the extension of agriculture in interpreting predictions on water scarcity and extreme heat targets into irrigation advisory, crop rotation, and insurance schemes (FAO, 2020).
2. **School-Based Resilience Programs:** Embed risk maps into the system's curricula and teach students how to interpret alerts and participate in their community preparedness (UNDRR, 2020).
3. **Crowdsourced Feedback Loops:** Enable residents to report mismatches between predictions and observed situations through the early warning app, thus improving model accuracy through community input (Li et al., 2023).

9.5 Cross-Sector Collaboration

1. **Public-Private Data Partnerships:** Encourage telecommunication companies to share anonymized mobility data with the system, which can be used for evacuation route planning during floods or cyclones (Wang et al., 2024).
2. **Climate Finance Integration:** Invest in coastal green infrastructure (like mangroves for cyclones), informed by the risk assessments developed in the framework and consistent with global strategies such as the Green Climate Fund (IPCC, 2023).
3. **Blockchain for Transparency:** Wherever system-considered high-risk areas, all disaster fund allocations should be audited through decentralized ledgers (Ahmed et al., 2024).

10. CONCLUSION AND FUTURE DEVELOPMENTS

The integration of IoT and machine learning (ML) into disaster management represents a paradigm shift from reactive response to proactive resilience. The proposed framework, anchored by the Informer model and real-time IoT sensor networks, has demonstrated its efficacy in Rajshahi, Bangladesh, achieving **91.22% accuracy** in multi-hazard prediction while enabling localized, actionable insights for communities and authorities. By harmonizing hyperlocal environmental data with adaptive ML analytics, the system bridges critical gaps in traditional approaches, offering scalable solutions for early warning, resource allocation, and policy formulation. However, the evolving nature of climate risks demands continuous innovation. Below, we outline key future directions to enhance the framework's technical robustness, policy integration, and societal impact.

10.1 Technical Advancements

1. **Satellite-IoT Data Fusion**: Integrating satellite imagery (e.g., Sentinel-1 SAR for flood mapping) with ground-based IoT data could enhance spatial resolution and prediction accuracy, particularly in remote regions with sparse sensor coverage (Wang et al., 2024).
2. **Explainable AI (XAI) Integration**: Embedding XAI techniques, such as SHAP (SHapley Additive exPlanations), would improve transparency in hazard predictions, fostering trust among policymakers and communities (Zhang et al., 2023).
3. **Reinforcement Learning for Adaptive Calibration**: Deploying reinforcement learning agents to automate sensor recalibration and model fine-tuning could further reduce latency and human intervention (Li et al., 2023).

10.2 Policy Integration

1. **Early Warning Mobile Apps**: Developing government-backed apps that translate the system's risk scores (e.g., flood_target, cyclone_target) into localized alerts and evacuation routes would democratize access to critical information. For instance, integrating with India's *UMANG* platform or Bangladesh's *Cyclone Preparedness Programme* could streamline disaster response (UNDRR, 2020).
2. **AI-Powered Policy Dashboards**: Centralized dashboards, updated in real time with the framework's outputs, could guide dynamic policy adjustments—such as revising building codes in high-risk seismic zones or allocating drought relief funds based on water_scarcity_target values (Ahmed et al., 2024).

3. **Blockchain for Accountability**: Implementing blockchain-based audit trails would ensure transparent resource allocation, particularly for funds directed to regions flagged by the system's risk assessments (Brian et al., 2020).

10.3 Community-Centric Innovations

1. **Crowdsourced Data Platforms:** These platforms enable citizens to provide ground-truth data using mobile applications (for example, to report the depth of flooding or sightings of wildfires) to improve model accuracy and engagement with the community (FAO, 2020).
2. **Education in Disaster Resilience:** Introducing and embedding predictive scenarios into school curricula and workshops within communities could help foster preparedness for future climate-vulnerable areas(IPCC, 2023).
3. **Microinsurance Schemes:** Partnering with insurers to design parametric policies turning owing to the system's predictions, say automatic payouts triggered when water_scarcity_target >0.8 for farmers, thus minimizing socio-economic loss (UNDRR, 2020).

10.4 Global Scalability

1. **Low-Cost Sensor Kits:** Modular solar-powered IoT kits (like Raspberry Pi-based nodes with DHT-22 and BMP280 sensors) would allow low-cost replication in low-resource settings across sub-Saharan Africa and Southeast Asia (Kumar et al., 2022).
2. **Cross-Border Data Sharing:** Agreement on the international sharing of data under frameworks, such as the Sendai Framework, would complement efforts for regional preparedness for transboundary disasters like cyclones and riverine floods (IPCC, 2023).
3. **Climate-Resilient Infrastructure Bonds:** Risk assessment under the system, to allow prioritizing investments in flood barriers, wildfire zones, and heat-prone urban planning, can leverage climate finance (Wang et al., 2024).

REFERENCES

Ahmed, A.. (2021). Real-Time Disaster Monitoring and Early Warning System using IoT. *IEEE Access : Practical Innovations, Open Solutions*, *9*, 44920–44934.

Akter, S., Haque, M. M., & Islam, M. R. (2021). IoT based real-time river water level monitoring and flood warning system for Bangladesh. ResearchGate. https://www.researchgate.net/publication/350760195

Alavi, S. H. S., Atani, R. E., & Qader, N. N. (2020). Internet of Things in Disaster Management: A Review. *International Journal of Advanced Computer Science and Applications*, *11*(3), 372–380.

Amini Hosseini, K., Alesheikh, A. A., & Javadpoor, M. (2020). A machine learning framework for multi-hazards modeling and mapping in a mountainous area. *Scientific Reports*, *10*(1), 1–12. DOI: 10.1038/s41598-020-70457-3 PMID: 31913322

Awaysheh, F. M., Alhussein, A., Al-Ali, A. R., & Alassery, H. A. (2020). IoT-based disaster management system for real-time flood detection. *Sustainable Computing : Informatics and Systems*, *28*, 100431.

Bhoi, S. K., & Turuk, P. K. "A Novel Energy Efficient Routing Protocol for MANET," in *Proc. of the Int. Conf. on Information Technology*, 2007, pp. 1–6.

Bindra, K. S., Yadav, R., & Rana, N. K. (2019). IoT-based Smart Disaster Management System for Flood Detection and Rescue. *Procedia Computer Science*, *152*, 106–113.

Bissyandé, T. F., Lo, D., Jiang, L., Réveillère, L., & Klein, J. (2016). Got issues? Who cares about it? A large scale investigation of issue trackers from the perspective of developers. *Empirical Software Engineering*, *21*(1), 150–188.

Gao, H., Wang, Y., Liu, J., & Liu, X. (2018). Real-time emergency rescue scheduling model for urban disaster management. *IEEE Access : Practical Innovations, Open Solutions*, *6*, 55284–55295.

Ghayvat, H., Mukhopadhyay, S., & Gui, X. (2015). Smart home-based ambient assisted living: Recognition of activities of daily living using 3D depth sensors. *IEEE Sensors Journal*, *15*(3), 1476–1483.

Ghosh, A., Datta, D., Saha, S. K., & Mukherjee, D. (2021). An IoT-cloud integrated disaster management framework using Artificial Intelligence. *Journal of Ambient Intelligence and Humanized Computing*, *12*(1), 991–1004.

Gupta, H., Jain, A., Singhal, R., & Kapoor, S. (2020). Analysis and optimization of deep learning models for disaster management. *Procedia Computer Science, 171*, 1600–1609.

Gupta, N., Roy, A., & Das, R. (2021). IoT-based Flood Monitoring and Alerting System using Machine Learning. *International Journal of Intelligent Engineering and Systems, 14*(1), 225–234.

Haque, A. M., & Kaiser, M. S. (2020). Smart Disaster Management System Using IoT and Cloud. *Lecture Notes in Networks and Systems, 123*, 233–241.

Harsono, T. W. R., Pramudito, A. A., & Nugroho, S. A. (2021). IoT and Deep Learning-based Flood Monitoring and Early Warning System. *Journal of King Saud University. Computer and Information Sciences.*

Hiremath, P. S., & Joshi, B. D. (2019). Content Based Image Retrieval using Color, Texture and Shape features. *International Journal of Advanced Computer Science and Applications, 10*(1), 18–24.

Islam, M. M., Hasan, M., & Rahman, M. M. (2019). A machine learning-based prediction and analysis of flood affected households: A case study of floods in Bangladesh. *International Journal of Disaster Risk Reduction, 40*, 101276. DOI: 10.1016/j.ijdrr.2019.101276

Kamal, S. A., & Alam, M. S. (2020). A Framework for Disaster Management in Smart Cities Using IoT. *IEEE Access : Practical Innovations, Open Solutions, 8*, 213897–213908.

Khan, M. A., & Khan, M. A. (2020). A robust framework for disaster management using hybrid cloud. *Journal of Cloud Computing (Heidelberg, Germany), 9*(1), 1–14. DOI: 10.1186/2192-113X-1-1

Kodali, R. K., Yerrapragada, S., & Mandal, A. (2018). An IoT-based disaster management system. *Procedia Computer Science, 132*, 1258–1265.

Kumar, V. A., & Kumar, A. A. (2020). IoT-based disaster management system: A review. *International Journal of Disaster Risk Reduction, 50*, 101735.

Li, L.. (2018). Emergency response system for natural disasters based on the Internet of Things. *IEEE Access : Practical Innovations, Open Solutions, 6*, 21020–21032.

Manogaran, B., & Thota, C. (2019). Big Data Analytics for Healthcare and Disaster Management. *Journal of Organizational and End User Computing, 31*(4), 1–22.

Mishra, R. K., Giri, V. K., & Tripathi, R. C. (2020). Disaster Management using Cloud and Big Data: A Review. *Procedia Computer Science, 167*, 1340–1350.

Ragel, R. G., Caldera, C. S., & Wijesundara, L. A. (2020). A review of the role of machine learning in IoT-based disaster management systems. *Computers & Electrical Engineering*, *89*, 106906.

Rahman, M. A., Hossain, M. A., & Kaiser, M. S. (2021). Deep Learning for Disaster Response: A Systematic Review. *IEEE Access : Practical Innovations, Open Solutions*, *9*, 144295–144312.

Raj, A. C., Anil, K. S., & Sastry, V. N. (2020). IoT based disaster monitoring and management system for real-time applications. *Procedia Computer Science*, *171*, 1738–1747.

Rana, S., Goyal, P., & Bansal, P. (2021). IoT and Machine Learning-Based Disaster Management System. *Journal of King Saud University. Computer and Information Sciences*.

Roy, A., Tazyeen, S. M. F., Sarker, S. R., & Hossain, M. S. (2021). An IoT based real-time disaster monitoring system using deep learning. *Computers, Materials & Continua*, *69*(2), 1745–1762.

Saha, A., Ahmed, M., & Rahman, A. (2023). Flood forecasting by using machine learning: A study leveraging historic climatic records of Bangladesh. *Water (Basel)*, *15*(5), 887. DOI: 10.3390/w15050887

Sani, O. G., Azizi, A., & Malek, M. A. (2020). Assessing and mapping multi-hazard risk susceptibility using a machine learning technique. *Scientific Reports*, *10*(1), 19620. DOI: 10.1038/s41598-020-76257-0 PMID: 33184353

Sharma, A., Singh, M., & Chauhan, V. (2020). Big data and machine learning in disaster management: A survey. *Procedia Computer Science*, *167*, 561–570.

Shi, Y., Sun, C., Yu, Y., & Chen, G. (2020). A hybrid machine learning model for flood prediction. *Applied Sciences (Basel, Switzerland)*, *10*(3), 1016.

Sultana, M. T. N., Ahmed, F. U., & Islam, M. R. (2021). A review on disaster management using big data analytics and machine learning. *Journal of King Saud University. Computer and Information Sciences*.

Suto, K., & Oie, Y. (2019). Anomaly detection for disaster management using mobile social networking services. *Sensors (Basel)*, *19*(6), 1362. PMID: 30893769

Tripathy, A. K., & Kumar, S. (2021). IoT and cloud-based disaster management system: A review. *Materials Today: Proceedings*, *47*, 1908–1913.

Tripathy, A. K., Sahoo, A., Mohanty, R., & Satapathy, S. (2021). An integrated approach to disaster management using IoT and machine learning. *International Journal of Information Management*, *57*, 102196.

Zhang, H. Y., Khan, M. K., & Wang, L. (2021). Blockchain-based disaster management system with situation awareness and decision support. *Computers & Electrical Engineering*, *93*, 107217.

Chapter 11
Integrating AI, Robotics, and Augmented Reality in Disaster Management

Ruby Jindal
https://orcid.org/0000-0002-8589-5549
School of Basic and Applied Sciences, K.R. Mangalam University, Gurugram, India

Archana Tripathi
Department of Physics, Deshbandhu College, University of Delhi, New Delhi, India

Harleen Kaur
Department of Applied Sciences, Baba Banda Singh Bahadur Engineering College, Fatehgarh Sahib, India

Sunita Sharma
School of Basic Sciences, Galgotias University, Greater Noida, India

ABSTRACT

This study investigates the transformative role of Artificial Intelligence (AI), Robotics, and Augmented Reality (AR) in enhancing disaster management across preparedness, response, and recovery phases. Using qualitative analysis and real-world case studies, the research highlights how AI improves early warning systems and decision support, robotics enables efficient and safe search-and-rescue operations, and AR enhances situational awareness for emergency responders. The findings demonstrate that these technologies significantly reduce response time, improve resource allocation, and enhance operational coordination during disasters. The study underscores the need for ethical frameworks, inclusive deployment strategies,

DOI: 10.4018/979-8-3373-1892-9.ch011

and further research to ensure equitable and effective integration of these technologies in global disaster management systems.

1. INTRODUCTION

Disasters, both natural and human-induced, pose significant threats to communities, economies, and ecosystems. In recent years, the increasing frequency and severity of disasters—ranging from hurricanes and wildfires to industrial accidents and pandemics—have highlighted the need for more efficient and adaptive disaster management strategies. Traditional disaster management approaches, while effective to an extent, often suffer from limitations such as slow response times, inadequate communication channels, and insufficient real-time data. These gaps in disaster response efforts can lead to delayed interventions, increased casualties, and extensive damage to infrastructure and the environment.

The evolution of technology, particularly in the fields of AI, robotics, and AR, has introduced new possibilities for mitigating disaster impacts. AI-driven predictive analytics can anticipate potential threats with remarkable accuracy, allowing authorities to take preventive measures before disasters strike. Robotics, equipped with advanced sensors and autonomous capabilities, are now being deployed for search and rescue operations, helping to locate survivors in hazardous environments and deliver essential aid. Meanwhile, AR enhances situational awareness by providing real-time data overlays and immersive training environments for emergency responders.

The integration of these technologies into disaster management represents a paradigm shift in how we prepare for, respond to, and recover from crises given in figure 1. AI-powered models can analyze vast amounts of historical and real-time data to detect early warning signs of disasters. Machine learning algorithms refine these predictions over time, improving their accuracy and reliability. Robotics, on the other hand, reduce the risks faced by human responders by operating in dangerous conditions, from collapsed buildings to flood zones (Bekele et al., 2022). AR bridges the gap between data and real-world application by visualizing complex information in an intuitive, actionable manner.

Recent literature has emphasized the limitations of conventional disaster management approaches, particularly in terms of real-time data collection, resource coordination, and situational awareness. For instance, Sun et al. (2020) highlight the fragmented nature of existing systems and the slow integration of digital technologies, which often results in delayed emergency responses and inefficient logistics. Xu and Xue (2024) further identify the lack of predictive capabilities and insufficient automation as critical gaps that hinder proactive disaster mitigation. The integration of AI has shown promise in overcoming these deficiencies through predictive analytics and automated decision-making (Lamsal & Kumar, 2020), while robotics has enabled safer and faster search-and-rescue missions in hazardous environments (Lindqvist et al., 2022). Augmented Reality, as discussed by Xu and Moreu (2021), contributes by enhancing operational awareness and training for first responders. Together, these technologies offer a multidimensional upgrade to traditional practices, providing more adaptable, data-driven, and resilient disaster management systems.

Governments, humanitarian organizations, and the private sector are increasingly recognizing the potential of AI, robotics, and AR in building resilient disaster management systems. Countries such as Japan, the United States, and Germany are already implementing AI-driven early warning systems and robotic rescue units, demonstrating the life-saving benefits of these technologies. The fusion of AI, robotics, and AR not only enhances disaster response but also improves post-disaster recovery efforts by streamlining logistics, restoring communication networks, and supporting infrastructure rebuilding (Abid et al., 2021).

Despite these advancements, challenges remain (C. Xu & Xue, 2024). Ethical concerns regarding AI decision-making, high deployment costs, and interoperability issues between different technologies need to be addressed. Additionally, developing nations, which often face the highest disaster risks, may struggle with the financial and technical barriers to adopting these cutting-edge solutions.

It is important to study AI, robotics, and AR in disaster management now because disasters are becoming more frequent and dangerous due to climate change and growing cities. Traditional methods often cannot respond quickly or effectively enough to save lives and reduce damage. Current systems may be slow, disconnected, or lacking real-time information. New technologies like AI can predict disasters earlier, robots can safely search for survivors, and AR can help emergency teams see and understand situations better. These tools can make disaster response faster, safer, and more organized. As risks increase around the world, it is necessary to explore how these technologies can improve the way we handle disasters today and in the future.

This chapter will explore these aspects in detail, highlighting both the opportunities and obstacles in leveraging AI, robotics, and AR for disaster management.

As we move further into the 21st century, the role of technology in disaster resilience will continue to expand. By fostering collaboration between governments, researchers, and technology developers, we can create a future where disasters are not only better managed but also significantly mitigated. The following sections will delve into the specific contributions of AI, robotics, and AR, providing case studies and real-world applications that illustrate their transformative impact on disaster management.

Figure 1. Disaster Management Cycle

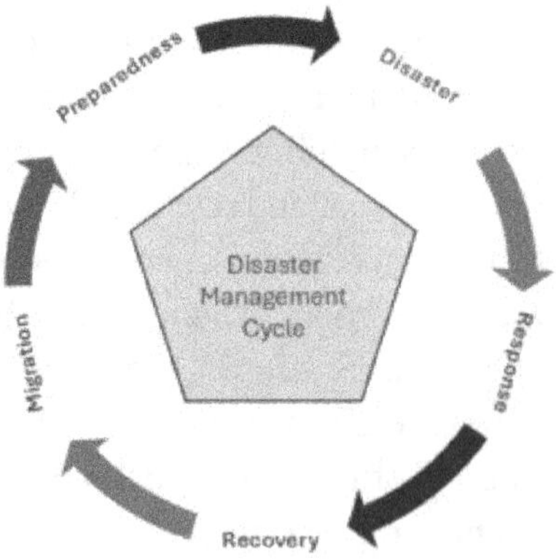

1.1 Methodology

This study follows a qualitative research approach to understand how Artificial Intelligence (AI), robotics, and Augmented Reality (AR) are being used in disaster management. The information was collected from various reliable sources, including published research papers, government reports, disaster response case studies, and official documentation from technology companies involved in disaster-related projects. The study focuses on real-life examples where these technologies have been used effectively, such as Google's Flood Forecasting system in India, the use of search-and-rescue robots during the Fukushima nuclear disaster in Japan, and AR-based training programs used to prepare emergency responders.

These case studies were chosen because they show clear and practical applications of the technologies in actual disaster situations. The collected data was carefully reviewed to find common patterns and key lessons—for example, how these tools helped reduce response time, improved coordination, or made rescue efforts safer. When available, numbers and statistics (like how many people were warned, or how much faster teams responded) were also included to support the findings. This approach helps us understand not just what the technologies can do, but how they actually work in the field and what benefits they bring. By combining real-world examples with expert insights, the study gives a deeper understanding of how these tools are changing disaster management today.

2. AI IN DISASTER MANAGEMENT

2.1 Predictive Analytics and Early Warning Systems

AI-powered predictive models analyze historical disaster data, weather patterns, and seismic activity to forecast potential threats as shown in figure 2. Machine learning algorithms improve the accuracy of predictions, enabling authorities to implement early warning systems for hurricanes, floods, and wildfires. These models leverage diverse datasets, including satellite imagery, geospatial data, and real-time sensor readings, to enhance situational awareness and disaster readiness.

One of the key benefits of AI-based predictive analytics is its ability to identify patterns and correlations that might be overlooked using traditional statistical methods. AI-driven systems such as IBM's Watson and Google's DeepMind are being used to analyze climate change trends, predict extreme weather conditions, and assess vulnerabilities in urban and rural areas. Such predictive insights allow governments, emergency services, and communities to take proactive measures, such as reinforcing infrastructure, pre-positioning emergency supplies, and issuing timely evacuation alerts.

Moreover, AI-powered flood forecasting models, such as Google's Flood Forecasting Initiative, utilize deep learning techniques to predict river water levels and potential flood zones days in advance. These forecasts are shared with local authorities and communities via mobile alerts, helping to mitigate flood-related damages and casualties. The integration of AI-driven analytics with Geographic Information Systems (GIS) further enhances the visualization and communication of risk data, facilitating more effective emergency planning and decision-making (Lamsal & Kumar, 2020).

Figure 2. Role of AI in Disaster Management

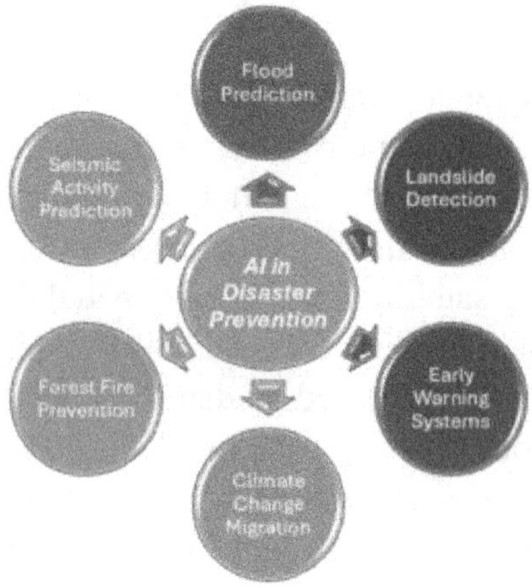

2.2 AI-driven Decision Support Systems

In addition to predictive analytics, AI plays a crucial role in decision support during disaster scenarios. AI-based decision support systems (DSS)(shown in figure 3) process vast amounts of real-time data, enabling emergency managers to make informed and timely decisions. These systems assess multiple factors such as population density, infrastructure resilience, road networks, and weather conditions to recommend optimal evacuation routes, resource allocation strategies, and emergency response priorities (Soori et al., 2024).

For example, AI-powered decision support platforms like IBM's Crisis Communications provide real-time insights by aggregating data from various sources, including social media, satellite imagery, and IoT sensors. By analyzing and interpreting this data, AI can detect crisis hotspots, assess the severity of incidents, and predict the needs of affected communities (Kostopoulos et al., 2024).

Furthermore, AI-driven chatbots and virtual assistants are increasingly being used to enhance communication during emergencies. These intelligent agents provide real-time guidance to individuals seeking help, directing them to the nearest shelters, medical facilities, or safe zones. During the COVID-19 pandemic, AI chatbots played a significant role in disseminating health information, tracking infections, and assisting in contact tracing efforts. Similarly, AI-powered social media monitoring tools help authorities detect misinformation and rumors, ensuring that only accurate and verified information is relayed to the public.

Figure 3. Decision support systems (DSS) process

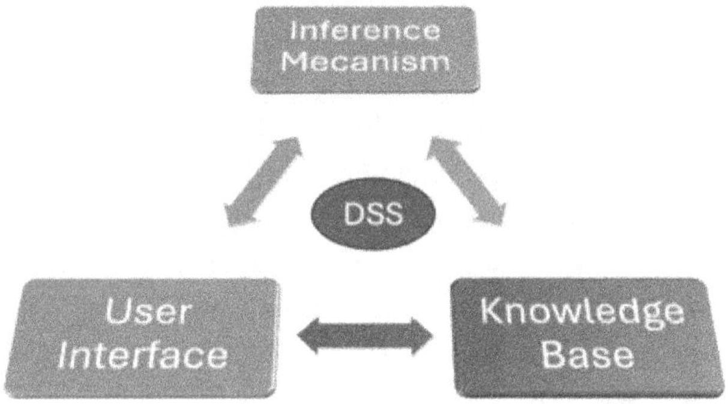

2.3 AI-powered Image and Speech Recognition for Disaster Assessment

Another crucial application of AI in disaster management is image and speech recognition. AI algorithms analyze satellite images, drone footage, and surveillance videos to assess disaster damage and identify areas in need of urgent assistance. Computer vision techniques help distinguish between collapsed buildings, flooded streets, or blocked roads, providing emergency responders with actionable intelligence (Sun et al., 2020).

For instance, after natural disasters like earthquakes or hurricanes, AI-powered image recognition tools can rapidly assess infrastructure damage by comparing pre- and post-disaster imagery. This technology was employed in the aftermath of the 2017 Hurricane Maria in Puerto Rico, where AI-assisted damage assessment enabled a more efficient response by pinpointing the most severely affected areas (DeVries et al., 2018).

In addition, AI-driven speech recognition technology plays a vital role in processing emergency calls and radio transmissions during crises. By analyzing distress calls in multiple languages and dialects, AI systems can prioritize urgent cases and dispatch emergency services accordingly. This capability is especially valuable in large-scale disasters where call centers are overwhelmed with distress signals.

2.4 Case Study: AI in Flood Prediction

Countries like India and the USA use AI-driven flood forecasting systems that integrate satellite imagery, sensor data, and hydrological models to predict water levels and mitigate flood risks effectively. Google's AI-powered flood forecasting initiative in India has successfully reduced flood-related fatalities by providing accurate early warnings to millions of people (Abdullahi et al., 2018).

In India, where monsoon-induced floods affect millions annually, AI-driven flood forecasting has been a game-changer. Google's Flood Forecasting Initiative, in partnership with local governments, utilizes deep learning algorithms to predict flood levels up to 48 hours in advance. By analyzing historical river flow data, topographical features, and real-time weather conditions, these AI models generate precise flood maps, which are then disseminated through Google Alerts, mobile apps, and emergency services.

Similarly, the National Oceanic and Atmospheric Administration (NOAA) in the United States employs AI-enhanced hydrological models to predict and monitor flood patterns. By leveraging AI's ability to analyze high-resolution satellite imagery, NOAA can provide more accurate flood warnings, allowing communities to take preventive measures such as reinforcing embankments, relocating vulnerable populations, and securing critical infrastructure.

AI's role in flood prediction extends beyond forecasting—it also aids in post-disaster recovery. AI-driven image analysis tools help assess the extent of flood damage by comparing satellite images taken before and after a disaster. This rapid assessment enables governments and humanitarian organizations to allocate resources efficiently and prioritize recovery efforts. Additionally, AI-based decision support systems optimize the distribution of relief supplies, ensuring that food, water, and medical aid reach the most affected areas promptly.

Overall, AI's integration into disaster management enhances resilience by enabling proactive planning, real-time decision-making, and efficient post-disaster recovery. As AI continues to evolve, its capabilities in disaster forecasting, emergency response, and crisis management will become increasingly sophisticated, paving the way for a future where disasters are mitigated more effectively and lives are saved through intelligent, data-driven solutions.

3. ROBOTICS IN DISASTER RESPONSE

3.1 Autonomous Search and Rescue Robots

Robotic systems equipped with AI navigate hazardous environments, identifying survivors in collapsed buildings, fire-stricken areas, or flooded regions. Drones equipped with thermal imaging assist rescue teams in locating people trapped under debris (Lindqvist et al., 2022). These autonomous robots are equipped with advanced sensors, including LiDAR, GPS, and infrared cameras, to map affected areas and detect human presence in challenging terrains (Zhang & Herrmann, 2024).

Rescue robots come in various forms, such as ground-based robots, aerial drones, and underwater autonomous vehicles. Ground-based robots like Boston Dynamics' Spot (shown in figure 4) can traverse difficult terrains, such as rubble or unstable surfaces, while aerial drones such as DJI's Mavic Enterprise can quickly scan large disaster zones. Underwater robots, such as the REMUS autonomous vehicle, assist in maritime search and rescue missions following tsunamis or shipwrecks.

AI-powered robotics also integrate natural language processing (NLP) capabilities, allowing them to communicate with survivors. For example, voice recognition software enables robots to respond to cries for help, ensuring a faster and more targeted response. Some robots are even equipped with robotic arms capable of clearing debris or providing medical supplies to those in need before human rescuers arrive.

Figure 4. Boston Dynamics' Spot Robot built by Marc Raibert

PAU BARRENA/GETTY IMAGES https://www.wired.com/story/boston-dynamics-institute-robots-marc-raibert/

3.2 Robotic Logistics and Medical Assistance

Autonomous robots deliver emergency supplies such as food, water, and medical aid in disaster-hit areas. AI-enabled robotic arms perform remote surgeries and medical diagnostics when human intervention is challenging. These robotic logistics systems improve supply chain efficiency in post-disaster scenarios by minimizing human exposure to hazardous conditions and ensuring timely aid distribution (Cruz et al., 2024; Deo & Anjankar, 2023).

Medical assistance robots, such as the Tele Robotic Surgery System, allow doctors to perform remote surgeries on critically injured victims in inaccessible areas. These robotic systems use high-definition cameras and AI-driven precision to conduct delicate procedures while being operated by medical professionals from distant locations. In addition, autonomous delivery drones from companies like Zipline have been deployed in disaster-affected regions to deliver life-saving medical supplies such as blood, vaccines, and essential drugs. A team of Johns Hopkins University researchers designed a, the Smart Tissue Autonomous Robot, that has performed laparoscopic surgery on the soft tissue of a pig without the guiding hand of a human (shown in figure 6).

Another significant advancement in robotic logistics is the development of exoskeletons for first responders. These AI-assisted wearable robotic suits enhance human endurance and strength, allowing emergency personnel to lift heavy debris, navigate unstable terrain, and perform rescues more efficiently. Companies such as Sarcos Robotics are pioneering exoskeleton technology for disaster response teams, enabling enhanced mobility and reduced fatigue during prolonged rescue operations.

Figure 5. A robot performing first laparoscopic operation without the guidance of a surgeon's hand

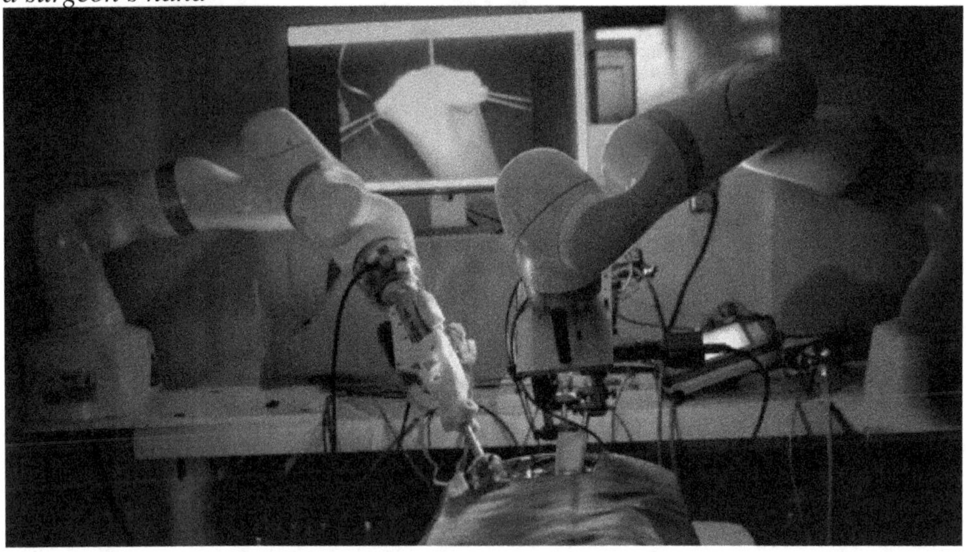

https://hub.jhu.edu/2022/01/26/star-robot-performs-intestinal-surgery/

3.3 Advanced Robotics for Hazardous Environments

Robots are increasingly used to access dangerous locations that are too risky for human responders (Gentile et al., 2023). For instance, chemical spills, nuclear disasters, and fire-ravaged buildings pose severe health risks to rescue teams. AI-powered robots equipped with hazardous material sensors can detect toxic gases, radiation levels, and structural instability in real time, relaying crucial information to human decision-makers (Trevelyan et al., 2008).

A notable example is the use of robots in nuclear disaster zones, such as the Fukushima Daiichi nuclear plant crisis in Japan. Following the 2011 tsunami, AI-driven robotic units, including the PackBot (Yamauchi, 2004) and Quince robots, were deployed to assess radiation levels, inspect damage, and perform cleanup operations in areas unsafe for human workers. Similarly, firefighting robots like the Thermite RS3 are being used to combat large-scale fires, reducing firefighter exposure to extreme heat and hazardous smoke.

Figure 6. IRobot PackBot military ground robot

(Yamauchi, 2004)

3.4 Human-Robot Collaboration in Disaster Response

While autonomous robots play a crucial role in disaster management, human-robot collaboration is also a growing trend. AI-driven robotics are designed to complement human decision-making rather than replace human responders. For instance, semi-autonomous robots can be controlled remotely by emergency teams, allowing for a more precise and coordinated response.

Wearable robotics, such as smart helmets with augmented reality (AR) integration, provide first responders with critical data overlays, including structural weaknesses in buildings, potential hazards, and real-time video feeds from drone scouts. These AR-assisted robotic systems ensure that human teams are equipped with the best possible situational awareness to optimize rescue missions (Wilk-Jakubowski et al., 2022).

Moreover, robotic swarms—groups of small, AI-powered robots that work together—are being developed to efficiently search disaster zones. These robotic networks communicate with each other, using AI-driven algorithms to distribute search tasks, identify survivors, and relay real-time data to control centers. The use of swarm robotics in earthquake or landslide scenarios significantly speeds up search and rescue efforts, reducing the time it takes to locate and assist survivors.

3.5 Case Study: Drones in Earthquake Rescue Operations

Following earthquakes in Japan and Nepal, drones played a crucial role in aerial reconnaissance, identifying structurally unsafe buildings and guiding rescue teams to survivors. After the devastating 2015 Nepal earthquake, AI-driven drones mapped disaster zones, delivered emergency aid, and monitored aftershock activity, significantly improving response times and coordination efforts.

In the wake of earthquakes, drones provide real-time imaging and thermal detection to locate survivors under collapsed structures. High-resolution cameras and AI-driven analysis allow for rapid assessment of infrastructure damage, helping emergency teams prioritize rescue operations. In Nepal, AI-powered drones mapped inaccessible mountainous regions, allowing responders to deliver medical supplies to remote villages.

Drones also play a vital role in post-disaster recovery by assessing damage to infrastructure, such as roads, bridges, and power lines. AI-driven aerial surveillance assists engineers in determining the best reconstruction strategies and resource allocation. In recent years, drone-assisted disaster response has become a standard practice in earthquake-prone regions, demonstrating the effectiveness of AI and robotics in improving crisis response and resilience.

3.6 Future Trends in Disaster Robotics

The future of robotics in disaster management is rapidly evolving, with advancements in AI, machine learning, and automation enhancing robotic capabilities. Future developments include:

- **AI-Powered Autonomous Rescue Pods:** Self-navigating rescue pods capable of deploying themselves in disaster-stricken areas to assist survivors with basic medical care and shelter.
- **Self-Repairing Robots:** Machines equipped with AI-driven self-repair mechanisms, allowing them to function longer in hazardous conditions without human intervention.
- **Bio-Inspired Robotics:** Inspired by nature, robots designed to mimic animal behaviors—such as snake-like robots for navigating debris or swarm robotics based on insect colonies for coordinated search and rescue.

- **Quantum Computing in Robotics:** Enhanced computational power for real-time data processing, improving the efficiency and decision-making capabilities of robotic disaster responders.

As robotics technology continues to advance, the integration of AI, automation, and human-robot collaboration will redefine the way disasters are managed, leading to faster, safer, and more efficient rescue and recovery operations.

4. AUGMENTED REALITY FOR SITUATIONAL AWARENESS

Augmented Reality (AR) has emerged as a crucial tool in disaster management, providing real-time data visualization and enhanced situational awareness for emergency responders. By overlaying digital information onto physical environments, AR improves decision-making and facilitates efficient crisis response.

4.1 Enhancing First Responders' Awareness

AR technology enables first responders to assess disaster-affected areas quickly and accurately. Smart helmets and AR glasses, such as the Microsoft HoloLens, provide firefighters, paramedics, and search-and-rescue teams with real-time data, including structural stability, heat maps, and potential hazards. These AR-assisted devices improve navigation in low-visibility conditions, helping responders avoid risks and prioritize rescue efforts.

For example, AR can project virtual escape routes on a firefighter's visor, guiding them through smoke-filled buildings. Similarly, AR-equipped drones can overlay damage assessments on live video feeds, allowing emergency teams to focus their efforts on high-risk zones. Additionally, AR interfaces can integrate biometric data, such as heart rate and oxygen levels, to monitor responders' health during rescue operations.

4.2 AR in Disaster Training and Simulations

Disaster preparedness relies heavily on training simulations. AR-based training platforms allow emergency responders to practice complex rescue missions in realistic virtual environments. These simulations provide hands-on experience in handling floods, earthquakes, fires, and chemical spills without real-world risks (Khanal et al., 2022).

AR enhances these training sessions by incorporating interactive elements, such as virtual victims requiring medical assistance or collapsing structures requiring strategic evacuations. Emergency personnel can refine their decision-making skills and coordination, ensuring a more effective response during actual disasters. Furthermore, AR-based training modules can be updated dynamically to include new disaster response strategies based on recent events and research.

4.3 AR-Assisted Infrastructure Damage Assessment

Post-disaster recovery involves assessing structural damage to buildings, bridges, and roads. AR applications, integrated with AI and drone imaging, create 3D models of affected areas, allowing engineers and disaster management authorities to evaluate infrastructure integrity remotely.

These AR-powered assessments help officials identify which structures require immediate repairs and which ones need demolition. By overlaying past structural blueprints onto real-time images, AR assists in determining the safest approach for reconstruction efforts. Additionally, AR technology can help detect hidden damages, such as internal cracks or weakened foundations, which may not be visible through traditional inspection methods (J. Xu & Moreu, 2021).

4.4 Public Awareness and Evacuation Guidance

AR applications are also beneficial for civilians during disasters. Mobile AR apps provide real-time evacuation routes, hazard alerts, and emergency instructions based on geolocation. In earthquake-prone regions, AR-enabled smartphones can guide users to the nearest safe zones, while in flood situations, AR overlays can indicate the depth of water levels and potential escape routes.

AR also plays a role in public awareness campaigns, educating communities about disaster preparedness. Interactive AR simulations teach individuals how to respond during emergencies, making disaster education more engaging and impactful. Schools and community centers can leverage AR applications to conduct disaster drills, ensuring that individuals are better equipped to handle real-life emergencies.

Furthermore, AR can facilitate crowd control and emergency communication during large-scale disasters. By integrating AR with public infrastructure, authorities can project real-time instructions on roads, buildings, and transportation systems to guide evacuations effectively.

4.5 Future Trends in AR for Disaster Management

As AR technology advances, its role in disaster management will expand significantly. Future trends include:

- **AI-Powered AR Assistants:** Virtual assistants capable of providing real-time disaster response guidance through AR interfaces. These AI-driven assistants can analyze live disaster data and suggest optimal escape routes or medical aid locations.
- **AR-Integrated Smart Cities:** Smart infrastructure that dynamically overlays emergency data in urban environments to guide evacuations. Future cities will incorporate AR-enabled warning systems that can instantly display emergency routes on roads, buildings, and transportation hubs.
- **AR-Enabled Autonomous Drones:** Drones equipped with AR visualization for more precise disaster assessment and relief distribution. These drones can identify survivors and deliver aid while projecting real-time AR overlays for ground responders (Bhat et al., 2024).
- **Wearable AR for Disaster Survivors:** AR wearables designed to provide survivors with situational awareness, communication channels, and resource locations. These wearables could include holographic interfaces displaying food and medical aid distribution points in real time.
- **AR for Psychological Support:** Disaster survivors often experience trauma and stress. AR applications may be developed to offer psychological support, using interactive experiences to help individuals cope with post-disaster anxiety and mental health issues.

As AR continues to integrate with AI and robotics, disaster response strategies will become more efficient, reducing casualties and economic losses. The fusion of these technologies will revolutionize disaster management, creating safer and more resilient communities worldwide.

5. CHALLENGES AND ETHICAL CONSIDERATIONS IN AI, ROBOTICS, AND AR FOR DISASTER MANAGEMENT

Despite the promising advancements in AI, robotics, and AR for disaster management, there are significant challenges and ethical considerations that must be addressed to ensure these technologies are effectively and responsibly implemented.

5.1 Technological Limitations and Reliability Issues

One of the primary challenges of deploying AI, robotics, and AR in disaster scenarios is ensuring their reliability under extreme conditions. Natural disasters often disrupt power grids, communication networks, and satellite systems, which are critical for the functioning of these technologies. For example:

- AI-driven predictive models require real-time data, but if sensors and satellites fail due to extreme weather, the accuracy of forecasts may decline.
- Robots and drones used for search and rescue missions must be designed to withstand harsh environments, including high temperatures, strong winds, and water exposure.
- AR systems need stable internet and GPS signals to provide real-time overlays and guidance, which may be compromised in disaster zones.

Future advancements in energy-efficient AI, autonomous robotic navigation, and offline AR capabilities will be essential to overcoming these limitations.

5.2 Ethical Concerns in AI Decision-Making

AI-driven disaster response raises ethical questions, particularly in life-and-death decision-making scenarios. For instance:

- **Bias in AI Predictions:** AI models rely on historical data, which may be biased due to incomplete or region-specific information. This can lead to disparities in disaster warnings and resource allocation, disproportionately affecting vulnerable communities.
- **Automated Prioritization of Rescue Efforts:** If AI systems are programmed to prioritize rescuing certain groups over others based on demographic data, ethical dilemmas may arise regarding fairness and human oversight.
- **Autonomous Disaster Response by Robots:** The increasing role of robotic systems in rescue operations raises concerns about accountability. If a robotic system makes an incorrect decision that results in harm, determining liability—whether it lies with developers, operators, or government agencies—remains a challenge.

To govern the responsible use of AI and robotics in disaster scenarios, ethical frameworks such as the IEEE Ethically Aligned Design (EAD) and the AI4People framework offer valuable guidance. These frameworks emphasize principles like transparency, accountability, privacy, fairness, and human oversight. In the context of disaster response, one key risk is algorithmic bias, where AI systems trained on non-representative datasets may prioritize responses that unintentionally exclude marginalized populations. Another concern is the autonomy of robotic systems in making life-critical decisions without sufficient human intervention. To mitigate these risks, AI and robotics systems should be designed with explainability in mind, allowing responders and decision-makers to understand how conclusions were reached. Human-in-the-loop (HITL) models, where humans maintain control over final decisions, are crucial for ensuring accountability. Additionally, impact assessments and ethical audits should be mandated before deployment to evaluate the societal consequences of these technologies in disaster-prone or politically sensitive regions. By embedding such ethical frameworks into disaster response protocols, stakeholders can ensure that technological solutions are effective, equitable, and trustworthy.

Developing transparent AI models with clear decision-making protocols and human oversight mechanisms will be critical in addressing these ethical concerns (Femi Osasona et al., 2024).

5.3 Privacy and Data Security in AR and AI Systems

As AI and AR technologies become more integrated into disaster management, concerns surrounding privacy and data security are growing. These technologies rely on vast amounts of data collected from various sources, including satellite imagery, drones, IoT devices, and personal mobile devices. While this data is crucial for effective disaster response, it also raises significant ethical and security concerns.

Data Collection and Consent: One of the key concerns is the collection of personal and sensitive data without informed consent. AI-powered surveillance systems and AR applications used for disaster monitoring often gather real-time data from individuals and affected communities. Without proper data governance frameworks, there is a risk of unauthorized data usage, leading to potential privacy violations. It is essential to develop clear policies that ensure transparency in data collection and secure user consent before utilizing personal information.

Cybersecurity Threats in AI and AR Systems: AI and AR systems used in disaster management are vulnerable to cyber threats, including data breaches, hacking attempts, and ransomware attacks. Malicious actors could exploit security loopholes to manipulate disaster response systems, leading to misinformation and delays in emergency interventions. Strengthening cybersecurity protocols, implementing encryption techniques, and conducting regular security audits are critical measures to protect sensitive disaster-related data from cyber threats.

Anonymization and Data Protection Measures: To mitigate privacy risks, AI-driven systems should incorporate anonymization techniques that strip personally identifiable information (PII) from datasets before processing. Blockchain technology can also be leveraged to ensure secure and tamper-proof data storage, allowing only authorized personnel to access sensitive information. Additionally, disaster management organizations must adhere to international data protection regulations, such as the General Data Protection Regulation (GDPR), to ensure compliance with privacy laws.

Balancing Surveillance and Privacy: While AI-enhanced surveillance systems and AR applications improve situational awareness in disaster-stricken areas, excessive surveillance may infringe on civil liberties. Striking a balance between effective disaster monitoring and individual privacy rights is crucial. Implementing ethical AI guidelines and ensuring oversight by regulatory bodies can help maintain this balance. Furthermore, decentralized data-sharing models can be explored to give individuals greater control over their data while still allowing authorities to access critical disaster-related insights.

Building Public Trust in AI and AR Technologies: Public trust is a vital component of effective disaster management technologies. Transparent policies regarding data usage, clear communication about privacy protections, and involving communities in decision-making processes can help build confidence in AI and AR applications. Educational initiatives aimed at informing the public about how these technologies enhance disaster preparedness while safeguarding privacy can further promote acceptance and cooperation.

Future Directions for Privacy and Security in Disaster Technologies: The future of AI and AR in disaster management will require ongoing advancements in privacy-preserving technologies. Federated learning, which allows AI models to be trained without directly accessing raw data, could enhance privacy while maintaining the effectiveness of predictive models. AI-driven anomaly detection can help identify security threats in real time, preventing cyber-attacks and unauthorized access. Strengthening global collaborations on data security standards and ethical AI governance will ensure that privacy remains a top priority as disaster response technologies continue to evolve.

By addressing privacy and data security challenges in AI and AR systems, disaster management can leverage these technologies responsibly, ensuring ethical implementation while maximizing their potential for saving lives and reducing disaster impact. (Aldoseri et al., 2023; Lami et al., 2024).

5.4 Cost and Accessibility of Advanced Disaster Management Technologies

The deployment of AI, robotics, and AR in disaster management is often limited by financial constraints, particularly in developing nations. Advanced disaster response systems require significant investment in:

- Infrastructure, including high-speed internet, data centers, and cloud computing facilities.
- Skilled workforce capable of maintaining and operating AI models, robotics, and AR applications.
- Procurement and maintenance of drones, robotic responders, and AR hardware.

Many low-income countries lack the resources to implement these technologies, resulting in a technology divide in disaster preparedness and response. To address this gap, international collaboration, funding from humanitarian organizations, and cost-effective technology solutions must be explored.

5.5 Public Acceptance and Human Adaptation to AI and Robotics

Despite their potential benefits, AI, robotics, and AR in disaster management may face resistance from the public and emergency responders due to:

- **Mistrust in AI Predictions:** Some disaster management authorities and communities may hesitate to rely on AI-driven predictions over traditional meteorological methods and human expertise.
- **Concerns About Job Displacement:** The introduction of AI-powered disaster response systems may lead to fears of job losses among emergency responders, even though these technologies are meant to augment human capabilities rather than replace them.
- **Complexity of AR Interfaces:** While AR has the potential to improve situational awareness, emergency responders must be adequately trained to use these tools effectively without causing cognitive overload during high-stress situations.

Public education campaigns, user-friendly technology interfaces, and integration of AI and robotics as supportive tools rather than replacements for human responders will help in addressing these concerns.

5.6 Environmental and Ethical Impact of AI, Robotics, and AR

The deployment of high-tech disaster management solutions also raises environmental and ethical concerns, such as:

- **Carbon Footprint of AI and Robotics:** The energy-intensive nature of AI models, data centers, and robotic operations contributes to carbon emissions. Sustainable AI development practices and renewable energy sources for data processing must be prioritized.
- **Electronic Waste Management:** The production and disposal of robotic and AR hardware can lead to e-waste challenges. Establishing recycling programs and eco-friendly manufacturing methods will be essential.
- **Ethical Use of Drones and Surveillance Technologies:** The use of AI-driven drones for disaster response must be regulated to prevent unauthorized surveillance and ensure they are used strictly for humanitarian purposes.

By integrating sustainability considerations into disaster management technology development, we can create solutions that are both effective and environmentally responsible.

6. FUTURE DIRECTIONS AND INNOVATIONS IN AI, ROBOTICS, AND AR FOR DISASTER MANAGEMENT

The future of disaster management will be shaped by continuous advancements in AI, robotics, and AR, driving the development of more effective, resilient, and adaptive disaster response strategies. As these technologies evolve, they will become more sophisticated, integrated, and capable of addressing challenges that current systems struggle to overcome.

6.1 AI-Driven Predictive Models and Real-Time Analytics

Artificial Intelligence (AI)-driven predictive models and real-time analytics are transforming disaster management by enabling early warnings, optimizing response efforts, and improving decision-making. These advancements rely on vast datasets collected from multiple sources, such as satellites, weather sensors, seismic monitors, and social media platforms, to identify trends and anomalies that signal potential disasters.

One of the most critical applications of AI in disaster management is early warning systems. Machine learning algorithms process historical and real-time data to predict earthquakes, hurricanes, floods, and wildfires with increasing accuracy. For example, AI-powered seismic activity models analyze minor tremors and pressure shifts along fault lines to anticipate earthquakes, allowing authorities to take preemptive action. Similarly, AI-enhanced meteorological models assess atmospheric conditions to provide early warnings for hurricanes and cyclones, reducing casualties and property damage.

Real-time analytics plays a crucial role in enhancing situational awareness during disasters. AI-powered platforms aggregate data from diverse sources, including emergency calls, drone footage, and satellite imagery, to generate live updates for disaster response teams. By employing natural language processing (NLP) techniques, AI can also analyze social media posts and emergency requests to identify affected areas and the severity of incidents. This real-time intelligence helps authorities allocate resources more effectively, ensuring that rescue operations are conducted efficiently.

Another key innovation is AI-driven simulation models. These models enable disaster response agencies to conduct predictive scenario planning and stress-test various response strategies before an actual disaster occurs. By simulating multiple disaster scenarios, emergency planners can identify vulnerabilities, optimize evacuation plans, and develop more effective resource distribution strategies.

AI is also playing a significant role in optimizing logistics and resource allocation. Machine learning algorithms analyze transportation networks, infrastructure data, and availability of relief supplies to develop the most efficient routes for delivering aid and deploying emergency teams. This minimizes response times and maximizes the effectiveness of disaster relief efforts.

Despite these advancements, AI-driven predictive models face several challenges. Data accuracy, model biases, and integration issues with legacy systems remain hurdles to fully realizing AI's potential in disaster management. Continuous improvements in data collection, algorithm refinement, and regulatory oversight are essential to ensure that AI-driven analytics remain reliable, transparent, and ethically sound.

As AI continues to evolve, future advancements will likely incorporate even more sophisticated models that leverage quantum computing, edge AI, and blockchain for enhanced predictive accuracy, security, and decentralized disaster management systems. These developments will further strengthen global disaster resilience, enabling a more proactive and adaptive approach to crisis management

6.2 Robotics for Autonomous Search, Rescue, and Relief Operations

Robotics is revolutionizing search, rescue, and relief operations by introducing autonomous systems that enhance efficiency, reduce response times, and minimize risks to human responders. These robotic systems include aerial drones, ground-based robots, and underwater autonomous vehicles, all of which play critical roles in different disaster scenarios.

- **Aerial Drones for Search and Reconnaissance**: Aerial drones, equipped with high-resolution cameras, thermal imaging, and LiDAR sensors, provide a real-time assessment of disaster-affected areas. These drones can navigate through inaccessible regions, locate survivors trapped under debris, and map hazardous zones for rescue teams. AI-powered drones can autonomously analyze terrain, detect movement patterns, and prioritize high-risk areas, enabling more effective coordination of relief efforts. For example, after earthquakes or hurricanes, drones can quickly scan large areas, reducing the time required to assess damages and deploy emergency services.
- **Ground-Based Robots for Rescue Operations**: Ground-based robots equipped with robotic arms, advanced sensors, and AI-driven navigation capabilities are being developed to assist in physically demanding rescue operations. These robots can crawl through rubble, lift debris, and deliver emergency supplies to survivors. Some advanced humanoid robots can interact with victims, provide immediate medical assessments, and relay critical information to first responders. Autonomous all-terrain robots are particularly useful in flood-affected areas, where they can transport food, water, and medical kits while navigating unstable surfaces.
- **Swarm Robotics for Large-Scale Operations**: Swarm robotics, inspired by collective intelligence in nature, involves deploying multiple robots that communicate and collaborate to cover vast areas efficiently. These robotic swarms can spread out over a disaster zone, identify hazards, and relay information to emergency teams. The ability of these robots to self-organize and adapt to changing environments makes them highly effective in unpredictable disaster scenarios, such as wildfires and landslides.
- **Underwater and Amphibious Robots**: Disaster situations involving floods, tsunamis, or hurricanes require specialized robotic systems that can operate in aquatic environments. Underwater autonomous vehicles (UAVs) are designed to assess submerged infrastructure, locate missing persons, and perform underwater repairs. Amphibious robots, capable of transitioning be-

tween land and water, are being developed for disaster response efforts in coastal regions and flood-prone areas.

- **Robotics in Relief Distribution**: Beyond search and rescue, robotics also plays a crucial role in the efficient distribution of relief supplies. Autonomous delivery robots and drones are being used to transport essential items such as food, water, and medical aid to affected populations, particularly in remote and inaccessible regions. This automation helps to accelerate relief efforts while reducing dependency on human logistics personnel.
- **Challenges and Future Prospects**: Despite their potential, robotic systems for disaster response face several challenges. Power supply limitations, operational durability in extreme environments, and autonomous decision-making constraints remain key hurdles. Additionally, regulatory and ethical considerations, such as the deployment of robots in sensitive areas and concerns about job displacement, need to be addressed.

Future advancements in robotics will focus on integrating AI-driven decision-making, enhanced machine learning models for real-time adaptation, and improved energy efficiency to prolong operational capabilities. The continued collaboration between governments, private sector innovators, and humanitarian organizations will be crucial in refining these robotic technologies and ensuring their seamless integration into global disaster management frameworks.

6.3 Augmented Reality for Enhanced Disaster Response and Training

Augmented Reality (AR) is transforming disaster response and training by providing real-time, interactive information to responders, improving decision-making, and enhancing situational awareness. AR overlays digital information onto the physical world, enabling emergency teams to visualize hazards, plan efficient rescue routes, and train for complex disaster scenarios more effectively.

- **Real-Time Information Overlay and Navigation**: One of the most significant applications of AR in disaster management is its ability to provide real-time overlays of critical information. AR-enabled smart glasses and headsets can display live maps, building layouts, and hazard locations directly within the field of vision of first responders. This enhances situational awareness and allows rescue teams to navigate hazardous areas without relying solely on external maps or devices. Additionally, AR can integrate with AI-powered analytics to provide recommendations on the safest and most efficient rescue paths.

- **AR for Search and Rescue Operations**: During search and rescue missions, AR enhances visibility in low-visibility environments, such as smoke-filled buildings, collapsed structures, or areas with poor lighting. By integrating with thermal imaging and LiDAR sensors, AR can highlight the locations of trapped victims, hazardous materials, and structural weaknesses, helping responders make quick, informed decisions. This capability is particularly useful in earthquake or fire scenarios, where every second counts in locating survivors.
- **Disaster Training and Simulation with AR**: AR-based training programs allow emergency responders and disaster management teams to prepare for crisis situations in a highly immersive and interactive environment. AR-powered simulations replicate real-world disaster scenarios, such as earthquakes, floods, and chemical spills, enabling personnel to practice response strategies without real-world risks. These training modules improve coordination, enhance muscle memory, and allow responders to experience dynamic disaster situations in a controlled setting.
- **Public Awareness and Preparedness**: Beyond professional responders, AR is also being used to educate the general public about disaster preparedness. AR mobile applications provide interactive guides on evacuation routes, first-aid procedures, and emergency protocols. Schools and community centers can leverage AR experiences to teach individuals how to react in different disaster scenarios, ultimately improving overall community resilience.
- **Medical Assistance and Triage with AR**: In post-disaster scenarios, medical teams can utilize AR to enhance patient triage and treatment. AR applications can provide step-by-step guidance for emergency procedures, overlay patient vitals in real-time, and assist remote doctors in guiding on-site responders through complex treatments. By streamlining the decision-making process, AR helps medical teams prioritize critical cases and provide faster, more effective care.
- **Future Prospects and Challenges**: Despite its potential, AR in disaster response faces several challenges, including hardware limitations, data processing requirements, and integration with existing emergency systems. Future advancements will focus on improving AR wearables, enhancing real-time data processing capabilities, and developing AI-powered AR assistants that can guide responders in high-pressure situations.

The continued evolution of AR, combined with AI and robotics, will revolutionize disaster response by improving coordination, efficiency, and overall preparedness. As AR technology advances, its role in enhancing emergency response and training will become increasingly vital in ensuring better outcomes in disaster situations.

6.4 AI-Enhanced Communication and Coordination in Crisis Response

Effective communication and coordination are crucial in disaster response, and AI-powered systems are revolutionizing how emergency teams collaborate and disseminate critical information. AI-enhanced communication tools ensure that responders receive real-time updates, coordinate efficiently, and optimize decision-making processes even in chaotic environments.

- **AI-Driven Emergency Communication Systems**: AI-based emergency communication platforms analyze vast amounts of real-time data from multiple sources, including social media, news reports, satellite feeds, and sensor networks. These platforms use natural language processing (NLP) to filter out misinformation and extract crucial details about affected areas. AI-powered chatbots and virtual assistants provide instant information to responders and the public, reducing panic and misinformation during crises.
- **Automated Alerts and Early Warning Systems**: AI enhances early warning systems by integrating meteorological data, seismic activity reports, and environmental sensors to predict disasters with greater accuracy. AI-driven algorithms automatically generate and distribute emergency alerts through multiple communication channels, including SMS, social media, and emergency broadcast systems. This ensures that affected communities receive timely and location-specific warnings, allowing them to take necessary precautions.
- **AI-Powered Decision Support for Emergency Teams**: In high-pressure situations, AI-driven decision support systems provide emergency responders with actionable insights. These systems process vast amounts of data from drones, IoT sensors, and surveillance cameras to identify risk-prone areas, track the movement of displaced populations, and suggest optimal resource allocation strategies. AI can also simulate different disaster scenarios, helping authorities develop contingency plans and improve crisis preparedness.
- **Real-Time Translation and Multilingual Support**: Disasters often impact diverse populations speaking different languages, which can create communication barriers. AI-powered translation tools and real-time speech recognition systems help emergency responders communicate effectively with affected communities. These AI-driven translation platforms convert emergency instructions, medical guidelines, and public safety announcements into multiple languages instantly, ensuring that crucial information reaches everyone.

- **AI-Enabled Network Resilience and Connectivity**: Maintaining reliable communication networks is critical in disaster-stricken areas where infrastructure may be damaged. AI optimizes network connectivity by predicting areas with high communication demands and dynamically reallocating bandwidth resources. AI-driven drones and satellite-based networks can provide temporary communication links, ensuring uninterrupted connectivity for rescue teams and medical personnel.
- **Coordination of Relief Efforts with AI**: AI streamlines coordination among government agencies, NGOs, and humanitarian organizations by managing logistics, tracking supply chains, and optimizing resource distribution. AI-driven systems analyze data on survivor locations, food and water shortages, and medical supply availability to direct relief efforts where they are most needed. Machine learning models predict future demands based on evolving disaster conditions, ensuring efficient allocation of aid and minimizing wastage.
- **Future Challenges and Opportunities**: Despite its potential, AI-enhanced communication in disaster response faces challenges such as cybersecurity threats, data privacy concerns, and the need for high-speed computing infrastructure. Future advancements will focus on improving AI's ability to interpret complex disaster scenarios, enhancing integration with existing emergency response frameworks, and ensuring ethical AI implementation in crisis management.

The ongoing integration of AI-driven communication tools into disaster management strategies will significantly enhance response effectiveness, minimize casualties, and improve overall resilience in the face of disasters.

6.5 Sustainable and Resilient Technology Development

Future disaster management technologies must prioritize sustainability to minimize environmental impact. Key innovations may include:

- **Eco-Friendly AI Models:** Development of energy-efficient AI algorithms to reduce the carbon footprint of data processing and predictive modeling.
- **Biodegradable and Recyclable Robotics:** The introduction of robots made from biodegradable materials and modular designs to facilitate easy recycling and reduce e-waste.
- **Solar-Powered Disaster Response Technologies:** Use of solar and alternative energy sources to power disaster response infrastructure, ensuring uninterrupted operations during power outages.

6.6 Integration of AI, Robotics, and AR into Smart Cities and Infrastructure

Future urban planning and disaster preparedness strategies will incorporate AI, robotics, and AR as core components of smart cities (Ortega-Fernández et al., 2020; Szpilko et al., 2023). Advancements in this area may include:

AI-Enabled Smart Infrastructure: Buildings and bridges equipped with AI-powered sensors will autonomously assess structural integrity post-disaster and alert authorities to potential hazards. These sensors will continuously monitor environmental conditions such as seismic activity, temperature fluctuations, and material fatigue, enabling proactive maintenance and early intervention. AI algorithms will process this data in real-time to predict structural vulnerabilities and suggest remedial measures before catastrophic failures occur.

Automated Emergency Response Systems: AI-driven emergency networks will predict disaster impacts, reroute traffic, and deploy robotic first responders before human teams arrive. By integrating AI-powered simulations and predictive modeling, smart cities can optimize resource allocation and response strategies in real time. Drones and autonomous vehicles equipped with AI will rapidly assess affected areas, delivering essential supplies and mapping out optimal rescue routes to minimize congestion and enhance response efficiency.

AR-Based Urban Evacuation Planning: Interactive AR interfaces will assist city planners in designing optimized evacuation routes and conducting real-time emergency drills. AR-powered simulations will allow city officials to visualize disaster scenarios and refine evacuation strategies based on different variables, such as population density and road accessibility. Public AR applications will also guide individuals during emergencies, providing step-by-step navigation to safe zones while overlaying real-time hazard information on mobile or AR headset displays.

Integration with IoT and Big Data Analytics: Smart cities will leverage IoT-connected devices and big data analytics to enhance disaster resilience. AI-driven predictive models will aggregate information from various sources, including social media, satellite imagery, and sensor networks, to identify emerging risks and inform decision-makers. This integrated approach will create an adaptive and self-regulating disaster management ecosystem capable of responding dynamically to evolving threats.

Cyber-Physical Resilience and Redundancy Systems: To prevent service disruptions, smart cities will incorporate decentralized, AI-driven emergency systems with redundant infrastructure. Edge computing and blockchain-based data management will ensure that disaster response systems remain operational even if central networks fail. AI-enhanced cybersecurity measures will also be critical in protecting smart city infrastructure from cyber threats that could compromise emergency response efforts.

By integrating AI, robotics, and AR into smart cities and infrastructure, urban centers can enhance disaster resilience, improve emergency response efficiency, and minimize human casualties. This forward-looking approach will play a pivotal role in shaping sustainable, disaster-ready cities of the future (Ortega-Fernández et al., 2020; Szpilko et al., 2023).

6.7 Ethical Considerations and Policy Development for Future Technologies

As AI, robotics, and AR become more deeply integrated into disaster management, ethical frameworks and policies must evolve to address emerging concerns:

- **Regulation of Autonomous AI Decision-Making:** Governments must establish clear guidelines for AI-driven disaster response systems to ensure ethical and unbiased decision-making.
- **Privacy Laws for AI and AR Data Usage:** Policies will need to balance the benefits of data-driven disaster management with the need to protect individual privacy and prevent unauthorized surveillance.
- **International Collaboration on Disaster Tech Ethics:** Establishing global standards for the ethical use of AI, robotics, and AR in disaster response will help prevent misuse and promote equitable access to technology.

6.8 Strategic Recommendations and Research Priorities

To fully realize the potential of AI, robotics, and AR in disaster management, several strategic actions and research priorities must be pursued. First, emerging technologies such as edge computing, quantum AI models, and bio-inspired robotics should be explored for their potential to increase real-time responsiveness and resilience in complex environments. Governments and research institutions should invest in the development of lightweight, energy-efficient robotics and offline-capable AR systems that can function in infrastructure-compromised disaster zones. Furthermore, there is a pressing need to design interoperable platforms that allow AI, AR, and robotic systems to seamlessly share data and coordinate actions in real-time. Future research should focus on ethically aligned autonomous decision-making, crowdsourced disaster intelligence, and inclusive technology deployment in low-resource settings. Collaboration between academia, industry, and disaster relief agencies is essential to pilot these innovations in real-world scenarios. These recommendations aim to bridge the gap between innovation and implementation, ensuring that future disaster technologies are not only advanced but also accessible, accountable, and adaptive to diverse global needs.

7. RESULTS AND DISCUSSION

This section presents the outcomes of the case studies analyzed in the study and discusses their broader implications in the context of AI, robotics, and AR integration into disaster management.

The Google Flood Forecasting Initiative in India has demonstrated how AI can predict high-risk flood zones and provide alerts to affected populations up to 48 hours in advance. This capability has significantly reduced loss of life and damage to property, especially in rural and vulnerable regions. The model's predictive accuracy and responsiveness highlight the practical benefits of AI in early warning systems, enabling authorities to mobilize evacuation plans and allocate emergency resources more effectively.

In the case of the Fukushima nuclear disaster, robotic systems played a vital role in assessing hazardous areas and conducting initial search and rescue missions. These robots operated in environments with high radiation, where human entry would have been fatal. Their deployment illustrated how robotics enhances safety, operational reach, and data collection in extreme disaster settings. The use of robotics also accelerated recovery timelines and reduced the number of responders exposed to danger.

AR applications were evaluated through emergency training programs where first responders used smart headsets and glasses to visualize structural risks and simulate evacuation procedures. The immersive nature of AR improved trainees' spatial awareness, decision-making, and reaction speed. These findings emphasize how AR enhances situational awareness and preparedness in both pre-disaster training and active response scenarios.

Together, these results show that the integration of AI, robotics, and AR not only improves individual operational components—such as forecasting, rescue, and coordination—but also fosters a more proactive, data-driven, and resilient disaster management system. The case studies provide strong evidence that these technologies can close critical gaps in traditional disaster response.

8. CONCLUSION

The integration of AI, robotics, and AR into disaster management marks a significant shift in how societies prepare for, respond to, and recover from natural and man-made disasters. These technologies provide unparalleled advantages in terms of speed, accuracy, and efficiency, leading to more effective disaster response and mitigation strategies.

AI-driven predictive models enable early warnings and precise disaster forecasting, minimizing loss of life and damage to infrastructure. These models continuously analyze large datasets, identifying patterns and anomalies that can indicate impending disasters. By leveraging machine learning algorithms and real-time data feeds from satellites, sensors, and social media, AI enhances situational awareness and decision-making capabilities. Additionally, AI-powered chatbots and virtual assistants can provide instant information to affected populations, ensuring timely dissemination of emergency alerts and evacuation procedures.

Robotics plays a crucial role in search and rescue missions, ensuring that human responders are not exposed to hazardous conditions. Autonomous drones equipped with thermal imaging and AI-based object recognition can locate survivors in collapsed buildings, flood zones, and wildfire-affected regions. Ground-based robots with dexterous manipulators can clear debris, deliver medical supplies, and perform critical rescue tasks in areas inaccessible to humans. Furthermore, advancements in swarm robotics, where multiple robots collaborate intelligently, can significantly enhance the efficiency of rescue operations by covering larger areas and executing tasks in a coordinated manner.

Augmented Reality enhances situational awareness, providing real-time information to first responders, decision-makers, and affected populations. AR-enabled wearable devices, such as smart glasses and headsets, can overlay critical data on a responder's field of view, highlighting escape routes, victim locations, and hazard zones. AR-based training simulations allow emergency personnel to undergo immersive disaster response exercises, preparing them for real-world scenarios with greater efficiency. Additionally, AR applications for the general public can facilitate disaster preparedness by providing interactive guides on emergency protocols, first aid techniques, and evacuation plans.

Despite these advancements, challenges remain, including ethical considerations, data privacy concerns, cost barriers, and the need for regulatory frameworks to ensure responsible deployment. The ongoing development of these technologies must balance innovation with safety, ensuring that AI, robotics, and AR are used in a way that benefits humanity while minimizing potential risks. Addressing bias in AI algorithms, ensuring transparency in automated decision-making, and implementing safeguards against cyber threats will be essential to building public trust in these technologies.

Looking ahead, continuous improvements in AI algorithms, robotics automation, and AR visualization will further strengthen disaster management capabilities. Future smart cities will integrate these technologies seamlessly into their infrastructure, making disaster preparedness an inherent aspect of urban planning. Intelligent surveillance systems with AI-driven anomaly detection will enable early threat identification, while robotic emergency response units will autonomously assess and mitigate disasters before they escalate. Additionally, international collaboration and policy development will be critical in standardizing the ethical use of these technologies across borders, ensuring equitable access to disaster management innovations worldwide.

In conclusion, AI, robotics, and AR are revolutionizing disaster management, making responses faster, smarter, and more coordinated. With further research, investment, and global cooperation, these technologies have the potential to save countless lives and build a more resilient world in the face of increasing environmental and humanitarian challenges. By fostering interdisciplinary collaboration and prioritizing human-centered technology development, the future of disaster management can be transformed into a highly proactive and adaptive system, capable of addressing the growing complexity of global disasters with unprecedented precision and effectiveness.

REFERENCES

Abdullahi, S. I., Habaebi, M. H., & Malik, N. A. (2018). Flood Disaster Warning System on the go. *2018 7th International Conference on Computer and Communication Engineering (ICCCE)*, 258–263. DOI: 10.1109/ICCCE.2018.8539253

Abid, S. K., Sulaiman, N., Chan, S. W., Nazir, U., Abid, M., Han, H., Ariza-Montes, A., & Vega-Muñoz, A. (2021). Toward an Integrated Disaster Management Approach: How Artificial Intelligence Can Boost Disaster Management. *Sustainability (Basel)*, *13*(22), 12560. DOI: 10.3390/su132212560

Aldoseri, A., Al-Khalifa, K. N., & Hamouda, A. M. (2023). Re-Thinking Data Strategy and Integration for Artificial Intelligence: Concepts, Opportunities, and Challenges. *Applied Sciences (Basel, Switzerland)*, *13*(12), 7082. DOI: 10.3390/app13127082

Bekele, T. W., Haile, A. T., Trigg, M. A., & Walsh, C. L. (2022). Evaluating a new method of remote sensing for flood mapping in the urban and peri-urban areas: Applied to Addis Ababa and the Akaki catchment in Ethiopia. *Natural Hazards Review*, *2*(2), 97–110. DOI: 10.1016/j.nhres.2022.03.001

Bhat, G. R., Dudhedia, M. A., Panchal, R. A., Shirke, Y. S., Angane, N. R., Khonde, S. R., Khedkar, S. P., Pansare, J. R., Bere, S. S., Wahul, R. M., & Gawande, S. H. (2024). Autonomous drones and their influence on standardization of rules and regulations for operating–A brief overview. *Results in Control and Optimization*, *14*, 100401. DOI: 10.1016/j.rico.2024.100401

Cruz, E. M. G. N. V., Oliveira, S., & Correia, A. (2024). Robotics Applications in the Hospital Domain: A Literature Review. *Applied System Innovation*, *7*(6), 125. DOI: 10.3390/asi7060125

Deo, N., & Anjankar, A. (2023). Artificial Intelligence With Robotics in Healthcare: A Narrative Review of Its Viability in India. *Cureus*, *15*(5), e39416. DOI: 10.7759/cureus.39416 PMID: 37362504

DeVries, P. M. R., Viégas, F., Wattenberg, M., & Meade, B. J. (2018). Deep learning of aftershock patterns following large earthquakes. *Nature*, *560*(7720), 632–634. DOI: 10.1038/s41586-018-0438-y PMID: 30158606

Gentile, C., Lunghi, G., Buonocore, L. R., Cordella, F., Di Castro, M., Masi, A., & Zollo, L. (2023). Manipulation Tasks in Hazardous Environments Using a Teleoperated Robot: A Case Study at CERN. *Sensors (Basel)*, *23*(4), 1979. DOI: 10.3390/s23041979 PMID: 36850591

Khanal, S., Medasetti, U. S., Mashal, M., Savage, B., & Khadka, R. (2022). Virtual and Augmented Reality in the Disaster Management Technology: A Literature Review of the Past 11 years. *Frontiers in Virtual Reality*, *3*, 843195. DOI: 10.3389/frvir.2022.843195

Kostopoulos, G., Davrazos, G., & Kotsiantis, S. (2024). Explainable Artificial Intelligence-Based Decision Support Systems: A Recent Review. *Electronics (Basel)*, *13*(14), 2842. DOI: 10.3390/electronics13142842

Lami, B., Hussein, S. M., Rajamanickam, R., & Emmanuel, G. K. (2024). The role of artificial intelligence (AI) in shaping data privacy. International Journal of Law and Management, (ahead-of-print).

Lamsal, R., & Kumar, T. V. V. (2020). Artificial Intelligence and Early Warning Systems. In Kumar, T. V. V., & Sud, K. (Eds.), *AI and Robotics in Disaster Studies* (pp. 13–32). Springer Nature Singapore., DOI: 10.1007/978-981-15-4291-6_2

Lindqvist, B., Karlsson, S., Koval, A., Tevetzidis, I., Haluška, J., Kanellakis, C., Agha-mohammadi, A., & Nikolakopoulos, G. (2022). Multimodality robotic systems: Integrated combined legged-aerial mobility for subterranean search-and-rescue. *Robotics and Autonomous Systems*, *154*, 104134. DOI: 10.1016/j.robot.2022.104134

Ortega-Fernández, A., Martín-Rojas, R., & García-Morales, V. J. (2020). Artificial Intelligence in the Urban Environment: Smart Cities as Models for Developing Innovation and Sustainability. *Sustainability (Basel)*, *12*(19), 7860. DOI: 10.3390/su12197860

Osasona, F., Amoo, O. O., Atadoga, A., Abrahams, T. O., Farayola, O. A., & Ayinla, B. S. (2024). REVIEWING THE ETHICAL IMPLICATIONS OF AI IN DECISION MAKING PROCESSES. *International Journal of Management & Entrepreneurship Research*, *6*(2), 322–335. DOI: 10.51594/ijmer.v6i2.773

Soori, M., Jough, F. K. G., Dastres, R., & Arezoo, B. (2024). AI-Based Decision Support Systems in Industry 4.0, A Review. *Journal of Economy and Technology*, *S2949948824000374*. Advance online publication. DOI: 10.1016/j.ject.2024.08.005

Sun, W., Bocchini, P., & Davison, B. D. (2020). Applications of artificial intelligence for disaster management. *Natural Hazards*, *103*(3), 2631–2689. DOI: 10.1007/s11069-020-04124-3

Szpilko, D., Naharro, F. J., Lăzăroiu, G., Nica, E., & Gallegos, A. D. L. T. (2023). Artificial Intelligence in the Smart City—A Literature Review. *Engineering Management in Production and Services*, *15*(4), 53–75. DOI: 10.2478/emj-2023-0028

Trevelyan, J. P., Kang, S.-C., & Hamel, W. R. (2008). Robotics in Hazardous Applications. In Siciliano, B., & Khatib, O. (Eds.), *Springer Handbook of Robotics* (pp. 1101–1126). Springer Berlin Heidelberg., DOI: 10.1007/978-3-540-30301-5_49

Wilk-Jakubowski, G., Harabin, R., & Ivanov, S. (2022). Robotics in crisis management: A review. *Technology in Society*, *68*, 101935. DOI: 10.1016/j.techsoc.2022.101935

Xu, C., & Xue, Z. (2024). Applications and challenges of artificial intelligence in the field of disaster prevention, reduction, and relief. *Natural Hazards Review*, *4*(1), 169–172. DOI: 10.1016/j.nhres.2023.11.011

Xu, J., & Moreu, F. (2021). A Review of Augmented Reality Applications in Civil Infrastructure During the 4th Industrial Revolution. *Frontiers in Built Environment*, *7*, 640732. DOI: 10.3389/fbuil.2021.640732

Yamauchi, B. M. (2004). In Gerhart, G. R., Shoemaker, C. M., & Gage, D. W. (Eds.), *PackBot: A versatile platform for military robotics* (pp. 228–237)., DOI: 10.1117/12.538328

Zhang, K., & Herrmann, G. (2024). Control practice for robotic applications in challenging environments. *Annual Reviews in Control*, *57*, 100956. DOI: 10.1016/j.arcontrol.2024.100956

Chapter 12
Transitioning to Electric Vehicles in India:
A Sustainable Path Forward

Priyanshi Gupta
https://orcid.org/0009-0003-6265-4464
Case Western Reserve University, USA

ABSTRACT

As the human race is becoming more conscious of the catastrophic state of environmental destruction they are growing more responsible towards its conservation. The need of the hour is to move towards sustainable development methods. One of the significant initiatives taken in recent years towards sustainability is to replace a conventional gasoline vehicle (CGV) with an electric vehicle (EV). It is considered to be a cleaner, greener and sustainable solution which will help reduce greenhouse gases emission. In this paper, a literature review is presented discussing about basic technology of EV and their role in promoting sustainability. Additionally, it addresses the challenges associated with EV manufacturing and the expanding hybrid vehicle market in India. The paper also outlines government policies and the future roadmap for EVs in India.

1. INTRODUCTION

The biggest environmental concern in recent times is environmental pollution, particularly in India, where several cities consistently rank among the most polluted in the world (Balakrishnan et al., 2019). According to WHO, air pollution contributes to severe respiratory and cardiovascular conditions, impacting quality of life and public health (Cohen et al., 2019). This is primarily due to transportation emissions, which

DOI: 10.4018/979-8-3373-1892-9.ch012

account for nearly 25% of the country's total air pollution (Ministry of Environment, Forest & Climate Change). Climate change and global warming are further aggravated by such emissions, posing a threat to ecological balance and human survival. These alarming realities have prompted researchers and policymakers to move towards greener technologies in the automobile sector. One such technology is the Electric Vehicle (EV), which runs on electricity produced by batteries and is considered a cleaner and more sustainable alternative to Conventional Gasoline Vehicles (CGVs) (Singla & Bansal, 2021). Beyond EVs, the exploration of solar-powered, hydrogen-based, and even water-powered vehicles offers promising alternatives that should be considered for long-term sustainability (Zhao et al., 2024).

The Paper is divided into three major segments as described below. The first part provides a literature overview about EV working technology, its environmental benefits and various production challenges from an Indian manufacturing prospective. The second section discusses the expanding market for hybrid cars and discusses the government policies and future roadmap of EVs in India.

1. Introduction
    - 1.1 Working Principle of EVs
    - 1.2 Impact on Environment
    - 1.3 Challenges in Production
2. Scenario in India
    - 2.1 Shift towards Sustainability
    - 2.2 Hybrid Vehicles
    - 2.3 Government Policies
    - 2.4 Future Roadmap
3. Conclusion

1.1 *Working Principle of EVs*

An EV works on the principle of converting chemical energy from the batteries to mechanical energy by rotating the shaft (Sarode & Eadke, 2020). The major components of an electric vehicle are shown in Fig.1. It consists of Battery Pack which is a collection of DC current generating rechargeable batteries. It acts as the fuel to the vehicle (Kumar & Padmanaban, 2019). It is further connected to a Controller which converts the DC current produced to AC current. Further, it is connected to an electric motor, which acts as the engine of the vehicle. (Cheng, 2009). It converts the alternating current into rotating motion of shaft using the principle of electro-magnetism. The rotating shaft is further connected to the transmission system which transmits the power to wheels of vehicle (Poornesh et al., 2020)

Figure 1. Schematic Diagram of an EV

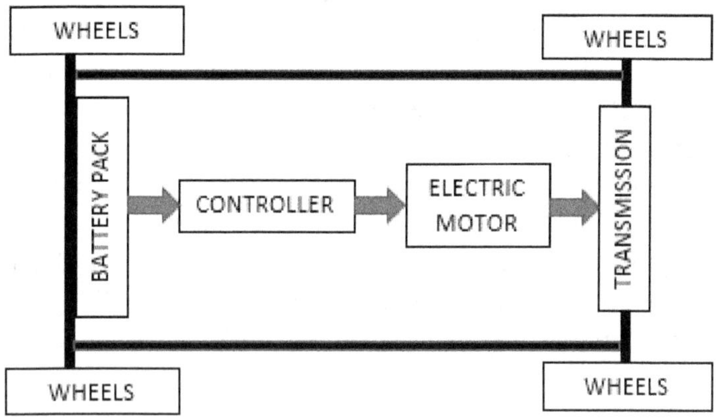

1.2 *Impact on Environment*

Based on its working phenomenon explained above, EVs contribute to some major environmental benefits which are discussed below-

A) Zero emission of greenhouse gases- EV does not have an exhaust system, therefore there are no tailpipe emissions which gives it a major environmental benefit compared to a CGV (Aijaz & Ahmad, 2022). Since, no non-renewable resource is used to generate the energy it reduces dependency on fossil fuels (Singh et al., 2021). Also, if the electricity provided to charge its batteries is generated using a renewable energy resource such a hydro, solar or wind if further reduces the carbon footprint and makes the product more sustainable (Hirz et al., 2022).
B) No Noise Pollution- Internal Combustion Engine in a CGV makes vibration and noise during operation because of changing pressures in cylinder during intake and exhaust process (Spessert et al., 2019). This adds to the noises coming from other sources especially in urban cities resulting in noise pollution (Maffei & Masullo, 2014). EVs are much quitter in operation compared to CGVs. The electric motor spin with barely audible humming sound (del et al., 2020).
C) Low maintenance- EVs require less maintenance compared to CGVs. Since, there are very few mechanical parts in an EV it demands less service and oiling than a CGV where Engine, gears etc. require high maintenance (Alanazi, 2023). This results in less wear and tear damage and longer life of an EV having a potential of 10-20 years before replacement (Mohamed et al., 2018).

D) Cheaper running cost- An EV feeds on electricity which is required to recharge its batteries (Wu et al., 2015). Electricity costs around 8-10 per KW and ~ 100-150 to fully charge an EV at a charging station which is much less compared to the cost of petrol which is ~ 100 per liter costing ~ 3000 to fill a 35 liter fuel tank of a CGV (Gujarathi et al.,2018).

E) Better driving experience- EV have simple operation and controls compared to CGV because electric motor can generate instant and maximum power across entire RPM range (Labeye et al, 2016). This eliminates the need to have set of gears to help regulate torque and speed (Sanguesa et al., 2021). This encourages larger section of people to drive the car with basic driving skills. Table 1 represents the summary of various benefits of EV over CGV.

Table 1. Benefits of EV compared to CGV.

Benefits	CGV	EV
Environmental Pollution	Tailpipe Emission of gaseous pollutants	No emission of gaseous pollutants
Noise	Vibration & operational noise of engine	No operational noise
Maintenance	High	Low
Running Cost	High	Low
Driving Experience	Complex operations- Professional driving skills required	Easy operations- Basic driving skills required

1.3 *Challenges in Production*

From the discussion below, it seems that EVs is the best sustainable alternative for CGV but with new technology comes new challenges. Some of them are discussed below-

A) Technology & Infrastructure

- **Range & Speed Limitations**- The EVs currently available in the Indian market have limited range and speed which restrict its usage for shorter travel distances (Abid et al., 2022).
- **Charging Time**- EV requires long charging hours for as long as 12 hours to fully charge the battery. (Olabi et al., 2022). This makes it unlikely to be preferred over a CGV.

- **Standardization Issue**- There is no standard electrical plug to charge the battery compared to a standard nozzle at a gas station for fuel input (Kumar et al., 2021)
- **Charging Infrastructure**- The major infrastructural challenge is to develop enough charging stations network and provide electricity with enough backup (Das et al., 2020). For a developing country like India, this poses a major hindrance (Goel et al., 2021).

B) Capital & Market

- **High Costs**- The electric cars available in India market costs around 14 Lacs which is much higher compared to an economical gasoline car costing around 5 Lacs (KV et al, 2022)
- **Lithium Import Dependency**- Higher price of EV is majorly due to import of lithium metal for manufacturing lithium batteries which is the main component of an EV. Lithium has its reserves in counties like Chile, Argentina, Australia (Kar et al., 2021).
- **Investment in Infrastructure**- In addition to this, setting up the infrastructure for the production of EVs requires a substantial investment, which poses a challenge and hindrance to investors given the limited purchasing power of Indian working class (Mushi et al., 2022).
- **Skilled Labor Shortage**- India still lacks in the skilled labor required for manufacturing lithium based batteries and have to depend on China for supply (Kambli, 2022).

C) Safety concerns

- **Fire Risk**- In case of short circuit, the heat can ignite the chemical reaction within the battery which may lead to fire (Lin, 2023). This poses a serious threat to the safety of the passengers (Sun et al., 2020).
- **Flammability**- Lithium batteries uses flammable liquid electrolyte making it more prone to fire incidents (Held et al., 2022). India, being a tropical having tremendously high heat and humidity chances of such accidents increases to a greater extend (Kethareswaran & Moulik, 2023).

D) Waste Management

- **E-waste disposal**- Another major issue is the management of e-waste generated from EVs (Preeti & Sayali, 2021). A sustainable method is required for disposal of batteries having rare earth metals such as cobalt, nickel, lithium etc. (Amusa et al., 2024).
- **Recycling Challenges** – Although 95% of a battery's materials can technically be recycled, India lacks the infrastructure and enforcement for efficient e-waste recycling (Leal et al., 2022; Gupta et al., 2022).

Improper disposal could lead to soil, air, and groundwater contamination, impacting public health and biodiversity (Chatterjee et al., 2021).

- **Policy Gaps** – While initiatives like the Battery Waste Management Rules, 2022 have been introduced, implementation and monitoring remain weak (Patel & Kumar, 2024). Stronger regulation and investment in circular economy models are necessary to manage this growing challenge sustainably (Bansal & Reena, 2023)

Various challenges in production of EVs is summarized in Fig.2.

Figure 2. Challenges in EV

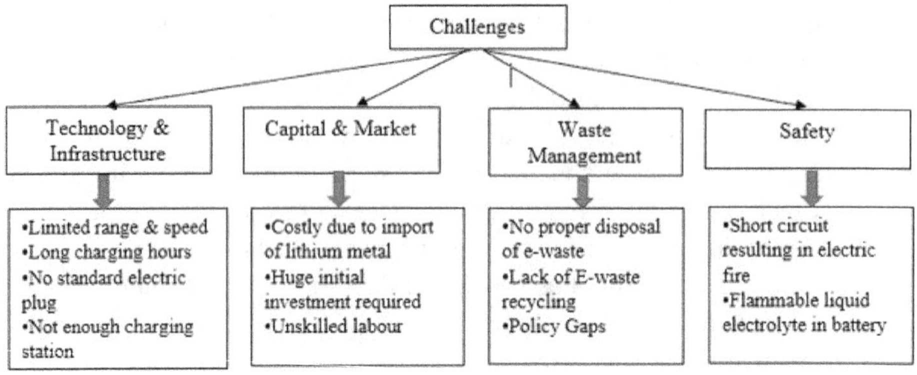

2. SCENARIO IN INDIA

The first electric car introduced in India was Reva, a two-seater electric vehicle launched by Mahindra Electric Mobility Ltd. (formerly known as Reva Electric Car Company) in 1994 (Mohanty & Kotak, 2017). Since then, the Indian electric vehicle market has gradually evolved, with significant growth observed particularly in the last decade. Currently, the EV landscape in India remains relatively small and is primarily dominated by two-wheelers (Bishnoi, 2024). Nevertheless, as the fastest-growing economy globally, India holds considerable potential to establish a strong and competitive base in the EV sector (Chaturvedi et al., 2022).

However, the feasibility of EV adoption across the country presents a mixed picture. While urban centers, especially Tier 1 cities, benefit from a growing network of charging stations, better infrastructure, and higher levels of consumer awareness, rural and semi-urban regions face substantial challenges (Garg, 2024). These include inconsistent electricity supply, limited access to public charging infrastructure, and a lack of trained personnel for EV servicing and maintenance (Dhankar et al., 2024). Additionally, the high initial purchase cost of electric vehicles remains a critical barrier for many consumers in lower-income and rural segments, where access to affordable financing options is limited.

There are also practical implementation hurdles such as inadequate road conditions in certain regions, and persistent concerns around vehicle range and battery reliability (Masurali & Surya, 2024).These issues are more pronounced in non-urban settings, where long-distance travel is common and emergency support infrastructure is sparse (Gaikwad & Tomar, 2024). Bridging this urban–rural divide is essential for India to achieve equitable and widespread EV adoption.

To explore these aspects further, Section 2.1 highlights successful EV implementations in India, while Section 2.2 discusses the expanding market for hybrid vehicles. Section 2.3 then examines how government initiatives are working to address some of these challenges and support the overall growth of the EV ecosystem in India.

2.1 Examples

A) E-Rickshaw- It is a three wheeler electric vehicle evolved from hand pulled rickshaw. It is majority manufactured in India and used for travelling distances in urban cities (Singh et al., 2021). It can be charged with household power and has become a popular option for door to door connectivity service (kokate at al., 2018). Some of the major benefits are-

- Robust design- having spacious seating capacity, clear entry & exit, separate compartment for driver, light in weight (kokate at al., 2018)
- Better income generating opportunity with les physical work, most of e-rickshaw owners were previously factory workers or daily laborers which have less job security and high physical activity (kumar et al, 2022).
- With the help of government initiatives such as Made in India, Deendayal scheme prices of e-rickshaws have fallen drastically making it affordable by the backward classes of society (kumar et al, 2022)

B) Metro- It is also known as the lifeline for the national capital of India. Metro has emerged to be one the biggest success in field of electric driven automobiles (Randhawa, 2012). It has not only helped reduced pollution and fuel consumption but changed the concept of public transport (Sharma et al., 2014). Considered to be one the most comfortable and smooth ride it has all the benefits that are discussed in section 1.2 and has overcome all the challenges as discussed in section 1.3. Doll & Balaban, 2013 discusses how metro network provides a sustainable ecosystem supporting other sustainable activities as well like-

- Rain water harvesting system installed to recharge ground water
- Solar Panels installed to generate electricity
- Afforestation in the Metro premises

2.2 Hybrid Vehicles

Hybrid technology has indeed seen significant adoption, particularly as a transitional solution towards fully electric vehicles (EVs) in India (Das & Bhat, 2022). By combining internal combustion engines (ICEs) with electric batteries, hybrids offer an appealing balance between the efficiency of EVs and the range reliability of conventional fuel engines (Ding et al., 2017). This combination allows for enhanced fuel efficiency, improved emissions performance, and potentially lower operational costs for consumers who may not have access to EV charging infrastructure.

Some major benefits of hybrid vehicles include:

- **No need for external charging:** Unlike full EVs, hybrids do not require charging stations, as their batteries recharge through regenerative braking, harnessing energy that would otherwise be lost (Cody et al., 2009).
- **Increased fuel efficiency and performance:** By utilizing both the battery and ICE, hybrids deliver better fuel efficiency and enhanced power, especially in start-stop city traffic, a common scenario in India (Rokadiya & Bandivadekar, 2016).
- **Optimized engine size with start/stop functionality:** The hybrid system allows for smaller, more efficient engines that automatically start and stop based on driving conditions, further contributing to fuel savings (Langford & Gillingham, 2023).

The recent launches of vehicles like the Maruti Suzuki Vitara Brezza and MG Hector are examples of this trend, showing that manufacturers are prioritizing a mix of technology options to meet the unique demands of the Indian market (Prajapati et al., 2014). These hybrid solutions represent a practical step toward the nation's goal of sustainable, lower-emission transportation, even as full EV adoption continues to grow (Hamamoto, 2019).

However, it is important to consider hybrid vehicles alongside other emerging clean technologies. Alternatives such as solar-powered vehicles and hydrogen fuel cell vehicles are gaining attention globally and could play a significant role in India's future mobility strategy. Solar-powered vehicles harness renewable energy directly, making them emission-free during operation, while hydrogen fuel cells produce only water vapor as a by-product, offering fast refueling and longer-range capabilities (Harichandan & Kar, 2023). Although still in nascent stages in India, these technologies are being actively researched and publicly endorsed by national leaders like Minister Nitin Gadkari (Prajapati et al., 2020). As India progresses toward a cleaner transportation ecosystem, evaluating and supporting a diverse mix of technological pathways including hybrids, EVs, solar, and hydrogen-based vehicles will be essential for achieving long-term sustainability and energy security.

2.3 Government Policies

In the past decade, Indian government have formed various policies to encourage and promote the growth of EV market in India. Niti Aayog, 2022 aims to achieve 100% e-mobility by the year 2030. Fig.3 represents the timeline of various initiatives taken by government

Figure 3. Initiatives by Government of India to promote EV

2.4 Future Roadmap

From sections 2.1 and 2.2, it is evident that India has the potential to become a key player in the electric vehicle (EV) market. According to the India Brand Equity Foundation, India's EV market could reach a value of $206 billion by 2030, provided steady growth is maintained. However, realizing this potential requires a collective effort from both the public and private sectors, with significant contributions from individuals, industries, and government.

A) Support from People

The role of people in this ecosystem, including EV manufacturers, investors, and buyers, is paramount to achieving the growth trajectory envisioned for India's EV market. While EVs are heralded as a sustainable alternative to traditional vehicles, the process of manufacturing them must also adhere to sustainable practices to truly fulfill the broader goal of sustainable development (Prakash, 2023).

One of the key strategies to achieving this is the adoption of **Circular Economy** principles. According to Ritcher (2022), the circular economy model can significantly reduce waste and optimize resource usage, aligning with the sustainability goals of the EV industry. Recycling e-waste and repurpose materials from discarded products can help reduce the dependence on raw materials and curb the environmental impact caused by the extraction of natural resources (Panchal et al., 2021). Moreover, reducing imports of these materials can bolster India's economic resilience while minimizing the environmental risks posed by activities like mining (Parthsarathy & Bulbule, 2017).

A significant portion of the EV industry's environmental impact stems from the manufacturing and disposal of **batteries**, which are the most resource-intensive component of EVs (Li-Cycle, 2022). However, advances in **battery recycling** technologies have made it possible to recover up to 95% of the materials in spent batteries (Picatoste, 2022), offering a promising solution to the industry's material supply concerns. Through these practices, manufacturers can lower their reliance on new raw materials, reduce pollution, and help drive India toward a more circular economy.

Another key solution to overcome the challenges of battery management is the implementation of **Battery Swapping** infrastructure. As proposed by Adu-Gyamfi et al. (2022), battery swapping involves providing consumers with charged batteries in exchange for depleted ones, reducing downtime for EVs and ensuring that EVs remain operational in regions lacking adequate charging infrastructure. This method can help address the critical challenge of long charging times, particularly in areas where charging stations are sparse (Kunar et al., 2024). By implementing shared technology for battery swapping, the industry can foster collaboration and innovation across various stakeholders.

In conclusion, the collective efforts of manufacturers, investors, and consumers must be centered on promoting sustainable manufacturing practices and adopting cutting-edge technologies like recycling and battery swapping (Financial Express, 2023) This will serve as the key to overcoming current production challenges and creating a sustainable and scalable EV ecosystem in India.

B) Support from Government

For India to achieve its ambitious goal of becoming a leader in the EV market, the government must play an active and supportive role. The government needs to adopt and enforce policies that align with India's long-term environmental goals, including the United Nations' target of **Net Zero emissions** by 2050 (Ansab & Kumar, 2024). The government's role is crucial in providing the necessary infrastructure, policy frameworks, and financial incentives that can drive the growth of the EV market.

Initiatives such as the **Production Linked Incentive (PLI)** scheme have been designed to incentivize local production of EV components and vehicles. According to Shrimali (2021), PLI will help reduce India's reliance on imported EV technologies and enable the domestic production of vehicles and components, which is essential for building a robust EV ecosystem. Furthermore, schemes like **FAME I & II** (Faster Adoption and Manufacturing of Hybrid and Electric Vehicles), which provide subsidies to buyers and manufacturers, are vital in driving the adoption of EVs and spurring investments in infrastructure (Dua et al., 2021). These schemes also contribute to creating jobs in both manufacturing and infrastructure development, making them a win-win for the economy and the environment.

However, for these policies to be effective, the government must establish strong feedback mechanisms and ensure active communication between policymakers and implementing agencies. As pointed out by Kumar et al. (2021), continuous feedback and the resolution of ground-level challenges are critical to achieving the desired outcomes. Proper implementation of these policies will not only facilitate the growth of the EV industry but will also help India meet its climate goals by significantly reducing carbon emissions.

C) Long-Term Replacements for EVs and Future Technologies

Looking beyond the current horizon of electric vehicles, there is a growing interest in exploring alternative propulsion technologies that could complement or eventually replace EVs. One such possibility is the development of **water-powered vehicles**, as discussed by India's Union Minister for Road Transport and Highways,

Mr. Nitin Gadkari (Rawat et al., 2024). Water-powered vehicles represent a future breakthrough in sustainable transportation, as they could significantly reduce the environmental footprint associated with traditional internal combustion engines and even electric vehicles, particularly in terms of raw material extraction and battery disposal (Shet & Moholkar, 2025).

Water-powered technology, if successfully developed and scaled, could offer several advantages over current electric battery technologies, such as faster refueling times and potentially lower costs in the long run (Ahsan & Dey, 2023). The concept is still in the early stages of research and development, but if it becomes commercially viable, it could play a pivotal role in meeting India's transportation and sustainability goals.

Moreover, future technologies such as **hydrogen fuel cells**, **solar-powered vehicles**, and innovations in **synthetic fuels** could further contribute to reducing dependency on fossil fuels and mitigating environmental impact (Jain & Verma, 2024). As the global transportation sector continues to evolve, India must keep an eye on these emerging technologies and ensure that it is well-positioned to adopt them as part of its broader sustainable development strategy.

3. CONCLUSION

The increasing environmental pollution caused by traditional vehicles has driven the shift toward green technology in the automobile sector, with Electric Vehicles (EVs) being a key sustainable solution. EVs operate on electric motors powered by rechargeable batteries, offering several environmental benefits, including zero emissions, reduced noise pollution, lower maintenance costs, and a smooth driving experience. These advantages significantly reduce air pollution and dependence on fossil fuels, making EVs essential for sustainable transportation.

However, the widespread adoption of EVs faces several challenges in India, including long charging times, insufficient charging infrastructure, high investment costs, and limited market reach. Additionally, concerns about e-waste management and battery safety need to be addressed to ensure the overall sustainability of EVs. While government policies such as FAME I & II have spurred some growth, challenges related to infrastructure, technology, and market penetration remain. Hybrid technology is also gaining popularity as an intermediary solution, blending traditional fuel with electric propulsion to ease the transition.

To ensure long-term sustainability, a diversified clean energy approach is required. This includes increasing the share of renewable energy sources and advancing energy storage technologies. A circular economy model, focusing on battery recycling and battery swapping, can reduce raw material dependency and minimize e-waste.

However, managing battery waste remains a critical challenge. Without effective recycling and second-life battery applications, the environmental benefits of EVs could be negated. Overcoming these challenges through collaboration between industry, government, and research institutions will be crucial in achieving India's ambitious emission reduction goals and fostering a greener future.

REFERENCES

Abid, M., Tabaa, M., Chakir, A., & Hachimi, H. (2022). Routing and charging of electric vehicles: Literature review. *Energy Reports*, *8*, 556–578. DOI: 10.1016/j.egyr.2022.07.089

Adu-Gyamfi, G., Song, H., Asamoah, A. N., Li, L., Nketiah, E., Obuobi, B., & Cudjoe, D. (2022). Towards sustainable vehicular transport: Empirical assessment of battery swap technology adoption in China. *Technological Forecasting and Social Change*, *184*, 121995. DOI: 10.1016/j.techfore.2022.121995

Ahsan, A., & Dey, P. K. (2023). Comparative analysis of hydrogen fuel cell and battery electric vehicles for sustainable transportation. *Journal of Clean Energy Technologies*, *11*(2), 45–55.

Aijaz, I., & Ahmad, A. (2022). Electric vehicles for environmental sustainability. Smart technologies for energy and environmental sustainability, 131-145.

Alanazi, F. (2023). Electric vehicles: Benefits, challenges, and potential solutions for widespread adaptation. *Applied Sciences (Basel, Switzerland)*, *13*(10), 6016. DOI: 10.3390/app13106016

Amusa, H. K., Sadiq, M., Alam, G., Alam, R., Siefan, A., Ibrahim, H., & Yildiz, B. (2024). Electric vehicle batteries waste management and recycling challenges: A comprehensive review of green technologies and future prospects. *Journal of Material Cycles and Waste Management*, *26*(4), 1–20. DOI: 10.1007/s10163-024-01982-y

Ansab, K. V., & Kumar, S. P. (2024). Influence of government financial incentives on electric car adoption: Empirical evidence from India. *South Asian Journal of Business Studies*, *13*(2), 226–243. DOI: 10.1108/SAJBS-03-2021-0088

Balakrishnan, K., Gupta, S. D., & Mathur, P. (2019). The impact of air pollution on deaths, disease burden, and life expectancy across the states of India: The Global Burden of Disease Study 2017. *The Lancet. Planetary Health*, *3*(1), e26–e39. DOI: 10.1016/S2542-5196(18)30261-4 PMID: 30528905

Bansal, R., & Meena, D. (2023). Policy and regulatory perspectives of waste battery management and recycling: A review and future research agendas. *Journal of Environmental Management*, *335*, 113227.

Bishnoi, M. R. (2024). A Study on Electric Vehicles in India Opportunities and Challenges. In Conference paper available at https://www. researchgate. net/publication/378737287

Chatterjee, P., Chakrabarti, S., & Gupta, A. (2021). Understanding the gap between formal and informal e-waste recycling facilities in India. *Environmental Health Perspectives, 129*(10), 105003.

Chaturvedi, B. K., Nautiyal, A., Kandpal, T. C., & Yaqoot, M. (2022). Projected transition to electric vehicles in India and its impact on stakeholders. *Energy for Sustainable Development, 66*, 189–200. DOI: 10.1016/j.esd.2021.12.006

Chawla, V. K., Chhabra, D., Gupta, P., & Naaz, S. (2021). Evaluation of green operations management by fuzzy analytical hierarchy process. *Materials Today: Proceedings, 38*, 274–279. DOI: 10.1016/j.matpr.2020.07.200

Cheng, K. W. E. (2009, May). Recent development on electric vehicles. In 2009 3rd International Conference on Power Electronics Systems and Applications (PESA) (pp. 1-5). IEEE.

Chhikara, R., Garg, R., Chhabra, S., Karnatak, U., & Agrawal, G. (2021). Factors affecting adoption of electric vehicles in India: An exploratory study. *Transportation Research Part D, Transport and Environment, 100*, 103084. DOI: 10.1016/j.trd.2021.103084

Cody, J., Göl, Ö., Nedic, Z., Nafalski, A., & Mohtar, A. (2009). Regenerative braking in an electric vehicle. Maszyny Elektryczne: zeszyty problemowe, 113-118.

Cohen, A. J., Brauer, M., Burnett, R., Anderson, H. R., Frostad, J., Estep, K., ... & Forouzanfar, M. H. (2019). *The impact of air pollution on deaths, disease burden, and life expectancy across the states of India: The Global Burden of Disease Study 2017. The Lancet Planetary Health*

Das, H. S., Rahman, M. M., Li, S., & Tan, C. W. (2020). Electric vehicles standards, charging infrastructure, and impact on grid integration: A technological review. *Renewable & Sustainable Energy Reviews, 120*, 109618. DOI: 10.1016/j.rser.2019.109618

Das, P. K., & Bhat, M. Y. (2022). Global electric vehicle adoption: Implementation and policy implications for India. *Environmental Science and Pollution Research International, 29*(27), 40612–40622. DOI: 10.1007/s11356-021-18211-w PMID: 35083674

del Carmen Pardo-Ferreira, M., Rubio-Romero, J. C., Galindo-Reyes, F. C., & Lopez-Arquillos, A. (2020). Work-related road safety: The impact of the low noise levels produced by electric vehicles according to experienced drivers. *Safety Science, 121*, 580–588. DOI: 10.1016/j.ssci.2019.02.021

Dhankhar, S., Pateer, S., Sandhu, V., & Kaur, H. (2024). Barriers to Electric Vehicle Adoption in India: A Comparative Review and Future Growth Prospects. *European Economic Letters*, *14*(3), 2007–2015.

Ding, N., Prasad, K., & Lie, T. T. (2017). The electric vehicle: A review. *International Journal of Electric and Hybrid Vehicles*, *9*(1), 49–66. DOI: 10.1504/IJEHV.2017.082816

Doll, C. N., & Balaban, O. (2013). A methodology for evaluating environmental co-benefits in the transport sector: Application to the Delhi metro. *Journal of Cleaner Production*, *58*, 61–73. DOI: 10.1016/j.jclepro.2013.07.006

Dua, R., Hardman, S., Bhatt, Y., & Suneja, D. (2021). Enablers and disablers to plug-in electric vehicle adoption in India: Insights from a survey of experts. *Energy Reports*, *7*, 3171–3188. DOI: 10.1016/j.egyr.2021.05.025

Financial Express. (2023). *Sustainable materials and eco-friendly innovations in EV manufacturing. Financial Express.* https://www.financialexpress.com/business/express-mobility-sustainable-materials-eco-friendly-innovations-in-ev-manufacturing-3317096

Gaikwad, R., & Tomar, A. (2024). Electric vehicles – Paving the road to widespread adoption: A review of technical and non-technical challenges. *International Journal of Research in Advanced Mobility*, *1*(2), 22–26.

Garg, R. K. (2024). Challenges and opportunities in electric vehicle charging infrastructure development. *Shodh Sagar Journal of Electric Vehicles*, *1*(1), 13–20.

Goel, S., Sharma, R., & Rathore, A. K. (2021). A review on barrier and challenges of electric vehicle in India and vehicle to grid optimisation. Transportation engineering, 4, 100057.

Gopal, A. R., Witt, M., Sheppard, C., & Harris, A. (2015). Battery Electric Vehicles can reduce greenhouse has emissions and make renewable energy cheaper in India.

Gujarathi, P. K., Shah, V. A., & Lokhande, M. M. (2018). Electric vehicles in India: Market analysis with consumer perspective, policies and issues. *Journal of Green Engineering*, *8*(1), 17–36. DOI: 10.13052/jge1904-4720.813

Gupta, P., Chawla, V., Jain, V., & Angra, S. (2022). Green operations management for sustainable development: An explicit analysis by using fuzzy best-worst method. *Decision Science Letters*, *11*(3), 357–366. DOI: 10.5267/j.dsl.2022.1.003

Hamamoto, M. (2019). An empirical study on the behavior of hybrid-electric vehicle purchasers. *Energy Policy*, *125*, 286–292. DOI: 10.1016/j.enpol.2018.10.042

Harichandan, S., & Kar, S. K. (2023). An empirical study on motivation to adopt hydrogen fuel cell vehicles in India: Policy implications for stakeholders. *Journal of Cleaner Production*, *408*, 137198. DOI: 10.1016/j.jclepro.2023.137198

Held, M., Tuchschmid, M., Zennegg, M., Figi, R., Schreiner, C., Mellert, L. D., & Nachef, L. (2022). Thermal runaway and fire of electric vehicle lithium-ion battery and contamination of infrastructure facility. *Renewable & Sustainable Energy Reviews*, *165*, 112474. DOI: 10.1016/j.rser.2022.112474

Hirz, M., Brunner, H., & Nguyen, T. T. (2022). Greenhouse Gas Emissions of Electric Cars-A Comprehensive Evaluation. Tehnički glasnik, 16(2), 280-287.

Jain, R., & Verma, S. (2024). Hydrogen fuel cell vehicles: Future prospects in India's sustainable transportation ecosystem. *Energy & Environmental Science*, *12*(3), 300–312.

K v, S., Michael, L. K., Hungund, S. S., & Fernandes, M.KV. (2022). Factors influencing adoption of electric vehicles–A case in India. *Cogent Engineering*, *9*(1), 2085375. DOI: 10.1080/23311916.2022.2085375

Kambli, R. O. Electric Vehicles in India: Future and Challenges.

Kar, B., Pal, T. K., & Bhattacharyya, S. (2021, February). Study of the Reality on Electric Vehicle in Indian Scenario. [). IOP Publishing.]. *IOP Conference Series. Materials Science and Engineering*, *1080*(1), 012003. DOI: 10.1088/1757-899X/1080/1/012003

Kethareswaran, V., & Moulik, S. (2023). Electric Vehicles and the Burning Question: Reasons, Risks, Ramifications and Remedies—An Indian Perspective. *Fire Technology*, *59*(5), 2189–2201. DOI: 10.1007/s10694-023-01453-0

Kohli, S. (2024). Electric vehicle demand incentives in India.

Kokate, V. L., Bankar, D. S., Vispi, R. H., & Karandikar, P. B. (2018, July). E-Rickshaw Present Past And Future With Reference to Current Transportation in India. In International Conference on Recent Innovations in Electrical, Electronics & Communication Engineering-(ICRIEECE), Kalinga Institute of Industrial Technology (KIIT), Bhubaneswar, India (Vol. 1, No. 01, pp. 1-15).

Kumar, A., Thomas, J., Wadhwa, S. S., Mishra, A., & Dasgupta, S. (2022). The last mile connectivity: Experience of passengers and drivers of E-rickshaws and cycle rickshaws in an Indian city. *Case Studies on Transport Policy*, *10*(2), 948–953. DOI: 10.1016/j.cstp.2022.03.008

Kumar, H., Bharule, S. P., & Kuntewar, M. (2024). Policy implications of battery swapping systems for three-wheelers in Tier II cities in India. [IJERT]. *International Journal of Engineering Research & Technology (Ahmedabad), 13*(3).

Kumar, P., Singh, R. K., Paul, J., & Sinha, O. (2021). Analyzing challenges for sustainable supply chain of electric vehicle batteries using a hybrid approach of Delphi and Best-Worst Method. *Resources, Conservation and Recycling, 175*, 105879. DOI: 10.1016/j.resconrec.2021.105879

Kumar, R., Lamba, K., & Raman, A. (2021). Role of zero emission vehicles in sustainable transformation of the Indian automobile industry. *Research in Transportation Economics, 90*, 101064. DOI: 10.1016/j.retrec.2021.101064

Kumar, R., & Padmanaban, S. (2019). Electric vehicles for India: overview and challenges. IEEE India info, 14(2), 139-142.

Labeye, E., Hugot, M., Brusque, C., & Regan, M. A. (2016). The electric vehicle: A new driving experience involving specific skills and rules. *Transportation Research Part F: Traffic Psychology and Behaviour, 37*, 27–40. DOI: 10.1016/j.trf.2015.11.008

Langford, R. P., & Gillingham, K. (2023). Quantifying the benefits of the introduction of the hybrid electric vehicle. *International Journal of Industrial Organization, 87*, 102904. DOI: 10.1016/j.ijindorg.2022.102904

Leal, V. M., Ribeiro, J. S., Coelho, E. L. D., & Freitas, M. B. J. G. (2022). Recycling of spent lithium-ion batteries as a sustainable solution to obtain raw materials for different applications. *Journal of Energy Chemistry*.

Li-Cycle. (2022). *Lithium battery recycling efficiency and recovery rates*. Retrieved from https://www.auto-innovations.net/news/62824-li-cycle-lithium-battery-recycling-efficiency-and-recovery-rates

Lin, T. (2023). Electric Vehicle Safety Issues and the Solutions. Highlights in Science. *Engineering and Technology, 46*, 91–96.

Maffei, L., & Masullo, M. (2014). Electric vehicles and urban noise control policies. *Archives of Acoustics, 39*(3), 333–341. DOI: 10.2478/aoa-2014-0038

Masurali, A., & Surya, P. (2024). Driving sustainable mobility: A study of electric vehicle adoption in rural India. *Proceedings of the Design Society: Design Conference, 4*, 1145–1154.

Ministry of Environment. Forest & Climate Change <https://moef.gov.in/moef/environment/pollution/>

Mohamed, M., Tamil Arasan, G., & Sivakumar, G. (2018). Study on electric vehicles in India opportunities and challenges. *International Journal of Scientific Research in Environmental Science and Toxicology*, •••, 5.

Mohanty, P., & Kotak, Y. (2017). Electric vehicles: Status and roadmap for India. In *Electric vehicles: Prospects and challenges* (pp. 387–414). Elsevier. DOI: 10.1016/B978-0-12-803021-9.00011-2

Munshi, T., Dhar, S., & Painuly, J. (2022). Understanding barriers to electric vehicle adoption for personal mobility: A case study of middle income in-service residents in Hyderabad city, India. *Energy Policy*, *167*, 112956. DOI: 10.1016/j.enpol.2022.112956

Olabi, A. G., Abdelkareem, M. A., Wilberforce, T., Alkhalidi, A., Salameh, T., Abo-Khalil, A. G., & Sayed, E. T. (2022). Battery Electric Vehicles: Progress, Power Electronic Converters, Strength (S), Weakness (W), Opportunity (O), and Threats (T). International Journal of Thermofluids, 100212.

Panchal, R., Singh, A., & Diwan, H. (2021). Economic potential of recycling e-waste in India and its impact on import of materials. *Resources Policy*, *74*, 102264. DOI: 10.1016/j.resourpol.2021.102264

Parthsarathy, P., & Bulbule, K. A. (2017). E-Waste Recycling: Best option for Mineral Substitution and Environmental Protection. *Journal of the Geological Society of India*, *89*(2), 221. DOI: 10.1007/s12594-017-0589-8

Patel, S., & Kumar, R. (2024). Analysing the draft Liquid Waste Management Rules 2024: Salient features and way forward. *National Law School of India University, Centre for Environmental Law and Policy*.

Picatoste, A., Justel, D., & Mendoza, J. M. F. (2022). Exploring the applicability of circular design criteria for electric vehicle batteries. *Procedia CIRP*, *109*, 107–112. DOI: 10.1016/j.procir.2022.05.222

Poornesh, K., Nivya, K. P., & Sireesha, K. (2020, September). A comparative study on electric vehicle and internal combustion engine vehicles. In *2020 International Conference on Smart Electronics and Communication (ICOSEC)* (pp. 1179-1183). IEEE. DOI: 10.1109/ICOSEC49089.2020.9215386

Prajapati, K. C., Patel, R., & Sagar, R. (2014). Hybrid vehicle: A study on technology. [IJERT]. *International Journal of Engineering Research & Technology (Ahmedabad)*, *3*(12), 8.

Prajapati, M. L., Gundaliya, P. J., & Sheth, A. (2020). Case study: Solar electric vehicles in India. *IOP Conference Series. Materials Science and Engineering, 937*(1), 012033. DOI: 10.1088/1757-899X/937/1/012033

Prakash, J. (2023). Study of environmental sustainability and green manufacturing practices in the Indian automobile industry. *International Journal for Research Publication and Seminar, 14*(2).

Preeti, M., & Sayali, A. (2021). Scientometric analysis of research on end-of-life electronic waste and electric vehicle battery waste. *Journal of Scientometric Research, 10*(1), 37–46. DOI: 10.5530/jscires.10.1.5

Randhawa, P. (2012). Delhi metro rail: Beyond mass transit. *Economic and Political Weekly*, •••, 25–29.

Rawat, A., Garg, C. P., & Sinha, P. (2024). Analysis of the key hydrogen fuel vehicles adoption barriers to reduce carbon emissions under net zero target in emerging markets. *Energy Policy, 184*, 113847. DOI: 10.1016/j.enpol.2023.113847

Richter, J. L. (2022). A circular economy approach is needed for electric vehicles. *Nature Electronics, 5*(1), 5–7. DOI: 10.1038/s41928-021-00711-9

Rokadiya, S., & Bandivadekar, A. (2016). *Hybrid and electric vehicles in India: current scenario and market incentives. The International Council on Clean Transportation*. ICCT.

Sanguesa, J. A., Torres-Sanz, V., Garrido, P., Martinez, F. J., & Marquez-Barja, J. M. (2021). A review on electric vehicles: Technologies and challenges. *Smart Cities, 4*(1), 372–404. DOI: 10.3390/smartcities4010022

Sarode, R., & Eadke, A. (2020). Electric Vehicle (Construction and Working principle). *Int. Res. J. Eng. Technol, 7*(10), 140–142.

Sharma, N., Singh, A., Dhyani, R., & Gaur, S. (2014). Emission reduction from MRTS projects–a case study of Delhi metro. *Atmospheric Pollution Research, 5*(4), 721–728. DOI: 10.5094/APR.2014.081

Shet, H. N., & Moholkar, V. S. (2025). Comparative life cycle analysis of hydrogen fuel cell electric vehicles and battery electric vehicles: An Indian perspective. *International Journal of Hydrogen Energy, 103*, 729–739. DOI: 10.1016/j.ijhydene.2025.01.254

Shrimali, G. (2021). Getting to India's electric vehicle targets cost-effectively: To subsidize or not, and how? *Energy Policy, 156*, 112384. DOI: 10.1016/j.enpol.2021.112384

Singh, R., Mishra, S., & Tripathi, K. (2021). Analysing acceptability of E-rickshaw as a public transport innovation in Delhi: A responsible innovation perspective. *Technological Forecasting and Social Change, 170*, 120908. DOI: 10.1016/j.techfore.2021.120908

Singh, S. P., Sharma, N., Chandrakant, S. A., & Singh, S. P. (2021). Electric Vehicles in India: A Literature Review.

Singh, V., Singh, V., & Vaibhav, S. (2021). Analysis of electric vehicle trends, development and policies in India. *Case Studies on Transport Policy, 9*(3), 1180–1197. DOI: 10.1016/j.cstp.2021.06.006

Singla, A., & Bansal, R. SUSTAINABILITY OF ELECTRIC VEHICLES.

Spessert, B. M., Fischer, M., & Kühn, B. (2019, September). Combustion Engine Noise-a Historical Review. In *INTER-NOISE and NOISE-CON Congress and Conference Proceedings* (Vol. 259, No. 9, pp. 347-358). Institute of Noise Control Engineering.

Srikanth, R. (2018). Role of electric mobility in a sustainable, and energy-secure future for India. *Current Science, 114*(4), 732–739. DOI: 10.18520/cs/v114/i04/732-739

Sun, P., Bisschop, R., Niu, H., & Huang, X. (2020). A review of battery fires in electric vehicles. *Fire Technology, 56*(4), 1361–1410. DOI: 10.1007/s10694-019-00944-3

Thakur, J., de Almeida, C. M. L., & Baskar, A. G. (2022). Electric vehicles batteries for a circular economy: Second life batteries as stationary storage. *Journal of Cleaner Production, 375*, 134066. DOI: 10.1016/j.jclepro.2022.134066

Wu, X., Freese, D., Cabrera, A., & Kitch, W. A. (2015). Electric vehicles' energy consumption measurement and estimation. *Transportation Research Part D, Transport and Environment, 34*, 52–67. DOI: 10.1016/j.trd.2014.10.007

Zhao, X., Zhang, L., & Li, H. (2024). Performance, emissions, and economic analyses of hydrogen fuel cell vehicles. *Renewable & Sustainable Energy Reviews, 168*, 112849.

Compilation of References

Abad-Segura, E., Gonzalez-Zamar, M. D., Rosa, A. L. D. L., & Morales Cevallos, M. B. (2020). Sustainability of educational technologies: An approach to augmented reality research. *Sustainability (Basel), 12*(10), 4091. DOI: 10.3390/su12104091

Abbasi, M., Abduli, M. A., Omidvar, B., & Baghvand, A. (2014). Results uncertainty of support vector machine and hybrid of wavelet transform-support vector machine models for solid waste generation forecasting. *Environmental Progress & Sustainable Energy*, *33*(1), 220–228. DOI: 10.1002/ep.11747

Abdallah, M., Talib, M. A., Feroz, S., Nasir, Q., Abdalla, H., & Mahfood, B. (2020). Artificial intelligence applications in solid waste management: A systematic research review. *Waste Management (New York, N.Y.), 109*, 231–246. DOI: 10.1016/j.wasman.2020.04.057

Abdullahi, S. I., Habaebi, M. H., & Malik, N. A. (2018). Flood Disaster Warning System on the go. *2018 7th International Conference on Computer and Communication Engineering (ICCCE)*, 258–263. DOI: 10.1109/ICCCE.2018.8539253

Abdul-Qawy, A. S., Pramod, P. J., Magesh, E., & Srinivasulu, T. (2015). The internet of things (iot): An overview. *International Journal of Engineering Research and Applications*, *5*(12), 71–82.

Abid, M., Tabaa, M., Chakir, A., & Hachimi, H. (2022). Routing and charging of electric vehicles: Literature review. *Energy Reports*, *8*, 556–578. DOI: 10.1016/j.egyr.2022.07.089

Abid, S. K., Sulaiman, N., Chan, S. W., Nazir, U., Abid, M., Han, H., Ariza-Montes, A., & Vega-Muñoz, A. (2021). Toward an Integrated Disaster Management Approach: How Artificial Intelligence Can Boost Disaster Management. *Sustainability (Basel), 13*(22), 12560. DOI: 10.3390/su132212560

Abioye, S. O., Oyedele, L. O., Akanbi, L., Ajayi, A., Delgado, J. M. D., Bilal, M., & Ahmed, A. (2021). Artificial intelligence in the construction industry: A review of present status, opportunities, and future challenges. *Journal of Building Engineering*, *44*, 10–32. DOI: 10.1016/j.jobe.2021.103299

Abubakar, I. R., Maniruzzaman, K. M., Dano, U. L., AlShihri, F. S., AlShammari, M. S., Ahmed, S. M. S., Al-Gehlani, W. A. G., & Alrawaf, T. I. (2022). Environmental sustainability impacts of solid waste management practices in the global South. *International Journal of Environmental Research and Public Health*, *19*(19), 12717. DOI: 10.3390/ijerph191912717 PMID: 36232017

Abutaha, F., & Dinler, C. (2024). Adopting Robotic Process Automation (RPA) in the Construction Industry. In *Complex AI Dynamics and Interactions in Management; 111* (pp. 273–310). IGI Global. DOI: 10.4018/979-8-3693-0712-0.ch013

Acosta Castellanos, P. M., & Queiruga-Dios, A. (2022). From environmental education to education for sustainable development in higher education: A systematic review. *International Journal of Sustainability in Higher Education*, *23*(3), 622–644. DOI: 10.1108/IJSHE-04-2021-0167

Adedeji, A., Ayodele, E., & Falaye, T. (2021). Barriers to the Adoption of Innovative Technologies in the Nigerian Construction Industry. *Journal of Construction Engineering and Management*, *147*(8), 1–12. DOI: 10.1061/(ASCE)CO.1943-7862.0002084

Adeosun, F. E., & Oke, A. E. (2024). Examining the awareness and usage of cyber-physical systems for construction projects in Nigeria. *Journal of Engineering. Defence Technology*, *22*(1), 281–294.

Adu-Gyamfi, G., Song, H., Asamoah, A. N., Li, L., Nketiah, E., Obuobi, B., & Cudjoe, D. (2022). Towards sustainable vehicular transport: Empirical assessment of battery swap technology adoption in China. *Technological Forecasting and Social Change*, *184*, 121995. DOI: 10.1016/j.techfore.2022.121995

Afolalu, S. A., Noiki, A. A., Ikumapayi, O. M., Ogundipe, A. T., & Oloyede, O. R. (2021). Development of smart waste bin for solid waste management. *International Journal of Sustainable Development and Planning*, *16*(8), 1449–1454. DOI: 10.18280/ijsdp.160805

Aghimien, D., Ikuabe, M., Aghimien, L. M., Aigbavboa, C., Ngcobo, N., & Yankah, J. (2024). PLS-SEM assessment of the impediments of robotics and automation deployment for effective construction health and safety. *Journal of Facilities Management*, *22*(3), 458–478. DOI: 10.1108/JFM-04-2022-0037

Agrawal, A., Gans, J., & Goldfarb, A. (2019). *The Economics of Artificial Intelligence: An Agenda*. University of Chicago Press. DOI: 10.7208/chicago/9780226613475.001.0001

Ahmed, A.A.A., Asadullah, A.B.M., (2020). Artificial intelligence and machine learning in waste management and recycling. Eng. Int. 8, 43–52. https://doi.org/.v8i1.498DOI: 10.18034/ei

Ahmed, R. (2024). Innovative Waste Management Solutions: A Global Perspective Challenges and Opportunities and the Bangladesh Context.

Ahmed, S. T., Reddy, R. V., & KM, K. R. (2025). Artificial Intelligence for Sustainable Waste Management: A Literature Review. *Artificial Intelligence for Sustainable Waste Management: A Literature Review (April 17, 2025)*.

Ahmed, A.. (2021). Real-Time Disaster Monitoring and Early Warning System using IoT. *IEEE Access : Practical Innovations, Open Solutions*, *9*, 44920–44934.

Ahsan, A., & Dey, P. K. (2023). Comparative analysis of hydrogen fuel cell and battery electric vehicles for sustainable transportation. *Journal of Clean Energy Technologies*, *11*(2), 45–55.

Ahvenniemi, H., Huovila, A., Pinto-Seppä, I., & Airaksinen, M. (2017). What are the differences between sustainable and smart cities? *Cities (London, England)*, *60*, 234–245. DOI: 10.1016/j.cities.2016.09.009

Aijaz, I., & Ahmad, A. (2022). Electric vehicles for environmental sustainability. Smart technologies for energy and environmental sustainability, 131-145.

Akande, I., Alagbe, O. T., Adebayo, O. A., Olaosun, S., & Oyedele, I. O. (2023). Prospects And Challenges Of Emerging Technologies In Architectural Design Practice Within The Nigerian Context. Journal of Namibian Studies, 33.

Akçayır, M., & Akçayır, G. (2017). Advantages and challenges associated with augmented reality for education: A systematic review of the literature. *Educational Research Review*, *20*, 1–11. DOI: 10.1016/j.edurev.2016.11.002

Akter, S., Haque, M. M., & Islam, M. R. (2021). IoT based real-time river water level monitoring and flood warning system for Bangladesh. ResearchGate. https://www.researchgate.net/publication/350760195

Al Duhayyim, M., Mohamed, H. G., Aljebreen, M., Nour, M. K., Mohamed, A., Abdelmageed, A. A., Yaseen, I., & Mohammed, G. P. (2022). Artificial ecosystem-based optimization with an improved deep learning model for IoT-assisted sustainable waste management. *Sustainability (Basel)*, *14*(18), 11704. DOI: 10.3390/su141811704

Alahi, M. E. E., Sukkuea, A., Tina, F. W., Nag, A., Kurdthongmee, W., Suwannarat, K., & Mukhopadhyay, S. C. (2023). Integration of IoT-enabled technologies and artificial intelligence (AI) for smart city scenario: Recent advancements and future trends. *Sensors (Basel)*, *23*(11), 5206. DOI: 10.3390/s23115206 PMID: 37299934

Alam, M. F., Katsikas, S., Beltramello, O., & Hadjiefthymiades, S. (2017). Augmented and virtual reality based monitoring and safety system: A prototype IoT platform. *Journal of Network and Computer Applications*, *89*, 109–119. DOI: 10.1016/j.jnca.2017.03.022

Alanazi, F. (2023). Electric vehicles: Benefits, challenges, and potential solutions for widespread adaptation. *Applied Sciences (Basel, Switzerland)*, *13*(10), 6016. DOI: 10.3390/app13106016

Al-Ansi, A. M., Jaboob, M., Garad, A., & Al-Ansi, A. (2023). Analyzing augmented reality (AR) and virtual reality (VR) recent development in education. *Social Sciences & Humanities Open*, *8*(1), 100532. DOI: 10.1016/j.ssaho.2023.100532

Alavi, S. H. S., Atani, R. E., & Qader, N. N. (2020). Internet of Things in Disaster Management: A Review. *International Journal of Advanced Computer Science and Applications*, *11*(3), 372–380.

Albahri, A. S., Khaleel, Y. L., Habeeb, M. A., Ismael, R. D., Hameed, Q. A., Deveci, M., Homod, R. Z., Albahri, O. S., Alamoodi, A. H., & Alzubaidi, L. (2024). A systematic review of trustworthy artificial intelligence applications in natural disasters. *Computers & Electrical Engineering*, *118*, 109409. DOI: 10.1016/j.compeleceng.2024.109409

Aldoseri, A., Al-Khalifa, K. N., & Hamouda, A. M. (2023). Re-Thinking Data Strategy and Integration for Artificial Intelligence: Concepts, Opportunities, and Challenges. *Applied Sciences (Basel, Switzerland)*, *13*(12), 7082. DOI: 10.3390/app13127082

Alhosani, K., & Alhashmi, S. M. (2024). Opportunities, challenges, and benefits of AI innovation in government services: A review. *Discover Artificial Intelligence*, *4*(1), 18. DOI: 10.1007/s44163-024-00111-w

Ali, O., Murray, P. A., Momin, M., Dwivedi, Y. K., & Malik, T. (2024). The effects of artificial intelligence applications in educational settings: Challenges and strategies. *Technological Forecasting and Social Change*, *199*, 123076. DOI: 10.1016/j.techfore.2023.123076

Ali, T., Irfan, M., Alwadie, A. S., & Glowacz, A. (2020). IoT-based smart waste bin monitoring and municipal solid waste management system for smart cities. *Arabian Journal for Science and Engineering*, *45*(12), 10185–10198. DOI: 10.1007/s13369-020-04637-w

Alotaibi, E., & Nassif, N. (2024). Artificial intelligence in environmental monitoring: In-depth analysis. *Discover Artificial Intelligence*, *4*(1), 84. DOI: 10.1007/s44163-024-00198-1

Altinpulluk, H. (2019). Determining the trends of using augmented reality in education between 2006-2016. *Education and Information Technologies*, *24*(2), 1089–1114. DOI: 10.1007/s10639-018-9806-3

Amaifeobu, O., Ogunbiyi, O., & Ayeni, O. (2023). The Challenges of Robotics in the Nigerian Construction Industry. *International Journal of Construction Research and Technology*, *15*(3), 456–471. DOI: 10.1016/j.jcrt.2023.05.007

Amini Hosseini, K., Alesheikh, A. A., & Javadpoor, M. (2020). A machine learning framework for multi-hazards modeling and mapping in a mountainous area. *Scientific Reports*, *10*(1), 1–12. DOI: 10.1038/s41598-020-70457-3 PMID: 31913322

Amusa, H. K., Sadiq, M., Alam, G., Alam, R., Siefan, A., Ibrahim, H., & Yildiz, B. (2024). Electric vehicle batteries waste management and recycling challenges: A comprehensive review of green technologies and future prospects. *Journal of Material Cycles and Waste Management*, *26*(4), 1–20. DOI: 10.1007/s10163-024-01982-y

Andeobu, L., Wibowo, S., & Grandhi, S. (2022). Artificial intelligence applications for sustainable solid waste management practices in Australia: A systematic review. *The Science of the Total Environment*, *834*, 155389. DOI: 10.1016/j.scitotenv.2022.155389 PMID: 35460765

Andújar, J. M., Mejías, A., & Márquez, M. A. (2011). Augmented reality for the improvement of remote laboratories: An augmented remote laboratory. *IEEE Transactions on Education*, *54*(3), 492–500. DOI: 10.1109/TE.2010.2085047

Anitha, K., Ghosal, I., Amala, J., & Hossain, I. (2025). Future Trends and Innovations in Environmental Governance for Sustainable Development. In *Intersecting Environmental Governance With Technological Advancements* (pp. 381–412). IGI Global Scientific Publishing.

Ansab, K. V., & Kumar, S. P. (2024). Influence of government financial incentives on electric car adoption: Empirical evidence from India. *South Asian Journal of Business Studies*, *13*(2), 226–243. DOI: 10.1108/SAJBS-03-2021-0088

Ansari, S. P., Odongo, J. O., Nomani, M. Z., Salahuddin, G., Faez Hasan, M., & Pallathadka, L. K. (2021). Evaluating the role of environment management system based on artificial intelligence. *Materials Today: Proceedings*.

Araujo, A. C., de Oliveira, J. R., & Lima, N. E. (2021). Augmented reality for biodiversity conservation: Challenges and opportunities in education. *Sustainability*, *13*(8), 4103.

Ardyansyah, A., & Rahayu, S. (2023). Development and implementation of augmented reality-based card game learning media with environmental literacy in improving students' understanding of carbon compounds. *Orbital*, *15*(2), 118–126. DOI: 10.17807/orbital.v15i2.17617

Arici, F., Yildirim, P., Caliklar, S., & Yilmaz, R. M. (2019). Research trends in the use of augmented reality in science education: Content and bibliometric mapping analysis. *Computers & Education*, *142*, 103647. DOI: 10.1016/j.compedu.2019.103647

Arvanitis, T. N., Petrou, A., Knight, J. F., Savas, S., Sotiriou, S., Gargalakos, M., & Gialouri, E. (2007). Human factors and qualitative pedagogical evaluation of a mobile augmented reality system for science education used by learners with physical disabilities. *Personal and Ubiquitous Computing*, *13*(3), 243–250. DOI: 10.1007/s00779-007-0187-7

Asif, M., Naeem, G., & Khalid, M. (2024). Digitalization for sustainable buildings: Technologies, applications, potential, and challenges. *Journal of Cleaner Production*, *450*, 141814. Advance online publication. DOI: 10.1016/j.jclepro.2024.141814

Atiku, S. O., Jeremiah, A., Semente, E., & Boateng, F. (Eds.). (2024). *Eco-Innovation and Sustainable Development in Industry 5.0*. IGI Global. DOI: 10.4018/979-8-3693-2219-2

Awaysheh, F. M., Alhussein, A., Al-Ali, A. R., & Alassery, H. A. (2020). IoT-based disaster management system for real-time flood detection. *Sustainable Computing : Informatics and Systems*, *28*, 100431.

Azghadi, M. R., Olsen, A., Wood, J., Saleh, A., Calvert, B., Granshaw, T., & Philippa, B. (2025). Precision robotic spot-spraying: Reducing herbicide use and enhancing environmental outcomes in sugarcane. *Computers and Electronics in Agriculture*, *235*, 110365. DOI: 10.1016/j.compag.2025.110365

Azuma, R., Baillot, Y., Behringer, R., Feiner, S., Julier, S., MacIntyre, B., 2001. Recent advances in augmented reality. IEEE Comput. Graph. Appl. 21 (6), 34e47.

Babalola, O., & Harinarain, N. (2024). Challenges in Adopting Construction Robotics in Sub-Saharan Africa. *Journal of Construction Engineering and Management, 145*(2), 215–228.

Bacca, J., Baldiris, S., Fabregat, R., & Graf, S., & Kinshuk. (2014). Augmented reality trends in education: A systematic review of research and applications. *Journal of Educational Technology & Society, 17*(4), 133–149.

Bademosi, F., & Issa, R. R. (2021) Factors influencing adoption and integration of construction robotics technology in the US. *Journal of Construction Engineering and Management*; 147(8), pp.04-21.

Bag, T. 2023. Socio-economic impacts of scientific-technological advancements. Int. J. Multidiscip. Educ. Res. 12. https://ijmer.s3.amazonaws. com/pdf/volume12/volume12-issue8(4)/13.pdf

Bag, S., & Pretorius, J. H. C. (2020). Relationships between industry 4.0, sustainable manufacturing and circular economy: Proposal of a research framework. *The International Journal of Organizational Analysis, 30*(4), 864–898. DOI: 10.1108/IJOA-04-2020-2120

Bakator, M., Ćoćkalo, D., Makitan, V., Stanisavljev, S., & Nikolić, M. (2024). The three pillars of tomorrow: How Marketing 5.0 builds on Industry 5.0 and impacts Society 5.0? *Heliyon, 10*(17), e36543. DOI: 10.1016/j.heliyon.2024.e36543

Balakrishnan, K., Gupta, S. D., & Mathur, P. (2019). The impact of air pollution on deaths, disease burden, and life expectancy across the states of India: The Global Burden of Disease Study 2017. *The Lancet. Planetary Health, 3*(1), e26–e39. DOI: 10.1016/S2542-5196(18)30261-4 PMID: 30528905

Balasubramanian, S., Shukla, V., Sethi, J. S., Islam, N., & Saloum, R. (2021). A readiness assessment framework for Blockchain adoption: A healthcare case study. *Technological Forecasting and Social Change, 165*, 120536. DOI: 10.1016/j.techfore.2020.120536

Bamakan, S. M. H., Malekinejad, P., & Ziaeian, M. (2022). Towards blockchain-based hospital waste management systems; applications and future trends. *Journal of Cleaner Production, 349*, 131440. DOI: 10.1016/j.jclepro.2022.131440

Bansal, R., & Meena, D. (2023). Policy and regulatory perspectives of waste battery management and recycling: A review and future research agendas. *Journal of Environmental Management, 335*, 113227.

Barata, J., & Kayser, I. (2024). How will the digital twin shape the future of industry 5.0? *Technovation, 134*, 103025. DOI: 10.1016/j.technovation.2024.103025

Barwood, M., Li, J., Pringle, T., & Rahimifard, S. (2015). Utilisation of Reconfigurable Recycling Systems for Improved Material Recovery from E-Waste. *Procedia CIRP*, *29*, 746–751. DOI: 10.1016/j.procir.2015.02.071

Bekele, T. W., Haile, A. T., Trigg, M. A., & Walsh, C. L. (2022). Evaluating a new method of remote sensing for flood mapping in the urban and peri-urban areas: Applied to Addis Ababa and the Akaki catchment in Ethiopia. *Natural Hazards Review*, *2*(2), 97–110. DOI: 10.1016/j.nhres.2022.03.001

Berger-Tal, O., & Lahoz-Monfort, J. J. (2018). Conservation technology: The next generation. *Conservation Letters*, *11*(6), e12458. DOI: 10.1111/conl.12458

Bernat, K. (2023). Post-consumer plastic waste management: From collection and sortation to mechanical recycling. *Energies*, *16*(8), 3504. DOI: 10.3390/en16083504

Bharti, S., Fatma, S., & Kumar, V. (2022). AI in Waste Management: The Savage of Environment. In Paul, P. K., Choudhury, A., Biswas, A., & Singh, B. K. (Eds.), *Environmental Informatics: Challenges and Solutions* (pp. 97–123)., DOI: 10.1007/978-981-19-2083-7_6

Bhat, F. A., & Parvez, S. (2024). Emerging Challenges in the Sustainable Manufacturing System: From Industry 4.0 to Industry 5.0. *Journal of The Institution of Engineers (India): Series C*, 1-15.

Bhat, G. R., Dudhedia, M. A., Panchal, R. A., Shirke, Y. S., Angane, N. R., Khonde, S. R., Khedkar, S. P., Pansare, J. R., Bere, S. S., Wahul, R. M., & Gawande, S. H. (2024). Autonomous drones and their influence on standardization of rules and regulations for operating–A brief overview. *Results in Control and Optimization*, *14*, 100401. DOI: 10.1016/j.rico.2024.100401

Bhattarai, U., Sapkota, R., Kshetri, S., Mo, C., Whiting, M. D., Zhang, Q., & Karkee, M. (2025). A vision-based robotic system for precision pollination of apples. *Computers and Electronics in Agriculture*, *234*, 110158. DOI: 10.1016/j.compag.2025.110158

Bhoi, S. K., & Turuk, P. K. "A Novel Energy Efficient Routing Protocol for MANET," in *Proc. of the Int. Conf. on Information Technology*, 2007, pp. 1–6.

Bianchini, A., Rossi, J., & Pellegrini, M. (2019). Overcoming the main barriers of circular economy implementation through a new visualization tool for circular business models. *Sustainability (Basel)*, *11*(23), 6614. DOI: 10.3390/su11236614

Bibri, S. E., & Krogstie, J. (2020). The emerging data–driven Smart City and its innovative applied solutions for sustainability: The cases of London and Barcelona. *Energy Informatics*, *3*(1), 5. DOI: 10.1186/s42162-020-00108-6

Bibri, S. E., Krogstie, J., Kaboli, A., & Alahi, A. (2024). Smarter eco-cities and their leading-edge artificial intelligence of things solutions for environmental sustainability: A comprehensive systematic review. *Environmental Science and Ecotechnology, 19*, 100330. DOI: 10.1016/j.ese.2023.100330 PMID: 38021367

Billinghurst, M., Clark, A., & Lee, G. (2015). A survey of augmented reality. *Foundations and Trends® in Human–Computer Interaction, 8*(2-3), 73–272.

Billinghurst, M., & Duenser, A. (2012). Augmented reality in the classroom. *Computer, 45*(7), 56–63. DOI: 10.1109/MC.2012.111

Bindra, K. S., Yadav, R., & Rana, N. K. (2019). IoT-based Smart Disaster Management System for Flood Detection and Rescue. *Procedia Computer Science, 152*, 106–113.

Bishnoi, M. R. (2024). A Study on Electric Vehicles in India Opportunities and Challenges. In Conference paper available at https://www. researchgate. net/publication/378737287

Bissyandé, T. F., Lo, D., Jiang, L., Réveillère, L., & Klein, J. (2016). Got issues? Who cares about it? A large scale investigation of issue trackers from the perspective of developers. *Empirical Software Engineering, 21*(1), 150–188.

Bliznina, N. (2024). *Redefining Economic Growth: the Review of Industry 4.0 and Industry 5.0 Progress*. Green and Low-Carbon Economy.

Bolón-Canedo, V., Morán-Fernández, L., Cancela, B., & Alonso-Betanzos, A. (2024). A review of green artificial intelligence: Towards a more sustainable future. *Neurocomputing, 599*, 128096. DOI: 10.1016/j.neucom.2024.128096

Borrello, M., Caracciolo, F., Lombardi, A., Pascucci, S., & Cembalo, L. (2017). Consumers' perspective on circular economy strategy for reducing food waste. *Sustainability (Basel), 9*(1), 141. DOI: 10.3390/su9010141

Boya, A., Akinradewo, O., Aigbavboa, C., Ebekozien, A., & Ramabodu, M. (2022) Bottlenecks to the Implementation of Automation and Robotics in the Construction Industry; In International Conference on Computing in Civil and Building Engineering pp. 139-149. Cham: Springer Nature Switzerland.

Boza, P., & Evgeniou, T. (2021). Artificial intelligence to support the integration of variable renewable energy sources to the power system. *Applied Energy, 290*, 116754. DOI: 10.1016/j.apenergy.2021.116754

Brintha, V. P., Rekha, R., Nandhini, J., Sreekaarthick, N., Ishwaryaa, B., & Rahul, R. (2020). Automatic Classification of Solid Waste Using Deep Learning, in: Kumar, L.A., Jayashree, L.S., & Manimegalai, R. (Eds.), *Proceedings of International Conference on Artificial Intelligence, Smart Grid and Smart City Applications*. Springer International Publishing, Cham, pp. 881–889. DOI: DOI: 10.1007/978-3-030-24051-6_83

Bronack, S. C. (2011). The role of immersive media in online education. *The Journal of Continuing Higher Education, 59*(2), 113–117. DOI: 10.1080/07377363.2011.583186

Brown, J. R., Doore, S. A., Dimmel, J. K., Giudice, N., & Giudice, N. A. (2023). Comparing Natural Language and Vibro-Audio Modalities for Inclusive STEM learning with blind and low vision users. In *Proceedings of the 25th International ACM SIGACCESS Conference on Computers and Accessibility*, pp. 1-17. DOI: DOI: 10.1145/3597638.3608429

Buchner, J., & Kerres, M. (2023). Media comparison studies dominate comparative research on augmented reality in education. *Computers & Education, 195*, 104711. DOI: 10.1016/j.compedu.2022.104711

Bui, T.-D., Tseng, J.-W., Tseng, M.-L., Wu, K.-J., Lim, M.K., (2023). Municipal solid waste management technological barriers: A hierarchical structure approach in Taiwan. Resour. Conserv. Recycl. 190, 106842 https://doi.org/. resconrec.2022. 106842DOI: 10.1016/j

Burns, H. (2021). A smart city masterplan, Kigali. *Urban planning for transitions*, 153-169.

Burton, E. P., Frazier, W., Annetta, L., Lamb, R., Cheng, R., & Chmiel, M. (2011). Modeling Augmented Reality Games with Preservice. *Journal of Technology and Teacher Education, 19*(3), 303–329.

Cabero, J., & Barroso, J. (2016). The educational possibilities of Augmented Reality. *Journal of New Approaches in Educational Research, 5*(1), 44–50. DOI: 10.7821/naer.2016.1.140

Cai, S., Ma, Z., Skibniewski, M. J., Bao, S., & Wang, H.; (2020). Construction Robotics for high-rise buildings: Development priorities and key challenges. *Journal of Construction Engineering and Management,* 146(8), pp. 04-20.

Calvin, K., Dasgupta, D., Krinner, G., Mukherji, A., Thorne, P. W., Trisos, C., Romero, J., Aldunce, P., Barrett, K., Blanco, G., Cheung, W. W. L., Connors, S., Denton, F., Diongue-Niang, A., Dodman, D., Garschagen, M., Geden, O., Hayward, B., Jones, C., . . . Péan, C. (2023). *IPCC, 2023: Climate Change 2023: Synthesis Report. Contribution of Working Groups I, II and III to the Sixth Assessment Report of the Intergovernmental Panel on Climate Change [Core Writing Team, H. Lee and J. Romero (eds.)]. IPCC, Geneva, Switzerland.* (First). Intergovernmental Panel on Climate Change (IPCC). DOI: 10.59327/IPCC/AR6-9789291691647

Camps-Valls, G., Fernández-Torres, M.-Á., Cohrs, K.-H., Höhl, A., Castelletti, A., Pacal, A., Robin, C., Martinuzzi, F., Papoutsis, I., Prapas, I., Pérez-Aracil, J., Weigel, K., Gonzalez-Calabuig, M., Reichstein, M., Rabel, M., Giuliani, M., Mahecha, M. D., Popescu, O.-I., Pellicer-Valero, O. J., & Williams, T. (2025). Artificial intelligence for modeling and understanding extreme weather and climate events. *Nature Communications*, *16*(1), 1919. DOI: 10.1038/s41467-025-56573-8 PMID: 39994190

Capraro, V., Lentsch, A., Acemoglu, D., Akgun, S., Akhmedova, A., Bilancini, E., Bonnefon, J.-F., Brañas-Garza, P., Butera, L., Douglas, K. M., Everett, J. A. C., Gigerenzer, G., Greenhow, C., Hashimoto, D. A., Holt-Lunstad, J., Jetten, J., Johnson, S., Kunz, W. H., Longoni, C., & Viale, R. (2024). The impact of generative artificial intelligence on socioeconomic inequalities and policy making. *PNAS Nexus*, *3*(6), pgae191. Advance online publication. DOI: 10.1093/pnasnexus/pgae191 PMID: 38864006

Cascante, M.Q., Garcıa, N.A. and Fold, N. (2022), "The role of external forces in the adoption of aquaculture innovations: an ex-ante case study of fish farming in Colombia's southern Amazonian region", Technological Forecasting and Social Change, Vol. 174, 121185, doi: . techfore.2021.121185.DOI: 10.1016/ j

Caudell, T. P., & Mizell, D. W. (1992). Augmented reality: An application of heads-up display technology to manual manufacturing processes. *Proceedings of the 25th Hawaii International Conference on System Sciences*, 2, 659–669. DOI: 10.1109/HICSS.1992.183317

Cha, G.-W., Kim, Y.-C., Moon, H.J. and Hong, W.-H. (2017), "New approach for forecasting demolition waste generation using chi-squared automatic interaction detection (CHAID) method", Journal of Cleaner Production, Vol. 168, pp. 375-385, doi: . 09.025.DOI: 10.1016/j.jclepro.2017

Chai, C. S., Koh, J. H. L., & Teo, Y. H. (2019). Fostering collaboration and creativity in education through AR and AI. *Journal of Educational Technology & Society*, *22*(2), 45–56.

Chang, H.Y., Wu, H.K., Hsu, Y.S., 2013. Integrating a mobile augmented reality activity to contextualize student learning of a socioscientific issue. Br. J. Educ. Technol. 44 (3), E95eE99.

Chan, H. Y., & Muralidharan, A. (2024). Care Robots for the Elderly: Legal, Ethical Considerations and Regulatory Strategies. In *Developments in Intellectual Property Strategy: The Impact of Artificial Intelligence, Robotics and New Technologies Cham* (pp. 129–156). Springer International Publishing. DOI: 10.1007/978-3-031-42576-9_6

Chatterjee, P., Chakrabarti, S., & Gupta, A. (2021). Understanding the gap between formal and informal e-waste recycling facilities in India. *Environmental Health Perspectives*, *129*(10), 105003.

Chaturvedi, B. K., Nautiyal, A., Kandpal, T. C., & Yaqoot, M. (2022). Projected transition to electric vehicles in India and its impact on stakeholders. *Energy for Sustainable Development*, *66*, 189–200. DOI: 10.1016/j.esd.2021.12.006

Chauhan, C., Parida, V. and Dhir, A. (2022), "Linking circular economy and digitalisation technologies: a systematic literature review of past achievements and future promises", Technological Forecasting and Social Change, Vol. 177, 121508, doi: . 121508.DOI: 10.1016/j.techfore.2022

Chavali, D., Baburajan, B., Gurusamy, A., Dhiman, V. K., & Katari, S. C. (2024). Regulating Artificial Intelligence: Developments And Challenges. *IJPS Journal*. DOI: 10.5281/zenodo.10898480

Chawla, V. K., Chhabra, D., Gupta, P., & Naaz, S. (2021). Evaluation of green operations management by fuzzy analytical hierarchy process. *Materials Today: Proceedings*, *38*, 274–279. DOI: 10.1016/j.matpr.2020.07.200

Chea, C. P., Bai, Y., Pan, X., Arashpour, M., & Xie, Y. (2020). An integrated review of automation and robotic technologies for structural prefabrication and construction. *Transportation Safety and Environment*, *2*(2), 81–96. DOI: 10.1093/tse/tdaa007

Chen, J., Begicheva, S., & Nazarov, D. (2024). The role of big data in advancing precision agriculture and ensuring food security. In BIO Web of Conferences (Vol. 121, p. 02013). EDP Sciences. DOI: 10.1051/bioconf/202412102013

Cheng, K. W. E. (2009, May). Recent development on electric vehicles. In 2009 3rd International Conference on Power Electronics Systems and Applications (PESA) (pp. 1-5). IEEE.

Cheng, K. H., & Tsai, C. C. (2020). A case study of immersive virtual field trips in an elementary classroom: Students' learning experience and teacher-student interaction behaviors. *Interactive Learning Environments*, *28*(4), 432–446.

Cheng, L., & Yu, T. (2019). A new generation of AI: A review and perspective on machine learning technologies applied to smart energy and electric power systems. *International Journal of Energy Research*, *43*(6), 1928–1973. DOI: 10.1002/er.4333

Chen, H., & Shen, J. (2017, October). A degradation-based sorting method for lithium- ion battery reuse. *PLoS One*, *12*(10), e0185922. DOI: 10.1371/journal.pone.0185922 PMID: 29023485

Chen, X., Chang-Richards, A. Y., Pelosi, A., Jia, Y., Shen, X., Siddiqui, M. K., & Yang, N. (2022). Implementation of technologies in the construction industry: A systematic review. *Engineering, Construction, and Architectural Management*, *29*(8), 3181–3209. DOI: 10.1108/ECAM-02-2021-0172

Chhikara, R., Garg, R., Chhabra, S., Karnatak, U., & Agrawal, G. (2021). Factors affecting adoption of electric vehicles in India: An exploratory study. *Transportation Research Part D, Transport and Environment*, *100*, 103084. DOI: 10.1016/j.trd.2021.103084

Chisom, O. N., Biu, P. W., Umoh, A. A., Obaedo, B. O., Adegbite, A. O., & Abatan, A. (2024). Reviewing the role of AI in environmental monitoring and conservation: A data-driven revolution for our planet. *World Journal of Advanced Research and Reviews*, *21*(1), 161–171. DOI: 10.30574/wjarr.2024.21.1.2720

Chutcheva, Y., Kuprianova, L., Seregina, A., & Kukushkin, S. (2022). Environmental management of companies in the oil and gas markets based on AI for sustainable development. *Frontiers in Environmental Science*, *10*, 952102. Advance online publication. DOI: 10.3389/fenvs.2022.952102

Ciucu-Durnoi, A. N., Delcea, C., Stănescu, A., Teodorescu, C. A., & Vargas, V. M. (2024). Beyond Industry 4.0: Tracing the Path to Industry 5.0 through Bibliometric Analysis. *Sustainability (Basel)*, *16*(12), 5251. DOI: 10.3390/su16125251

Cody, J., Göl, Ö., Nedic, Z., Nafalski, A., & Mohtar, A. (2009). Regenerative braking in an electric vehicle. Maszyny Elektryczne: zeszyty problemowe, 113-118.

Cohen, A. J., Brauer, M., Burnett, R., Anderson, H. R., Frostad, J., Estep, K., ... & Forouzanfar, M. H. (2019). *The impact of air pollution on deaths, disease burden, and life expectancy across the states of India: The Global Burden of Disease Study 2017. The Lancet Planetary Health*

Comninos, A., Muller, E. S., & Mutung'u, G. (2019). Artificial intelligence for sustainable human development. Association for Progressive Communications (APC), Article, 19. https://tarbiatbadani.farafile.ir/content/demo/202009/eea99b6b-c4 8b-46e7-8703-bce6845e64e1.pdf

Corrêa, N. K., Galvão, C., Santos, J. W., Del Pino, C., Pinto, E. P., Barbosa, C., Massmann, D., Mambrini, R., Galvão, L., Terem, E., & De Oliveira, N. (2023). Worldwide AI ethics: A review of 200 guidelines and recommendations for AI governance. *Patterns (New York, N.Y.)*, *4*(10), 100857. DOI: 10.1016/j.patter.2023.100857 PMID: 37876898

Cossio-Montefinale, L., Quiñinao, C., & Verschae, R. (2025). Orchard sweet cherry color distribution estimation from wireless sensor networks and video-based fruit detection. *Computers and Electronics in Agriculture*, *235*, 110334. DOI: 10.1016/j.compag.2025.110334

Creswell, J. W. (2014). *Research Design: Qualitative, Quantitative, and Mixed Methods Approaches* (4th ed.). SAGE Publications.

Cruz, E. M. G. N. V., Oliveira, S., & Correia, A. (2024). Robotics Applications in the Hospital Domain: A Literature Review. *Applied System Innovation*, *7*(6), 125. DOI: 10.3390/asi7060125

Dacre, N., Yan, J., Frei, R., Al-Mhdawi, M. K. S., & Dong, H. (2024). Advancing sustainable manufacturing: A systematic exploration of industry 5.0 supply chains for sustainability, human-centricity, and resilience. *Production Planning and Control*, •••, 1–30. DOI: 10.1080/09537287.2024.2380361

Daniela, L., Visvizi, A., Gutierrez-Braojos, C., & Lytras, M. (2018). Sustainable higher education and technology-enhanced learning (TEL). *Sustainability (Basel)*, *10*(11), 3883. DOI: 10.3390/su10113883

Das, D. K. (2025). Digital Technology and AI for Smart Sustainable Cities in the Global South: A Critical Review of Literature and Case Studies. *Urban Science (Basel, Switzerland)*, *9*(3), 72. DOI: 10.3390/urbansci9030072

Das, H. S., Rahman, M. M., Li, S., & Tan, C. W. (2020). Electric vehicles standards, charging infrastructure, and impact on grid integration: A technological review. *Renewable & Sustainable Energy Reviews*, *120*, 109618. DOI: 10.1016/j.rser.2019.109618

Das, P. K., & Bhat, M. Y. (2022). Global electric vehicle adoption: Implementation and policy implications for India. *Environmental Science and Pollution Research International, 29*(27), 40612–40622. DOI: 10.1007/s11356-021-18211-w PMID: 35083674

Das, S., Lee, S.-H., Kumar, P., Kim, K.-H., Lee, S. S., & Bhattacharya, S. S. (2019). Solid waste management: Scope and the challenge of sustainability. *Journal of Cleaner Production, 228*, 658–678. DOI: 10.1016/j.jclepro.2019.04.323

de Moraes Rossetto, A. G., Martins, T. C., Silva, L. A., Leithardt, D. R., Bermejo-Gil, B. M., & Leithardt, V. R. (2023). An analysis of the use of augmented reality and virtual reality as educational resources. *Computer Applications in Engineering Education, 31*(6), 1761–1775. DOI: 10.1002/cae.22671

de Soto, B. G., & Skibniewski, M. J. (2020). Future of robotics in construction. In *Construction 4.0* (pp. 289–306). Routledge. DOI: 10.1201/9780429398100-15

Dede, C. (2009). Immersive interfaces for engagement and learning. *Science, 323*(5910), 66–69. DOI: 10.1126/science.1167311

del Carmen Pardo-Ferreira, M., Rubio-Romero, J. C., Galindo-Reyes, F. C., & Lopez-Arquillos, A. (2020). Work-related road safety: The impact of the low noise levels produced by electric vehicles according to experienced drivers. *Safety Science, 121*, 580–588. DOI: 10.1016/j.ssci.2019.02.021

Delgado, J. M. D., Oyedele, L., Ajayi, A., Akanbi, L., Akinade, O., Bilal, M., & Owolabi, H. (2019). Robotics and automated systems in construction: Understanding industry-specific challenges for adoption. *Journal of Building Engineering, 26*, 10–18. DOI: 10.1016/j.jobe.2019.100868

Denis, J., & Florentin, D. (2024). *"Urban infrastructures' maturity and the age (s) of maintenance," in Handbook of Infrastructures and Cities*. Edward Elgar Publishing.

Deo, N., & Anjankar, A. (2023). Artificial Intelligence With Robotics in Healthcare: A Narrative Review of Its Viability in India. *Cureus, 15*(5), e39416. DOI: 10.7759/cureus.39416 PMID: 37362504

DeVries, P. M. R., Viégas, F., Wattenberg, M., & Meade, B. J. (2018). Deep learning of aftershock patterns following large earthquakes. *Nature, 560*(7720), 632–634. DOI: 10.1038/s41586-018-0438-y PMID: 30158606

Dhankhar, S., Pateer, S., Sandhu, V., & Kaur, H. (2024). Barriers to Electric Vehicle Adoption in India: A Comparative Review and Future Growth Prospects. *European Economic Letters, 14*(3), 2007–2015.

Di Vaio, A., Hassan, R., D'Amore, G., & Strologo, A. D. (2022). Digital technologies for sustainable waste management on-board ships: An analysis of best practices from the cruise industry. *IEEE Transactions on Engineering Management*.

Díaz-Rodríguez, N., Del Ser, J., Coeckelbergh, M., López De Prado, M., Herrera-Viedma, E., & Herrera, F. (2023). Connecting the dots in trustworthy Artificial Intelligence: From AI principles, ethics, and key requirements to responsible AI systems and regulation. *Information Fusion*, *99*, 101896. DOI: 10.1016/j.inffus.2023.101896

Ding, N., Prasad, K., & Lie, T. T. (2017). The electric vehicle: A review. *International Journal of Electric and Hybrid Vehicles*, *9*(1), 49–66. DOI: 10.1504/IJEHV.2017.082816

Doll, C. N., & Balaban, O. (2013). A methodology for evaluating environmental co-benefits in the transport sector: Application to the Delhi metro. *Journal of Cleaner Production*, *58*, 61–73. DOI: 10.1016/j.jclepro.2013.07.006

Doluweera, L. K. (2024). Industry 5.0 Technologies for Supply Chain Management through Triple Bottom Line Approach in Companies of Sri Lanka for Economic Growth and Socio-Environmental Protection.

Dong, Z., Chen, J., & Lu, W. (2022). Computer vision to recognize construction waste compositions: A novel boundary-aware transformer (BAT) model. *Journal of Environmental Management*, *305*, 114405. DOI: 10.1016/j.jenvman.2021.114405 PMID: 34995944

Dua, R., Hardman, S., Bhatt, Y., & Suneja, D. (2021). Enablers and disablers to plug-in electric vehicle adoption in India: Insights from a survey of experts. *Energy Reports*, *7*, 3171–3188. DOI: 10.1016/j.egyr.2021.05.025

Dunleavy, M. (2010). Persisient design challenges: Augmeflling reafiry for teaming with wireless mobile devices. San Diego, CA: Invitation Symposia at Society for Information Technology and Teacher Education (SITE).

Dunleavy, M., Dede, C., & Mitchell, R. (2009). Affordances and limitations of immersive participatory augmented reality simulations for teaching and learning. *Journal of Science Education and Technology*, *18*(1), 7–22. DOI: 10.1007/s10956-008-9119-1

Dwivedi, Y. K., Hughes, L., Kar, A. K., Baabdullah, A. M., Grover, P., Abbas, R., Andreini, D., Abumoghli, I., Barlette, Y., Bunker, D., Chandra Kruse, L., Constantiou, I., Davison, R. M., De', R., Dubey, R., Fenby-Taylor, H., Gupta, B., He, W., Kodama, M., & Wade, M. (2022). Climate change and COP26: Are digital technologies and information management part of the problem or the solution? An editorial reflection and call to action. *International Journal of Information Management, 63*, 102456. DOI: 10.1016/j.ijinfomgt.2021.102456

Ebekozien, A., & Aigbavboa, C. (2021). Technological Integration in the Construction Industry: Opportunities and Uncertainties. *Journal of Building Engineering, 40*, 1–10. DOI: 10.1016/j.jobe.2021.102773

Ejimofor, M. I., Aniagor, C. O., Oba, S. N., Menkiti, M. C., & Ugonabo, V. I. (2022). Artificial intelligence in the reduction and management of land pollution. In *Current Trends and Advances in Computer-Aided Intelligent Environmental Data Engineering* (pp. 319–333). Elsevier., DOI: 10.1016/B978-0-323-85597-6.00009-4

Ejohwomu, O., Adekunle, S. A., Aigbavboa, C. O., & Bukoye, O. T. (2021). *Construction Industry and the Fourth Industrial Revolution: Issues and Strategies. The Construction Industry: Global Trends, Job Burnout, and Safety Issues*. Nova Science Publishers.

El Sayed, N. A. M., Zayed, H. H., & Sharawy, M. I. (2011). ARSC: Augmented reality student card – an augmented reality solution for the education field. *Computers & Education, 56*(4), 1045–1061. DOI: 10.1016/j.compedu.2010.10.019

Emma Klein & Stewart Patrick. (2024). *Envisioning a Global Regime Complex to Govern Artificial Intelligence*. Carnegie Endowment for International Peace. https://carnegieendowment.org/research/2024/03/envisioning-a-global-regime-complex-to-govern-artificial-intelligence?lang=en

Environment, U. N. (2024a, May 13). *AI Solutions for the Environment | UNEP - UN Environment Programme*. https://www.unep.org/topics/digital-transformations/digital-accelerator-lab/ai-solutions-environment

Environment, U. N. (2024b, July 8). *Artificial Intelligence (AI) end-to-end: The Environmental Impact of the Full AI Lifecycle Needs to be Comprehensively Assessed | UNEP - UN Environment Programme*. https://www.unep.org/resources/report/artificial-intelligence-ai-end-end-environmental-impact-full-ai-lifecycle-needs-be

Enyedy, N., Danish, J. A., Delacruz, G., & Kumar, M. (2012). Learning physics through play in an augmented reality environment. *International Journal of Computer-Supported Collaborative Learning, 7*(3), 347–378. DOI: 10.1007/s11412-012-9150-3

European Commission (2019), Accelerating the Transition to the Circular Economy: Improving Access to Finance for Circular Economy Projects, Publications Office, LU.

European Commission. (2020). *Categorisation System for the Circular Economy*. European Commission.

European Parliament (2022), "Towards carbon neutrality through ambitious transformation of the EU energy system".

Faltein, S. A., & Sukdeo, N. I. (2024). Culture-driven quality enhancement: Uncovering the impact of robotics integration in the South African construction sector. *Ain Shams Engineering Journal*, *102*(6), 728. DOI: 10.1016/j.asej.2024.102728

Fang, B., Yu, J., Chen, Z., Osman, A. I., Farghali, M., Ihara, I., Hamza, E. H., Rooney, D. W., & Yap, P. S. (2023). Artificial intelligence for waste management in smart cities: A review. *Environmental Chemistry Letters*, *21*(4), 1959–1989. DOI: 10.1007/s10311-023-01604-3 PMID: 37362015

Farroni, T., Valori, I., & Carnevali, L. (2022). Multimedia Interventions for Neurodiversity: Leveraging Insights from Developmental Cognitive Neuroscience to Build an Innovative Practice. *Brain Sciences*, *12*(2), 147. DOI: 10.3390/brainsci12020147

Fathy, A. E. A. E., Abdel-Hamid, M., & Abdelhaleem, H. M. (2024). Developing a predictive maintenance model for the rehabilitation of roads' assets. *International Journal of Construction Management*, •••, 1–11.

Fazel, A., & Izadi, A. (2018). An interactive augmented reality tool for constructing free-form modular surfaces. *Automation in Construction*, *85*, 135–145. DOI: 10.1016/j.autcon.2017.10.015

Fernández-Robles, B. (2018). La utilización de objetos de aprendizaje de realidad aumentada en la enseñanza universitaria de educación primaria. *IJERI Int. J. Educ. Res. Innov.*, *9*, 90–104.

Ferrer-Torregrosa, J., Torralba, J., Jimenez, M. A., García, S., & Barcia, J. M. (2015). ARBOOK: Development and assessment of a tool based on augmented reality for anatomy. *Journal of Science Education and Technology*, *24*(1), 119–124. DOI: 10.1007/s10956-014-9526-4

Field, A. (2018). *Discovering Statistics Using IBM SPSS Statistics* (5th ed.). SAGE Publications.

Financial Express. (2023). *Sustainable materials and eco-friendly innovations in EV manufacturing. Financial Express.* https://www.financialexpress.com/business/express-mobility-sustainable-materials-eco-friendly-innovations-in-ev-manufacturing-3317096

Forward, T. W. (2024). Industry 5.0, Quality of Life, and Sustainable Adaptation to Climate Change. *Quality of Life and Climate Change: Impacts, Sustainable Adaptation, and Social-Ecological Resilience: Impacts, Sustainable Adaptation, and Social-Ecological Resilience*, 200.

Francisco, M., & Linnér, B.-O. (2023). AI and the governance of sustainable development. An idea analysis of the European Union, the United Nations, and the World Economic Forum. *Environmental Science & Policy, 150*, 103590. DOI: 10.1016/j.envsci.2023.103590

Gaikwad, R., & Tomar, A. (2024). Electric vehicles – Paving the road to widespread adoption: A review of technical and non-technical challenges. *International Journal of Research in Advanced Mobility, 1*(2), 22–26.

Gal, I.-A., Ciocîrlan, A.-C., & M˘arg˘aritescu, M. (2021). State machine-based hybrid position/ force control architecture for a waste management mobile robot with 5DOF manipulator. *Applied Sciences (Basel, Switzerland), 11*(9), 4222. DOI: 10.3390/app11094222

Gamberini, L., & Pluchino, P. (2024). Industry 5.0: A comprehensive insight into the future of work, social sustainability, sustainable development, and career. *Australian Journal of Career Development, 33*(1), 5–14. DOI: 10.1177/10384162241231118

Gao, H., Wang, Y., Liu, J., & Liu, X. (2018). Real-time emergency rescue scheduling model for urban disaster management. *IEEE Access : Practical Innovations, Open Solutions, 6*, 55284–55295.

García de Soto, B., Agustí-Juan, I., Joss, S., & Hunhevicz, J. (2022). Implications of Construction 4.0 to the workforce and organizational structures. *International Journal of Construction Management, 22*(2), 205–217. DOI: 10.1080/15623599.2019.1616414

Garg, R. K. (2024). Challenges and opportunities in electric vehicle charging infrastructure development. *Shodh Sagar Journal of Electric Vehicles, 1*(1), 13–20.

Garzón, J., & Acevedo, J. (2019). Meta-analysis of the impact of augmented reality on students' learning gains. *Educational Research Review, 27*, 244–260. 1

Garzón, J. (2021). An overview of twenty-five years of augmented reality in education. *Multimodal Technologies and Interaction, 5*(7), 37. DOI: 10.3390/mti5070037

Gbadegesin, O. A. (2023). Leveraging Artificial Intelligence (AI) in Strengthening the Legal Framework for Regulation of Wildlife and Forest Crimes in Nigeria. *Environmental Policy and Law*, *53*(4), 259–274. DOI: 10.3233/EPL-230011

Gentile, C., Lunghi, G., Buonocore, L. R., Cordella, F., Di Castro, M., Masi, A., & Zollo, L. (2023). Manipulation Tasks in Hazardous Environments Using a Teleoperated Robot: A Case Study at CERN. *Sensors (Basel)*, *23*(4), 1979. DOI: 10.3390/s23041979 PMID: 36850591

Gerhards, R., Spaeth, M., Heyn, A., & Saile, M. (2025). Sensor-Guided Mechanical Weed Control in Transplanted Lettuce and Cabbage. *Journal of Crop Health*, *77*(2), 67. DOI: 10.1007/s10343-025-01139-3

Geroimenko, V. (2020). *Augmented reality in education. Springer Series on Cultural Computing*. Springer International Publishing. DOI: 10.1007/978-3-030-42156-4

Geroimenko, V. (2020). *Augmented reality in tourism, museums and heritage: Emerging applications and future trends*. Springer.

Ghahramani, M., Zhou, M., Molter, A., & Pilla, F. (2022). IoT-based route recommendation for an intelligent waste management system. *IEEE Internet of Things Journal*, *9*(14), 11883–11892. DOI: 10.1109/JIOT.2021.3132126

Ghanbari, F., Kamalan, H., & Sarraf, A. (2023). Predicting solid waste generation based on the ensemble artificial intelligence models under uncertainty analysis. *Journal of Material Cycles and Waste Management*, *25*(2), 920–930. DOI: 10.1007/s10163-023-01589-9

Gharbia, M., Chang-Richards, A., Lu, Y., Zhong, R. Y., & Li, H. (2020). Robotic technologies for on-site building construction: A systematic review. *Journal of Building Engineering*, *32*, 101–115. DOI: 10.1016/j.jobe.2020.101584

Ghayvat, H., Mukhopadhyay, S., & Gui, X. (2015). Smart home-based ambient assisted living: Recognition of activities of daily living using 3D depth sensors. *IEEE Sensors Journal*, *15*(3), 1476–1483.

Ghazal, S., Munir, A., & Qureshi, W. S. (2024). Computer vision in smart agriculture and precision farming: Techniques and applications. *Artificial Intelligence in Agriculture*, *13*, 64–83. DOI: 10.1016/j.aiia.2024.06.004

Ghobakhloo, M., Fathi, M., Iranmanesh, M., Vilkas, M., Grybauskas, A., & Amran, A. (2024). Generative artificial intelligence in manufacturing: Opportunities for actualizing Industry 5.0 sustainability goals. *Journal of Manufacturing Technology Management*, *35*(9), 94–121. DOI: 10.1108/JMTM-12-2023-0530

Ghobakhloo, M., Iranmanesh, M., Fathi, M., Rejeb, A., Foroughi, B., & Nikbin, D. (2024). Beyond Industry 4.0: A systematic review of Industry 5.0 technologies and implications for social, environmental and economic sustainability. *Asia-Pacific Journal of Business Administration*. Advance online publication. DOI: 10.1108/APJBA-08-2023-0384

Ghobakhloo, M., Mahdiraji, H. A., Iranmanesh, M., & Jafari-Sadeghi, V. (2024). From Industry 4.0 digital manufacturing to Industry 5.0 digital society: A roadmap toward human-centric, sustainable, and resilient production. *Information Systems Frontiers*, •••, 1–33. DOI: 10.1007/s10796-024-10476-z

Ghosh, A., Datta, D., Saha, S. K., & Mukherjee, D. (2021). An IoT-cloud integrated disaster management framework using Artificial Intelligence. *Journal of Ambient Intelligence and Humanized Computing*, *12*(1), 991–1004.

Giannakidou, S., Radoglou-Grammatikis, P., Lagkas, T., Argyriou, V., Goudos, S., Markakis, E. K., & Sarigiannidis, P. (2024). Leveraging the power of internet of things and artificial intelligence in forest fire prevention, detection, and restoration: A comprehensive survey. *Internet of Things : Engineering Cyber Physical Human Systems*, *26*, 101171. DOI: 10.1016/j.iot.2024.101171

Gkini, E., Voyiatzis, I., & Sgouropoulou, C. 2021. Head-Mounted Display Systems as Visual Aids for the Visually Impaired: A Survey. In *25th Pan-Hellenic Conference on Informatics (PCI 2021)*, November 26–28, 2021, Volos, Greece. ACM, New York, NY, USA 5 Pages. DOI: 10.1145/3503823.3503883

Gkontzis, A. F., Kotsiantis, S., Feretzakis, G., & Verykios, V. S. (2024). Enhancing urban resilience: Smart city data analyses, forecasts, and digital twin techniques at the neighborhood level. *Future Internet*, *16*(2), 47. DOI: 10.3390/fi16020047

Goel, S., Sharma, R., & Rathore, A. K. (2021). A review on barrier and challenges of electric vehicle in India and vehicle to grid optimisation. Transportation engineering, 4, 100057.

Golbaz, S., Nabizadeh, R., & Sajadi, H. S. (2019). Comparative study of predicting hospital solid waste generation using multiple linear regression and artificial intelligence. *Journal of Environmental Health Science & Engineering*, *17*(1), 41–51. DOI: 10.1007/s40201-018-00324-z

Gomaa, M., Schade, S., Bao, D. W., & Xie, Y. M. (2023). Automation in rammed earth construction for industry 4.0: Precedent work, current progress, and prospect. *Journal of Cleaner Production*, *398*, 13–25. DOI: 10.1016/j.jclepro.2023.136569

Gonsalves, N., Akanmu, A., Shojaei, A., & Agee, P. (2024). Factors influencing the adoption of passive exoskeletons in the construction industry: Industry perspectives. *International Journal of Industrial Ergonomics*, *100*, 10–35. DOI: 10.1016/j.ergon.2024.103549

Gopal, A. R., Witt, M., Sheppard, C., & Harris, A. (2015). Battery Electric Vehicles can reduce greenhouse has emissions and make renewable energy cheaper in India.

Gopalakrishnan, P., (2019). Blockchain Based Waste Management 8.

Gujarathi, P. K., Shah, V. A., & Lokhande, M. M. (2018). Electric vehicles in India: Market analysis with consumer perspective, policies and issues. *Journal of Green Engineering*, *8*(1), 17–36. DOI: 10.13052/jge1904-4720.813

Guma, T. N. (2021). An Overview of Industrial Automation and its Implementation Approaches for Boosting Nigerian and Other Developing Economies. Federal University Wukari. *Trends in Science & Technology Journal*, *6*(1), 154–160.

Gundupalli, S. P., Hait, S., & Thakur, A. (2018, August). Classification of metallic and non-metallic fractions of e-waste using thermal imaging-based technique. *Process Safety and Environmental Protection*, *118*, 32–39. DOI: 10.1016/j.psep.2018.06.022

Gupta, H., Jain, A., Singhal, R., & Kapoor, S. (2020). Analysis and optimization of deep learning models for disaster management. *Procedia Computer Science*, *171*, 1600–1609.

Gupta, N., Roy, A., & Das, R. (2021). IoT-based Flood Monitoring and Alerting System using Machine Learning. *International Journal of Intelligent Engineering and Systems*, *14*(1), 225–234.

Gupta, P., Chawla, V., Jain, V., & Angra, S. (2022). Green operations management for sustainable development: An explicit analysis by using fuzzy best-worst method. *Decision Science Letters*, *11*(3), 357–366. DOI: 10.5267/j.dsl.2022.1.003

Gupta, R., Manna, A., Sharma, H., Pandit, P., Pan, S., & Singh, Y. (2024). Air Pollution Prediction Systems in City Highways using Artificial Intelligence. *2024 International Conference on Electrical Electronics and Computing Technologies (ICEECT)*, *1*, 1-6. DOI: 10.1109/ICEECT61758.2024.10739234

Hacker, P. (2024). Article: Sustainable AI Regulation. *Common Market Law Review*, *61*(2), 345–386. Advance online publication. DOI: 10.54648/COLA2024025

Haleem, A., Khan, S., Luthra, S., Varshney, H., Alam, M., & Khan, M. I. (2021). Supplier evaluation in the context of circular economy: A forward step for resilient business and environment concern. *Business Strategy and the Environment, 30*(4), 2119–2146. DOI: 10.1002/bse.2736

Hamamoto, M. (2019). An empirical study on the behavior of hybrid-electric vehicle purchasers. *Energy Policy, 125*, 286–292. DOI: 10.1016/j.enpol.2018.10.042

Hannan, M. A., Arebey, M., Begum, R. A., Basri, H., & Al Mamun, M. A. (2016). Content-based image retrieval system for solid waste bin level detection and performance evaluation. *Waste Management (New York, N.Y.), 50*, 10–19. DOI: 10.1016/j.wasman.2016.01.046

Haque, A. M., & Kaiser, M. S. (2020). Smart Disaster Management System Using IoT and Cloud. *Lecture Notes in Networks and Systems, 123*, 233–241.

Harichandan, S., & Kar, S. K. (2023). An empirical study on motivation to adopt hydrogen fuel cell vehicles in India: Policy implications for stakeholders. *Journal of Cleaner Production, 408*, 137198. DOI: 10.1016/j.jclepro.2023.137198

Harper, G., Sommerville, R., Kendrick, E., Driscoll, L., Slater, P., Stolkin, R., ... & Anderson, P. (2019). Recycling lithium-ion batteries from electric vehicles. *nature, 575*(7781), 75-86.

Harsono, T. W. R., Pramudito, A. A., & Nugroho, S. A. (2021). IoT and Deep Learning-based Flood Monitoring and Early Warning System. *Journal of King Saud University. Computer and Information Sciences.*

Hasan, M. R., Khan, M. A., & Wuest, T. (2024). Towards Industry 5.0: A Systematic Literature Review on Sustainable and Green Composite Materials Supply Chains. *arXiv preprint arXiv:2402.06100.*

Hassan, M. M. (2017). Monitoring land use/land cover change, urban growth dynamics and landscape pattern analysis in five fastest urbanized cities in Bangladesh. *Remote Sensing Applications: Society and Environment, 7*, 69–83. DOI: 10.1016/j.rsase.2017.07.001

Hassija, V., Chamola, V., Mahapatra, A., Singal, A., Goel, D., Huang, K., Scardapane, S., Spinelli, I., Mahmud, M., & Hussain, A. (2024). Interpreting Black-Box Models: A Review on Explainable Artificial Intelligence. *Cognitive Computation, 16*(1), 45–74. DOI: 10.1007/s12559-023-10179-8

Hawkins, C. V., & Wang, X. (2012). Sustainable development governance: Citizen participation and support networks in local sustainability initiatives. *Public Works Management & Policy, 17*(1), 7–29. DOI: 10.1177/1087724X11429045

He, L. (2023). Variable rate technologies for precision agriculture. In *Encyclopedia of Digital Agricultural Technologies* (pp. 1533–1542). Springer International Publishing. DOI: 10.1007/978-3-031-24861-0_34

Held, M., Tuchschmid, M., Zennegg, M., Figi, R., Schreiner, C., Mellert, L. D., & Nachef, L. (2022). Thermal runaway and fire of electric vehicle lithium-ion battery and contamination of infrastructure facility. *Renewable & Sustainable Energy Reviews, 165*, 112474. DOI: 10.1016/j.rser.2022.112474

Hernandez Bennetts, V., Lilienthal, A. J., Neumann, P. P., & Trincavelli, M. (2012). Mobile Robots for Localizing Gas Emission Sources on Landfill Sites: Is Bio-Inspiration the Way to Go? *Frontiers in Neuroengineering, 4*. Advance online publication. DOI: 10.3389/fneng.2011.00020 PMID: 22319493

Hewitt, N., Wood, S., & Wilson, B. (2022). Ecosystem education with augmented reality: A flexible tool for in-field learning. *The Professional Geographer, 75*(4), 577–590. DOI: 10.1080/00330124.2022.2134151

Hiremath, P. S., & Joshi, B. D. (2019). Content Based Image Retrieval using Color, Texture and Shape features. *International Journal of Advanced Computer Science and Applications, 10*(1), 18–24.

Hirz, M., Brunner, H., & Nguyen, T. T. (2022). Greenhouse Gas Emissions of Electric Cars-A Comprehensive Evaluation. Tehnički glasnik, 16(2), 280-287.

Hoeft, M., Kronsell, S., Manzoor, S., Johansson, F., Gustafson, A., von Haslingen, T., & Eriksson, K. (2022).. . *Construction Robotics in Infrastructure., 12*, 23–34.

Hoofnagle, C. J., Van Der Sloot, B., & Borgesius, F. Z. (2019). The European Union general data protection regulation: What it is and what it means. *Information & Communications Technology Law, 28*(1), 65–98. DOI: 10.1080/13600834.2019.1573501

Hoque, M. M., & Rahman, M. T. U. (2020). Landfill area estimation based on solid waste collection prediction using ANN model and final waste disposal options. *Journal of Cleaner Production, 256*, 120387. DOI: 10.1016/j.jclepro.2020.120387

Hossain, I., Haque, A. M., Kadir, A., & Rana, M. S. (2025). Achieving sustainable development goals through knowledge management in Industry 4.0: A critical review from the perspective of Bangladesh. *Artificial Intelligence and Machine Learning Applications for Sustainable Development*, 82-102.

Hossain, I., Mahmudul Haque, A. K. M., & Akram Ullah, S. M. (2024b). Socio-economic Dimensions of Climate Change in Urban Bangladesh: A Focus on the Initiatives of Local Governing Agencies. In *Climate Crisis, Social Responses and Sustainability: Socio-ecological Study on Global Perspectives* (pp. 293-316). Cham: Springer Nature Switzerland. DOI: 10.1007/978-3-031-58261-5_13

Hossain, I., Haque, A. K. M., Ullah, S. M., & Salehin, A. (2025). Assessing the Role of Urban Governance in Promoting Sustainable Cities and Communities: Evidence from Four City Corporations in Bangladesh. *International Journal of Community Well-being*, •••, 1–28. DOI: 10.1007/s42413-025-00247-2

Hossain, I., Haque, A. M., Kılıç, Z., Ullah, S. A., Azrour, M., & Mabrouki, J. (2024b). Exploring Household Waste Management Practices and IoT Adoption in Barisal City Corporation. In *Smart Internet of Things for Environment and Healthcare* (pp. 27–45). Springer Nature Switzerland. DOI: 10.1007/978-3-031-70102-3_2

Hossain, I., Haque, A. M., & Ullah, S. A. (2023). Role of Government Institutions in Promoting Sustainable Development in Bangladesh: An Environmental Governance Perspective. *Journal of Current Social and Political Issues*, *1*(2), 42–53. DOI: 10.15575/jcspi.v1i2.485

Hossain, I., Haque, A. M., & Ullah, S. A. (2024). Policy evaluation and performance assessment for sustainable urbanization: A study of selected city corporations in Bangladesh. *Frontiers in Sustainable Cities*, *6*, 1377310. DOI: 10.3389/frsc.2024.1377310

Hossain, I., Haque, A. M., & Ullah, S. A. (2024a). Assessing Sustainable Waste Management Practices in Rajshahi City Corporation: An Analysis for Local Government Enhancement using IoT and AI Technology. *Environmental Science and Pollution Research International*. Advance online publication. DOI: 10.1007/s11356-024-33171-7

Hossain, I., Rana, M. S., Haque, A. K. M., & Al Masud, A. (2024a). Urban household adaptation to extreme heatwaves: Health impacts, socio-economic disparities and sustainable strategies in Rajshahi. *Discover Sustainability*, *5*(1), 1–18. DOI: 10.1007/s43621-024-00697-2

Hossen, M. S., Haque, A. M., Hossain, I., Haque, M. N., & Hossain, M. K. (2024). Towards comprehensive urban sustainability: Navigating predominant urban challenges and assessing their severity differential in Bangladeshi city corporations. *Urbanization. Sustainability Science*, *1*(1), 1–17. DOI: 10.1108/USS-10-2023-0009

Hsu, E., Barmak, K., West, A. C., & Park, A.-H. A. (2019). Advancements in the treatment and processing of electronic waste with sustainability: A review of metal extraction and recovery technologies. *Green Chemistry*, *21*(5), 919–936. DOI: 10.1039/C8GC03688H

Huang, C., Zhao, Y., & Zhang, M. (2022). Smart urban sustainability: Integrating AR, AI, and IoT for environmental awareness. *Sustainable Cities and Society*, *84*, 104030.

Huang, Q., Someya, K., Dewi, R. K., & Komatsuzaki, M. (2025). No-tillage intercropping with a robotic mower: Advancing a high productivity, low-carbon and energy-efficient organic farming system. *Energy*, *324*, 135851. DOI: 10.1016/j.energy.2025.135851

Huang, Z., Mao, C., Wang, J., & Sadick, A. M. (2022);Huang, Z., Mao, C., Wang, J., & Sadick, A.-M. (2022, November 24). Understanding the key takeaway of construction robots towards construction automation. *Engineering, Construction, and Architectural Management*, *29*(9), 3664–3688. DOI: 10.1108/ECAM-03-2021-0267

Hussain, A., Draz, U., Ali, T., Tariq, S., Irfan, M., Glowacz, A., Antonino Daviu, J. A., Yasin, S., & Rahman, S. (2020). Waste management and prediction of air pollutants using IoT and machine learning approach. *Energies*, *13*(15), 3930. DOI: 10.3390/en13153930

Ibáñez, M. B., & Delgado-Kloos, C. (2018). Augmented reality for STEM learning: A systematic review. *Computers & Education*, *123*, 109–123. DOI: 10.1016/j.compedu.2018.05.002

Ibekwe, K. I., Umoh, A. A., Nwokediegwu, Z. Q. S., Etukudoh, E. A., Ilojianya, V. I., & Adefemi, A. (2024). Energy efficiency in industrial sectors: A review of technologies and policy measures. *Engineering Science & Technology Journal*, *5*(1), 169–184.

IBM. (2015). *IBM Expands Green Horizons Initiative Globally To Address Pressing Environmental and Pollution Challenges*. IBM UK Newsroom. https://uk.newsroom.ibm.com/2015-Dec-09-IBM-Expands-Green-Horizons-Initiative-Globally-To-Address-Pressing-Environmental-and-Pollution-Challenges

Ibrahim, K., Simpeh, F., & Adebowale, O. J. (2024). Awareness and adoption of wearable technologies for health and safety management in the Nigerian construction industry. *Frontiers in Engineering and Built Environment*, *4*(1), 15–28. DOI: 10.1108/FEBE-11-2022-0041

Ikram, L., Afaf, D., & Latifa, E. (2024, May). Beyond Industry 5.0: Tracing the Past, Present, and Future of Digital Industrial Transformations. In *2024 4th International Conference on Innovative Research in Applied Science, Engineering and Technology (IRASET)* (pp. 1-7). IEEE.

Ikuabe, M., Obikwelu, O., & Ebekozien, A. (2022). Robotic Advancements in the Fourth Industrial Revolution: A Nigerian Construction Perspective. *Journal of Sustainable Construction and Engineering*, *10*(4), 223–240. DOI: 10.1007/s40999-022-00794-w

Ikuabe, M., Obikwelu, O., & Ebekozien, A. (2023). Benefits of Robotics in Nigerian Construction: A Comprehensive Analysis. *International Journal of Construction Innovation*, *18*(1), 15–30. DOI: 10.1108/IJCI.2023.0015

Ilesanmi, O.O., Moyanga, D.T., Oke, A.E. and Aliu, J. (2024), "Unraveling the awareness dynamics of smart building features and technologies in the Nigerian construction sector", *International Journal of Building Pathology and Adaptation*, Vol. ahead-of-print No. ahead-of-print. DOI: 10.1108/IJBPA-04-2024-0073

Iliev, D. (2022). Electronic waste management: A review of the limiting factors and robotic solutions. *Engineering 4.0 and the Internet of Everything*, 75.

Inoue, Y. (2020). Satellite-and drone-based remote sensing of crops and soils for smart farming–a review. *Soil Science and Plant Nutrition*, *66*(6), 798–810. DOI: 10.1080/00380768.2020.1738899

Islam, F. S. (2016). Solid waste management system in Dhaka City of Bangladesh. *Journal of Modern Science and Technology*, *4*(1), 192–209.

Islam, M. M., Hasan, M., & Rahman, M. M. (2019). A machine learning-based prediction and analysis of flood affected households: A case study of floods in Bangladesh. *International Journal of Disaster Risk Reduction*, *40*, 101276. DOI: 10.1016/j.ijdrr.2019.101276

Issa, T., Isaias, P., & Issa, T. (2017). *Sustainability, Green IT and Education Strategies in the Twenty-First Century*. Springer International Publishing. DOI: 10.1007/978-3-319-57070-9

Itamiya, T., Tohara, H., & Nasuda, Y. 2019, March. Augmented Reality Floods and Smoke Smartphone App Disaster Scope utilizing Real-time Occlusion. In 2019 IEEE Conference on Virtual Reality and 3D User Interfaces (VR)(pp. 1397-1397). IEEE. DOI: 10.1109/VR.2019.8798269

Jain, R., & Verma, S. (2024). Hydrogen fuel cell vehicles: Future prospects in India's sustainable transportation ecosystem. *Energy & Environmental Science, 12*(3), 300–312.

Janga, B., Asamani, G., Sun, Z., & Cristea, N. (2023). A Review of Practical AI for Remote Sensing in Earth Sciences. *Remote Sensing (Basel), 15*(16), 4112. DOI: 10.3390/rs15164112

Javaid, M., Haleem, A., Singh, R. P., & Suman, R. (2021). Substantial capabilities of robotics in enhancing industry 4.0 implementation. *Cognitive Robotics, 1*, 58–75. DOI: 10.1016/j.cogr.2021.06.001

Jewell, C. 2014. Augmented Reality Could Have Sustainability Applications. https:// www.thefifthestate.com.au/innovation/design/augmented-reality-could -havesustainability-applications/61208. (Accessed 13 May 2018).

Jin, Q., Chen, H., & Hu, F. (2024). Proposal of Industry 5.0-Enabled Sustainability of Product–Service Systems and Its Quantitative Multi-Criteria Decision-Making Method. *Processes (Basel, Switzerland), 12*(3), 473. DOI: 10.3390/pr12030473

Jogarao, M., Lakshmanna, B., & Naidu, S. (2024). *AI-Enabled circular economy management for sustainable smart cities: integrating artificial intelligence in resource optimization and waste reduction. Smart Cities and Circular Economy.* Emerald Publishing Limited.

John, B., Adekunle, S. A., & Aigbavboa, C. (2022). Construction Industry and the Fourth Industrial Revolution: The Key Impediments in Developing Countries. Proc. *Int. Struct. Eng. Constr, 9*(1), 3–14. DOI: 10.14455/ISEC.2022.9(1).CON-29

K v, S., Michael, L. K., Hungund, S. S., & Fernandes, M.KV. (2022). Factors influencing adoption of electric vehicles–A case in India. *Cogent Engineering, 9*(1), 2085375. DOI: 10.1080/23311916.2022.2085375

Kalirajan, M., Mageshen, V. R., Aswitha, K., & Saranya, M. (2024). AI-Based Regulation of Water Supply and Pest Management in Farming. In *Artificial Intelligence Techniques in Smart Agriculture* (pp. 195–215). Springer Nature Singapore. DOI: 10.1007/978-981-97-5878-4_12

Kamal, S. A., & Alam, M. S. (2020). A Framework for Disaster Management in Smart Cities Using IoT. *IEEE Access : Practical Innovations, Open Solutions, 8*, 213897–213908.

Kamarainen, A. M., Metcalf, S., Grotzer, T., Browne, A., Mazzuca, D., Tutwiler, M. S., & Dede, C. (2013). EcoMOBILE: Integrating augmented reality and probeware with environmental education field trips. *Computers & Education*, *68*, 545–556. DOI: 10.1016/j.compedu.2013.02.018

Kambli, R. O. Electric Vehicles in India: Future and Challenges.

Kannan, D., Amiri, A. S., Shaayesteh, M. T., Nasr, A. K., & Mina, H. (2024). Unveiling barriers to the integration of blockchain-based circular economy and Industry 5.0 in manufacturing industries: A strategic prioritization approach. *Business Strategy and the Environment*, *33*(8), 7855–7886. DOI: 10.1002/bse.3886

Kapetas, D., Christakakis, P., Goglia, V., & Pechlivani, E. M. (2025, January). AI-based robotic trap for real-time insect detection, monitoring and population prediction. In 2025 IEEE 6th International Conference on Image Processing, Applications and Systems (IPAS) (pp. 1-7). IEEE.

Kar, B., Pal, T. K., & Bhattacharyya, S. (2021, February). Study of the Reality on Electric Vehicle in Indian Scenario. []). IOP Publishing.]. *IOP Conference Series. Materials Science and Engineering*, *1080*(1), 012003. DOI: 10.1088/1757-899X/1080/1/012003

Katari, V., Goyal, S., Nigam, V., Jana, M., Maitra, A., Sharma, H., & Sahu, K. K. (2024). A Confluence of Emerging Technologies Like IoT, Edge & Cloud Computing, Blockchain, Industry 4.0 & 5.0, AI & ML toward the Realization of Eco-Friendly Supercapacitors. In *Eco-Friendly Supercapacitors: Design and Future Perspectives in Sustainable and Green Energy Storage Devices* (pp. 163-204). American Chemical Society.

Kato, H., & Billinghurst, M. (1999). Marker tracking and HMD calibration for a video-based augmented reality conferencing system. *Proceedings of the 2nd International Workshop on Augmented Reality*, 85–94. DOI: 10.1109/IWAR.1999.803809

Kerawalla, L., Luckin, R., Seljeflot, S., & Woolard, A. (2006). "Making it real": Exploring the potential of augmented reality for teaching primary school science. *Virtual Reality (Waltham Cross)*, *10*(3), 163–174. DOI: 10.1007/s10055-006-0036-4

Kethareswaran, V., & Moulik, S. (2023). Electric Vehicles and the Burning Question: Reasons, Risks, Ramifications and Remedies—An Indian Perspective. *Fire Technology*, *59*(5), 2189–2201. DOI: 10.1007/s10694-023-01453-0

Khaleel, M., Murtaza, N., Mueen, Q. H., Ahmad, S. A., & Qadri, S. F. (2024). Use of AI in conservation and for understanding climate change. In *A Biologist s Guide to Artificial Intelligence* (pp. 201–240). Elsevier., DOI: 10.1016/B978-0-443-24001-0.00013-0

Khanal, S., Medasetti, U. S., Mashal, M., Savage, B., & Khadka, R. (2022). Virtual and Augmented Reality in the Disaster Management Technology: A Literature Review of the Past 11 years. *Frontiers in Virtual Reality*, *3*, 843195. DOI: 10.3389/frvir.2022.843195

Khan, M. A., & Khan, M. A. (2020). A robust framework for disaster management using hybrid cloud. *Journal of Cloud Computing (Heidelberg, Germany)*, *9*(1), 1–14. DOI: 10.1186/2192-113X-1-1

Khan, M. I., Yasmeen, T., Khan, M., Hadi, N. U., Asif, M., Farooq, M., & Al-Ghamdi, S. G. (2025). Integrating industry 4.0 for enhanced sustainability: Pathways and prospects. *Sustainable Production and Consumption*, *54*, 149–189. DOI: 10.1016/j.spc.2024.12.012

Khoshfetrat, R., Sarvari, H., Chan, D. W. M., & Rakhshanifar, M. (2022, September 10). Critical risk factors for implementing building information modeling (BIM): A Delphi-based survey. *International Journal of Construction Management*, *22*(12), 2375–2384. DOI: 10.1080/15623599.2020.1788759

Kissi, E., Aigbavboa, C., & Kuoribo, E. (2023). Emerging technologies in the construction industry: Challenges and strategies in Ghana. *Construction Innovation*, *23*(2), 383–405. DOI: 10.1108/CI-11-2021-0215

Klopfer, E., & Squire, K. (2008). Environmental detectives: The development of an augmented reality platform for environmental simulations. *Educational Technology Research and Development*, *56*(2), 203–228. DOI: 10.1007/s11423-007-9037-6

Kodali, R. K., Yerrapragada, S., & Mandal, A. (2018). An IoT-based disaster management system. *Procedia Computer Science*, *132*, 1258–1265.

Kohli, S. (2024). Electric vehicle demand incentives in India.

Kokate, V. L., Bankar, D. S., Vispi, R. H., & Karandikar, P. B. (2018, July). E-Rickshaw Present Past And Future With Reference to Current Transportation in India. In International Conference on Recent Innovations in Electrical, Electronics & Communication Engineering-(ICRIEECE), Kalinga Institute of Industrial Technology (KIIT), Bhubaneswar, India (Vol. 1, No. 01, pp. 1-15).

Kolb, D. A. (1984). *Experiential learning: Experience as the source of learning and development*. Prentice Hall.

König, P. D., Wurster, S., & Siewert, M. B. (2023). Sustainability challenges of artificial intelligence and Citizens' regulatory preferences. *Government Information Quarterly*, *40*(4), 101863. DOI: 10.1016/j.giq.2023.101863

Konrad Meier & Roger Spichiger. (2024). *The EU AI Act: What it means for your business*. https://www.ey.com/en_ch/insights/forensic-integrity-services/the-eu-ai-act-what-it-means-for-your-business

Konya, A., & Nematzadeh, P. (2024). Recent applications of AI to environmental disciplines: A review. *The Science of the Total Environment*, *906*, 167705. DOI: 10.1016/j.scitotenv.2023.167705 PMID: 37820816

Kostenko, V. V., Bykanova, A. Y., & Tolstonogov, A. Y. (2019). Underwater robotics complex for inspection and laser cleaning of ships from biofouling. *IOP Conference Series. Earth and Environmental Science*, *272*(2), 1–7. DOI: 10.1088/1755-1315/272/2/022103

Kostopoulos, G., Davrazos, G., & Kotsiantis, S. (2024). Explainable Artificial Intelligence-Based Decision Support Systems: A Recent Review. *Electronics (Basel)*, *13*(14), 2842. DOI: 10.3390/electronics13142842

Kouhizadeh, M., Sarkis, J., & Zhu, Q. (2019). At the nexus of blockchain technology, the circular economy, and product deletion. *Applied Sciences (Basel, Switzerland)*, *9*(8), 1712. DOI: 10.3390/app9081712

Krueger, M. W. (1977). Responsive environments. *Proceedings of the National Computer Conference*, 267–277.

Kuhlen, T., & Dohle, C. (1995). Virtual reality for physically disabled people. *Computers in Biology and Medicine*, *25*(2), 205–211. DOI: 10.1016/0010-4825(94)00039-S

Kumar, R., & Padmanaban, S. (2019). Electric vehicles for India: overview and challenges. IEEE India info, 14(2), 139-142.

Kumar, S., & Katoria, D. (2013). Air Pollution and its Control Measures.

Kumar, A., Thomas, J., Wadhwa, S. S., Mishra, A., & Dasgupta, S. (2022). The last mile connectivity: Experience of passengers and drivers of E-rickshaws and cycle rickshaws in an Indian city. *Case Studies on Transport Policy*, *10*(2), 948–953. DOI: 10.1016/j.cstp.2022.03.008

Kumar, H., Bharule, S. P., & Kuntewar, M. (2024). Policy implications of battery swapping systems for three-wheelers in Tier II cities in India. [IJERT]. *International Journal of Engineering Research & Technology (Ahmedabad)*, *13*(3).

Kumar, N. M., Mohammed, M. A., Abdulkareem, K. H., Damasevicius, R., Mostafa, S. A., Maashi, M. S., & Chopra, S. S. (2021). Artificial intelligence-based solution for sorting COVID related medical waste streams and supporting data-driven decisions for smart circular economy practice. *Process Safety and Environmental Protection, 152*, 482–494. DOI: 10.1016/j.psep.2021.06.026

Kumar, P., Singh, R. K., Paul, J., & Sinha, O. (2021). Analyzing challenges for sustainable supply chain of electric vehicle batteries using a hybrid approach of Delphi and Best-Worst Method. *Resources, Conservation and Recycling, 175*, 105879. DOI: 10.1016/j.resconrec.2021.105879

Kumar, R., Lamba, K., & Raman, A. (2021). Role of zero emission vehicles in sustainable transformation of the Indian automobile industry. *Research in Transportation Economics, 90*, 101064. DOI: 10.1016/j.retrec.2021.101064

Kumar, V. A., & Kumar, A. A. (2020). IoT-based disaster management system: A review. *International Journal of Disaster Risk Reduction, 50*, 101735.

Kumar, V., Sharma, K. V., Kedam, N., Patel, A., Kate, T. R., & Rathnayake, U. (2024). A comprehensive review on smart and sustainable agriculture using IoT technologies. *Smart Agricultural Technology, 8*, 100487. DOI: 10.1016/j.atech.2024.100487

Kuppusamy, S., Meivelu, M., Praburaman, L., Mujahid Alam, M., Al-Sehemi, A. G., & K, A. (2024). Integrating AI in food contaminant analysis: Enhancing quality and environmental protection. *Journal of Hazardous Materials Advances, 16*, 100509. DOI: 10.1016/j.hazadv.2024.100509

Labeye, E., Hugot, M., Brusque, C., & Regan, M. A. (2016). The electric vehicle: A new driving experience involving specific skills and rules. *Transportation Research Part F: Traffic Psychology and Behaviour, 37*, 27–40. DOI: 10.1016/j.trf.2015.11.008

Lakhouit, A. (2025). Revolutionizing Urban Solid Waste Management with AI and IoT: A review of smart solutions for waste collection, sorting, and recycling. *Results in Engineering, 25*, 104018. DOI: 10.1016/j.rineng.2025.104018

Lami, B., Hussein, S. M., Rajamanickam, R., & Emmanuel, G. K. (2024). The role of artificial intelligence (AI) in shaping data privacy. International Journal of Law and Management, (ahead-of-print).

Lamsal, R., & Kumar, T. V. V. (2020). Artificial Intelligence and Early Warning Systems. In Kumar, T. V. V., & Sud, K. (Eds.), *AI and Robotics in Disaster Studies* (pp. 13–32). Springer Nature Singapore., DOI: 10.1007/978-981-15-4291-6_2

Langford, R. P., & Gillingham, K. (2023). Quantifying the benefits of the introduction of the hybrid electric vehicle. *International Journal of Industrial Organization*, *87*, 102904. DOI: 10.1016/j.ijindorg.2022.102904

Lanzalonga, F., Oppioli, M., Dal Mas, F., & Secinaro, S. (2023). Drones in Venice: Exploring business model applications for disruptive mobility and stakeholders' value proposition. *Journal of Cleaner Production*, *423*, 138764. DOI: 10.1016/j.jclepro.2023.138764

Lawan, R., Uumar, H. S., & Auwalu, A. (2023). EFFECTS OF AIR POLLUTION ON HUMAN HEALTH: A REVIEW. *International Research Journal of Modernization in Engineering Technology and Science*. Advance online publication. DOI: 10.56726/IRJMETS33055

Leal, V. M., Ribeiro, J. S., Coelho, E. L. D., & Freitas, M. B. J. G. (2022). Recycling of spent lithium-ion batteries as a sustainable solution to obtain raw materials for different applications. *Journal of Energy Chemistry*.

Lee, S. W.-Y., Tsai, C.-C., Wu, Y.-T., Tsai, M.-J., Liu, T.-C., Hwang, F.-K., Lai, C.-H., Liang, J.-C., Wu, H.-C., & Chang, C.-Y. (2011). Internet-based science learning: A review of journal publications. *International Journal of Science Education*, *33*(14), 1893–1925. DOI: 10.1080/09500693.2010.536998

Leesakul, N., Oostveen, A.-M., Eimontaite, I., Wilson, M. L., & Hyde, R. (2022, March 11). Workplace 4.0: Exploring the implications of technology adoption in digital manufacturing on a sustainable workforce. *Sustainability (Basel)*, *14*(6), 3311. DOI: 10.3390/su14063311

Lehtokunnas, T., Mattila, M., & Mesiranta, N. (2020). Towards a circular economy in food consumption: Food waste reduction practices as ethical work. *Journal of Consumer Culture*, *22*(1), 227–245. DOI: 10.1177/1469540520926252

Li, X., Jahani, A., Kokol, P., cnYaolongChen, Guo, Q., Ren, M., Wu, S., Sun, Y., Wang, J., Wang, Q., Ma, Y., & Song, X. Applications of artificial intelligence in the field of air pollution: A bibliometric analysis.

Licardo, J. T., Domjan, M., & Orehovački, T. (2024). Intelligent robotics—A systematic review of emerging technologies and trends. *Electronics (Basel)*, *13*(3), 542. DOI: 10.3390/electronics13030542

Li-Cycle. (2022). *Lithium battery recycling efficiency and recovery rates*. Retrieved from https://www.auto-innovations.net/news/62824-li-cycle-lithium-battery-recycling-efficiency-and-recovery-rates

Li, L.. (2018). Emergency response system for natural disasters based on the Internet of Things. *IEEE Access : Practical Innovations, Open Solutions*, *6*, 21020–21032.

Li, L., Yuan, P., Tang, Y., Cooper, G., Thurlbeck, S., Cheung, C. M., Manu, P., Yunusa-Kaltungo, A., & Weightman, A. (2024). The potential of construction robotics to reduce airborne virus transmission in the construction industry in the UK and China. *Heliyon*, *13*(9), 12–21. DOI: 10.1016/j.heliyon.2024.e29697 PMID: 38694123

Lindqvist, B., Karlsson, S., Koval, A., Tevetzidis, I., Haluška, J., Kanellakis, C., Agha-mohammadi, A., & Nikolakopoulos, G. (2022). Multimodality robotic systems: Integrated combined legged-aerial mobility for subterranean search-and-rescue. *Robotics and Autonomous Systems*, *154*, 104134. DOI: 10.1016/j.robot.2022.104134

Lin, K., Zhao, Y., Kuo, J.-H., Deng, H., Cui, F., Zhang, Z., Zhang, M., Zhao, C., Gao, X., Zhou, T., & Wang, T. (2022). Toward smarter management and recovery of municipal solid waste: A critical review on deep learning approaches. *Journal of Cleaner Production*, *346*, 130943. DOI: 10.1016/j.jclepro.2022.130943

Lin, T. (2023). Electric Vehicle Safety Issues and the Solutions. Highlights in Science. *Engineering and Technology*, *46*, 91–96.

Liu, Z., Liu, T., Liu, X., Wei, A., Wang, X., Yin, Y., & Li, Y. (2021). Research on optimization of healthcare waste management system based on green governance principle in the COVID-19 pandemic. *International Journal of Environmental Research and Public Health*, *18*(10), 5316. DOI: 10.3390/ijerph18105316 PMID: 34067709

Li, W., Xu, J., & Zheng, M. (2018). Green governance: New perspective from open innovation. *Sustainability (Basel)*, *10*(11), 3845. DOI: 10.3390/su10113845

Lo, H. W., Chan, H. W., Lin, J. W., & Lin, S. W. (2024). Evaluating the interrelationships of industrial 5.0 development factors using an integration approach of Fermatean Fuzzy Logic. *Journal of Operations Intelligence*, *2*(1), 95–113. DOI: 10.31181/jopi21202416

Lohani, N. (2024). *AI-based Environmental Sustainability: Transforming Conservation Efforts*. International Journal For Multidisciplinary Research.

Lopez-Medina, I. M., Álvarez-Nieto, C., Grose, J., Elsbernd, A., Huss, N., Huynen, M., & Richardson, J. (2019). Competencies on environmental health and pedagogical approaches in the nursing curriculum: A systematic review of the litera- ture. *Nurse Education in Practice*, *37*, 1–8. DOI: 10.1016/j.nepr.2019.04.004

Lu, W. (2019), "Big data analytics to identify illegal construction waste dumping: a Hong Kong study", Resources, Conservation and Recycling, Vol. 141, pp. 264-272, doi: . 10.039.DOI: 10.1016/j.resconrec.2018

Lu, J. W., Chang, N. B., & Liao, L. (2013). Environmental informatics for solid and hazardous waste management: Advances, challenges, and perspectives. *Critical Reviews in Environmental Science and Technology, 43*(15), 1557–1656. DOI: 10.1080/10643389.2012.671097

Lu, W., & Chen, J. (2022). Computer vision for solid waste sorting: A critical review of academic research. *Waste Management (New York, N.Y.), 142*, 29–43. DOI: 10.1016/j.wasman.2022.02.009

Lu, W., Chen, J., & Xue, F. (2022). Using computer vision to recognize composition of construction waste mixtures: A semantic segmentation approach. *Resources, Conservation and Recycling, 178*, 106022. DOI: 10.1016/j.resconrec.2021.106022

Maali Esfangareh, A. (2022), Adopting AI Techniques In Robotic Fabrication In Architecture: Intelligent Robotic Bricklaying Using Reinforcement Learning Algorithms; (Master's thesis, Middle East Technical University).

Macchione, F., Costabile, P., Costanzo, C., & De Santis, R. (2019). Moving to 3-D flood hazard maps for enhancing risk communication. *Environmental Modelling & Software, 111*, 510–522. DOI: 10.1016/j.envsoft.2018.11.005

Macdonald, S. J., & Clayton, J. (2013). Back to the Future, Disability and the Digital Divide. *Disability & Society, 28*(5), 702–718. DOI: 10.1080/09687599.2012.732538

Macrorie, R., Marvin, S., & While, A. (2021). Robotics and automation in the city: A research agenda. *Urban Geography, 42*(2), 197–217. DOI: 10.1080/02723638.2019.1698868

Maffei, L., & Masullo, M. (2014). Electric vehicles and urban noise control policies. *Archives of Acoustics, 39*(3), 333–341. DOI: 10.2478/aoa-2014-0038

Maffettone, G. C., Liguori, L., Palermo, E., Di Bernardo, M., & Porfiri, M. (2024). Mixed reality environment and high-dimensional continuification control for swarm robotics. *IEEE Transactions on Control Systems Technology*.

Mahbub, R. (2008), An investigation into the barriers to the implementation of Robotics technologies in the construction industry (Doctoral dissertation, Queensland University of Technology).

Majeed, Z. H., & Ali, H. A. (2020). A review of augmented reality in educational applications. *International Journal of Advanced Technology and Engineering Exploration*, *7*(62), 20–27. DOI: 10.19101/IJATEE.2019.650068

Makarova, I., Shubenkova, K., & Pashkevich, A. (2018), "The Concept of the decision support system to plan the reverse logistics in automotive industry", 2018 26th International Conference on Software, Telecommunications and Computer Networks (SoftCOM), IEEE, pp. 1-6. DOI: 10.23919/SOFTCOM.2018.8555760

Malhi, Y., Franklin, J., Seddon, N., Solan, M., Turner, M. G., Field, C. B., & Knowlton, N. (2020). Climate change and ecosystems: Threats, opportunities and solutions. *Philosophical Transactions of the Royal Society B: Biological Sciences, 375*(1794), 20190104. DOI: 10.1098/rstb.2019.0104

Mana, A. A., Allouhi, A., Hamrani, A., Rehman, S., El Jamaoui, I., & Jayachandran, K. (2024). Sustainable AI-based production agriculture: Exploring AI applications and implications in agricultural practices. *Smart Agricultural Technology*, *7*, 100416. DOI: 10.1016/j.atech.2024.100416

Mani, G., Viswanadhapalli, J. K., & Sriramalakshmi, P. (2021, November). AI powered IoT based real-time air pollution monitoring and forecasting. [). IOP Publishing.]. *Journal of Physics: Conference Series*, *2115*(1), 012016. DOI: 10.1088/1742-6596/2115/1/012016

Manirakiza, V., Mugabe, L., Nsabimana, A., & Nzayirambaho, M. (2019). City Profile: Kigali, Rwanda. *Environment & Urbanization Asia*, *10*(2), 290–307. DOI: 10.1177/0975425319867485

Manogaran, B., & Thota, C. (2019). Big Data Analytics for Healthcare and Disaster Management. *Journal of Organizational and End User Computing*, *31*(4), 1–22.

Mao, W.-L., Chen, W.-C., Wang, C.-T., & Lin, Y.-H. (2021). Recycling waste classification using optimized convolutional neural network. *Resources, Conservation and Recycling*, *164*, 105132. DOI: 10.1016/j.resconrec.2020.105132

Mappin, M. J., & Johnson, E. A. (2005). Changing perspectives of ecology and education in environmental education. In Johnson, E. A., & Mappin, M. J. (Eds.), *Environmental education and advocacy* (pp. 1–27). Cambridge University Press.

Marconi, M., Palmieri, G., Callegari, M., & Germani, M. (2019, February). Feasibility Study and Design of an Automatic System for Electronic Components Disassembly. *Journal of Manufacturing Science and Engineering*, *141*(2), 021011. DOI: 10.1115/1.4042006

Marrucci, L., Daddi, T., & Iraldo, F. (2021). The contribution of green human resource management to the circular economy and performance of environmental certified organisations. *Journal of Cleaner Production, 319*, 128859. DOI: 10.1016/j.jclepro.2021.128859

Martín-Gómez, A. M., Agote-Garrido, A., & Lama-Ruiz, J. R. (2024). A Framework for Sustainable Manufacturing: Integrating Industry 4.0 Technologies with Industry 5.0 Values. *Sustainability (Basel), 16*(4), 1364. DOI: 10.3390/su16041364

Martin-Gutierrez, J., Guinters, E., Perez-Lopez, D., 2012. Improving strategy of selflearning in engineering: laboratories with augmented reality. Procedia e Soc. Behav. Sci. 51, 832e839.

Martini, B., Bellisario, D., & Coletti, P. (2024). Human-centered and sustainable artificial intelligence in industry 5.0: Challenges and perspectives. *Sustainability (Basel), 16*(13), 5448. DOI: 10.3390/su16135448

Martin, S., Diaz, G., Sancristobal, E., Gil, R., Castro, M., & Peire, J. (2011). New technology trends in education: Seven years of forecasts and convergence. *Computers & Education, 57*(3), 1893–1906. DOI: 10.1016/j.compedu.2011.04.003

Ma, S., Zhang, Y., Liu, Y., Yang, H., Lv, J., & Ren, S. (2020). Data-driven sustainable intelligent manufacturing based on demand response for energy-intensive industries. *Journal of Cleaner Production, 274*, 123155. DOI: 10.1016/j.jclepro.2020.123155

Masood, A., & Ahmad, K. (2021). A review on emerging artificial intelligence (AI) techniques for air pollution forecasting: Fundamentals, application and performance. *Journal of Cleaner Production, 322*, 129072. DOI: 10.1016/j.jclepro.2021.129072

Masurali, A., & Surya, P. (2024). Driving sustainable mobility: A study of electric vehicle adoption in rural India. *Proceedings of the Design Society: Design Conference, 4*, 1145–1154.

McElwee, P., Turnout, E., Chiroleu-Assouline, M., Clapp, J., Isenhour, C., Jackson, T., Kelemen, E., Miller, D. C., Rusch, G., Spangenberg, J. H., Waldron, A., Baumgartner, R. J., Bleys, B., Howard, M. W., Mungatana, E., Ngo, H., Ring, I., & Santos, R. (2020). Ensuring a Post-COVID Economic Agenda Tackles Global Biodiversity Loss. *One Earth, 3*(4), 448–461. DOI: 10.1016/j.oneear.2020.09.011

Mckenzie, R., Cervello, A., Chabalala, V., Chiraira, I., Mathaha, T., Mayana, L., Njara, N., Tshuma, M. T., Kong, J., Cheng, L., Smith, B., & Mellado, B. (2024, November). A Cost- Effective Air Quality Monitoring System for the Global South. In *2024 IEEE International Humanitarian Technologies Conference (IHTC)* (pp. 1-7). IEEE. DOI: 10.1109/IHTC61819.2024.10855074

Medvedev, A., Fedchenkov, P., Zaslavsky, A., Anagnostopoulos, T., & Khoruzhnikov, S. (2015). Waste management as an IoT-enabled service in smart cities. In *Internet of Things, Smart Spaces, and Next Generation Networks and Systems: 15th International Conference, NEW2AN 2015, and 8th Conference, ruSMART 2015, St. Petersburg, Russia, August 26-28, 2015* [Springer International Publishing.]. *Proceedings, 15*, 104–115.

Meshram, A., Vanalkar, A., Meshram, K., Badar, A., Mehta, G., & Kaushik, V. (2024). Enhancing crop protection through smart autonomous pesticide spray bot in sustainable agriculture. *Engineering Research Express, 6*(4), 045563. DOI: 10.1088/2631-8695/ad9545

Middha, R., Srivastava, N., & Saxena, N. (2024). Chemical Principles in Waste Segregation and Recycling. In *Waste Management for Smart Cities* (pp. 1–46). Springer Nature Singapore.

Miller, D.. (2020). Real-time waste monitoring using AR and IoT: Applications in smart waste management. *Journal of Cleaner Production, 258*, 120758.

Mills, K.. (2020). Geospatial AR in climate education: Real-world data visualization for secondary learners. *Environmental Science Education, 45*(2), 103–118.

Ministry of Environment. Forest & Climate Change <https://moef.gov.in/moef/environment/pollution/>

Mishra, R. K., Giri, V. K., & Tripathi, R. C. (2020). Disaster Management using Cloud and Big Data: A Review. *Procedia Computer Science, 167*, 1340–1350.

Mistry P, Maes P, Chang L (2009) WUW – Wear Ur World – A Wearable Gestural Interface", ACM, CHI 2009, Boston, April 4-9.

Modgil, S., Gupta, S., Sivarajah, U., & Bhushan, B. (2021). Big data-enabled large-scale group decision making for circular economy: An emerging market context. *Technological Forecasting and Social Change, 166*, 120607. DOI: 10.1016/j.techfore.2021.120607

Mohamed, M., Tamil Arasan, G., & Sivakumar, G. (2018). Study on electric vehicles in India opportunities and challenges. *International Journal of Scientific Research in Environmental Science and Toxicology*, •••, 5.

Mohanty, P., & Kotak, Y. (2017). Electric vehicles: Status and roadmap for India. In *Electric vehicles: Prospects and challenges* (pp. 387–414). Elsevier. DOI: 10.1016/B978-0-12-803021-9.00011-2

Mohapatra, B., Mohapatra, S., & Mohapatra, S. (2024). *Automation in the Construction Industry, Process Automation Strategy in Services, Manufacturing and Construction*. Emerald Publishing Limited., DOI: 10.1108/978-1-80455-143-120231018

Monzambe, G. M., Mpofu, K., & Daniyan, I. A. (2019). Statistical analysis of determinant factors and framework development for the optimal and sustainable design of municipal solid waste management systems in the context of industry 4.0. *Procedia CIRP*, *84*, 245–250. DOI: 10.1016/j.procir.2019.04.182

Moores, J. L., Miller, R., & Nichols, M. (2012). Testing Survey Instruments for Reliability: Methods and Challenges. *Journal of Survey Methodology*, *34*(1), 45–62.

Morandín-Ahuerma, F. (2022). What is Artificial Intelligence? *International Journal of Research Publication and Reviews*.

Mourtzis, D., & Angelopoulos, J. (2024). Challenges and Opportunities for the Design and Development of Intelligent, Sustainable and Resilient Personalized Product-Service Systems Towards Industry 5.0. *Data-Driven Decision Making for Product Service Systems*, 27-62.

Mourtzis, D., Angelopoulos, J., & Panopoulos, N. (2022). A literature review of the challenges and opportunities of the transition from industry 4.0 to society 5.0. *Energies*, *15*(17), 6276. DOI: 10.3390/en15176276

Munoz-Cristobal, J. A., Jorrin-Abellan, I. M., Asensio-Perez, J. I., Martinez-Mones, A., Prieto, L. P., & Dimitriadis, Y. (2015). Supporting teacher orchestration in ubiquitous learning environments: A study in primary education. Learning Technologies. *IEEE Transactions on Learning Technologies*, *8*(1), 83–97. DOI: 10.1109/TLT.2014.2370634

Muñoz-Saavedra, L., Miró-Amarante, L., & Domínguez-Morales, M. (2020). Augmented and Virtual Reality Evolution and Future Tendency. *Applied Sciences (Basel, Switzerland)*, *10*(1), 322. DOI: 10.3390/app10010322

Munshi, T., Dhar, S., & Painuly, J. (2022). Understanding barriers to electric vehicle adoption for personal mobility: A case study of middle income in-service residents in Hyderabad city, India. *Energy Policy*, *167*, 112956. DOI: 10.1016/j.enpol.2022.112956

Musarat, M. A., Alaloul, W. S., Zainuddin, S. M. B., Qureshi, A. H., & Maqsoom, A.; (2024), Digitalization in malaysian construction industry: Awareness, challenges and opportunities. *Results in Engineering*; 102 p.013.

Nachiappan, B., Najmusher, H., Nagarajan, G., Rajkumar, N., Loganathan, D., & Gobinath, G. (2024). Exploring the Application of Drone Technology in the Construction Sector. *Salud. Ciencia y Tecnología-Serie de Conferencias*, *3*, 713–715. DOI: 10.56294/sctconf2024713

Nafiz, M. S., Das, S. S., Morol, M. K., Al Juabir, A., & Nandi, D. (2023, January). Convowaste: An automatic waste segregation machine using deep learning. In 2023 3rd International conference on robotics, electrical and signal processing techniques (ICREST) (pp. 181-186). IEEE.

Nagendraswamy, C., & Salis, A. (2021). A review article on artificial intelligence. *Annals of Biomedical Science and Engineering*.

Nahaei, V. S., & Naziri-Oskuei, F. (2021). Fuzzy clustering approach for marketing recycled products of tabriz municipality waste management organization. Int. J. Innov. Market. *Elem.*, *1*(1), 25–36. DOI: 10.59615/ijime.1.1.25

Nallaperuma, D., Nawaratne, R., Bandaragoda, T., Adikari, A., Nguyen, S., Kempitiya, T., De Silva, D., Alahakoon, D., & Pothuhera, D. (2019). Online Incremental Machine Learning Platform for Big Data-Driven Smart Traffic Management. *IEEE Transactions on Intelligent Transportation Systems*, *20*(12), 4679–4690. DOI: 10.1109/TITS.2019.2924883

Narayana, T. L., Venkatesh, C., Kiran, A., Kumar, A., Khan, S. B., Almusharraf, A., & Quasim, M. T. (2024). Advances in real time smart monitoring of environmental parameters using IoT and sensors. *Heliyon*, *10*(7), e28195. DOI: 10.1016/j.heliyon.2024.e28195 PMID: 38571667

Narkhede, G. B., Pasi, B. N., Rajhans, N., & Kulkarni, A. (2024). Industry 5.0 and sustainable manufacturing: A systematic literature review. *Benchmarking*.

Neo, E., Hasikin, K., Mokhtar, M., Lai, K., Azizan, M., Razak, S., & Hizaddin, H. (2022). Towards Integrated Air Pollution Monitoring and Health Impact Assessment Using Federated Learning: A Systematic Review. *Frontiers in Public Health*, *10*, 851553. Advance online publication. DOI: 10.3389/fpubh.2022.851553 PMID: 35664109

Ngcobo, N., & Akinradewo, O. (2023). Revolutionizing high-rise building construction: The role of robotics in South Africa's construction industry. *Journal of Construction Project Management and Innovation*, *13*(2), 93–102.

Nisi, V., Paredes, H., Jardim, M. A., & Nunes, N. (2019). Exploring AR for urban sustainability engagement. *International Journal of Human-Computer Studies*, *130*, 52–65.

Nižetić, S., Djilali, N., Papadopoulos, A., & Rodrigues, J. J. (2019). Smart technologies for promotion of energy efficiency, utilization of sustainable resources and waste management. *Journal of Cleaner Production, 231*, 565–591.

Nkollo, O., & Saidu, I. (2023). Appraising the Benefits and Challenges of Implementing Artificial Intelligence in The Nigerian Construction Industry. *International Journal of Environmental Design and Construction Management, 350*, 6.

Nwokediegwu, Z. Q. S., & Ugwuanyi, E. D. (2024). Implementing ai-driven waste management systems in underserved communities in the USA. *Eng. Sci. Technol. J.*, *5*(3), 794–802. DOI: 10.51594/estj.v5i3.903

Odirichukwu, J. C., Odirichukwu, S. P. C., Okolie, S. A., Njoku, O. A., Dimoji, C., Ndigwe, C., Ekuma, G.O., Onuoha, G., Igwe, F.A., Chika-Ugada, P.K. & Chima, T. Leveraging AI- Driven IoT Systems for Enhanced Air Quality Management in Nigeria: An Impact Examination towards Sustainable Development Goal.

Odulana, A. O., & Oyewobi, L. O. (2019). Effect of implementation of E-procurement on corrupt practices in the Nigerian construction industry. International Conference of Environmental Sciences, ICES. *1st International Conference of the Faculty of Environmental Sciences,* University of Ilorin, Nigeria.

Ojha, S., Geng, W., & Vasilakis, M. (2022). Eliminating Barriers to Robotics Adoption in the US Construction Industry. *Construction Management and Economics, 40*(7-8), 673–684. DOI: 10.1080/01446193.2022.2093743

Oke, A., Aigbavboa, C., & Ebekozien, A. (2023). Exploring Innovative Approaches to Robotics Adoption in the Nigerian Construction Industry. *Journal of Construction Technology and Management, 145*(5), 111–126. DOI: 10.1016/j.jctm.2023.04.015

Olabi, A. G., Abdelkareem, M. A., Wilberforce, T., Alkhalidi, A., Salameh, T., Abo-Khalil, A. G., & Sayed, E. T. (2022). Battery Electric Vehicles: Progress, Power Electronic Converters, Strength (S), Weakness (W), Opportunity (O), and Threats (T). International Journal of Thermofluids, 100212.

Olanipekun, A. O., Aje, I. O., & Fapohunda, J. A. (2015). Critical Success Factors for Construction Projects in Lagos State. *Journal of Project Management, 13*(3), 101–117.

Olaoye, F., Potter, K., & Doris, L. (2024). *Robotics and AI: Synergies and Impact on Automation*. EasyChair.

Olawade, D. B., Fapohunda, O., Wada, O. Z., Usman, S. O., Ige, A. O., Ajisafe, O., & Oladapo, B. I. (2024). *Smart waste management: a paradigm shift enabled by artificial intelligence, Waste Manage*. Bull.

Olawade, D. B., Fapohunda, O., Wada, O. Z., Usman, S. O., Ige, A. O., Ajisafe, O., & Oladapo, B. I. (2024). *Smart waste management: A paradigm shift enabled by artificial intelligence*. Waste Management Bulletin.

Olawade, D. B., Wada, O. Z., David-Olawade, A. C., Fapohunda, O., Ige, A. O., & Ling, J. (2024). Artificial intelligence potential for net zero sustainability: Current evidence and prospects. *Next Sustainability*, *4*, 100041. DOI: 10.1016/j.nxsust.2024.100041

Olawade, D. B., Wada, O. Z., Ige, A. O., Egbewole, B. I., Olojo, A., & Oladapo, B. I. (2024). Artificial intelligence in environmental monitoring: Advancements, challenges, and future directions. *Hygiene and Environmental Health Advances*, *12*, 100114. DOI: 10.1016/j.heha.2024.100114

Oluseye, O., & Aigbavboa, C. (2020)... *Practice-driven Difficulties and Need-driven Plausibility for the Utilization of Robotic and Autonomous Construction Systems in Nigeria.*, *30*, 1–15.

Onoda, H. (2020). Smart approaches to waste management for post-COVID-19 smart cities in Japan. *IET Smart Cities*, *2*(2), 89–94. DOI: 10.1049/iet-smc.2020.0051

Ooko, S., & Rweyemamu, E. (2024). *Monitoring and Predicting African Rural Household Air Pollution Using Internet of Things and Artificial Intelligence*. Pan-African Journal of Health and Environmental Science., DOI: 10.56893/ajhes2024v03i01.06

Ortega-Fernández, A., Martín-Rojas, R., & García-Morales, V. J. (2020). Artificial Intelligence in the Urban Environment: Smart Cities as Models for Developing Innovation and Sustainability. *Sustainability (Basel)*, *12*(19), 7860. DOI: 10.3390/su12197860

Osasona, F., Amoo, O. O., Atadoga, A., Abrahams, T. O., Farayola, O. A., & Ayinla, B. S. (2024). REVIEWING THE ETHICAL IMPLICATIONS OF AI IN DECISION MAKING PROCESSES. *International Journal of Management & Entrepreneurship Research*, *6*(2), 322–335. DOI: 10.51594/ijmer.v6i2.773

Ospina, R., & Noguchi, N. (2025). Real-time work progress estimation based on GIS remote monitoring system for agricultural robot vehicles. *Computers and Electronics in Agriculture*, *234*, 110313. DOI: 10.1016/j.compag.2025.110313

Ouariach, F. Z., & Raghav, A. (2025). Emerging Trends of IoT in Conservation of Biodiversity: A Futuristic Approach. *Machine Learning and Internet of Things in Fire Ecology*, 113-128.

Owojori, O. M., & Okoro, C. (2022). The private sector role as a key supporting stakeholder towards circular economy in the built environment: A scientometric and content analysis. *Buildings (Basel, Switzerland), 12*(5), 695. DOI: 10.3390/buildings12050695

Owolabi, K. A., Okorie, N. C., Yemi-Peters, O. E., Oyetola, S. O., Bello, T. O., & Oladokun, B. D. (2022). Readiness of academic librarians towards the use of robotic technologies in Nigerian university libraries. *Library Management, 43*(3/4), 296–305. DOI: 10.1108/LM-11-2021-0104

Oyedotun, T. D. T., & Moonsammy, S. (2021). Linking national policies to beneficiaries: Geospatial and statistical focus to waste and sanitation planning. *Environmental Challenges, 4*, 100142. DOI: 10.1016/j.envc.2021.100142

Oyejobi, D. O., Firoozi, A. A., Fern'andez, D. B., & Avudaiappan, S. (2024). Integrating circular economy principles into concrete technology: Enhancing sustainability through industrial waste utilization. *Results in Engineering, 24*, 102846. DOI: 10.1016/j.rineng.2024.102846

Ozmen Garibay, O., Winslow, B., Andolina, S., Antona, M., Bodenschatz, A., Coursaris, C., Falco, G., Fiore, S. M., Garibay, I., Grieman, K., Havens, J. C., Jirotka, M., Kacorri, H., Karwowski, W., Kider, J., Konstan, J., Koon, S., Lopez-Gonzalez, M., Maifeld-Carucci, I., & Xu, W. (2023). Six Human-Centered Artificial Intelligence Grand Challenges. *International Journal of Human-Computer Interaction, 39*(3), 391–437. DOI: 10.1080/10447318.2022.2153320

Padhiary, M., Roy, P., & Roy, D. (2025). *"The Future of Urban Connectivity: AI and IoT in Smart Cities," in Sustainable Smart Cities and the Future of Urban Development*. IGI Global Scientific Publishing.

Padhiary, M., Saha, D., Kumar, R., Sethi, L. N., & Kumar, A. (2024). Enhancing precision agriculture: A comprehensive review of machine learning and AI vision applications in all-terrain vehicle for farm automation. *Smart Agricultural Technology, 8*, 100483. DOI: 10.1016/j.atech.2024.100483

Paes, M. X., de Medeiros, G. A., Mancini, S. D., Ribeiro, F. M., & Puppim de Oliveira, J. A. (2019). Transition to circular economy in Brazil: A look at the municipal solid waste management in the state of S~ ao Paulo. *Management Decision, 59*(8), 1827–1840. DOI: 10.1108/MD-09-2018-1053

Pagallo, U. (2025). On Twelve Shades of Green: Assessing the Levels of Environmental Protection in the Artificial Intelligence Act. *Minds and Machines, 35*(1), 10. DOI: 10.1007/s11023-025-09713-4

Pagoropoulos, A., Pigosso, D.C. and McAloone, T.C. (2017), "The emergent role of digital technologies in the Circular Economy: a review", Procedia CIRP, Vol. 64, pp. 19-24, doi: . 02.047.DOI: 10.1016/j.procir.2017

Pallathadka, H., Mustafa, M., Sanchez, D. T., Sajja, G. S., Gour, S., & Naved, M. (2023). Impact of machine learning on management, healthcare and agriculture. *Materials Today: Proceedings*, *80*, 2803–2806. DOI: 10.1016/j.matpr.2021.07.042

Panait, C., Ljubenkov, D., & Alic, D. (2021). *Striking the balance between innovation and regulation in AI – is Europe leading the way or lagging behind? 1*, 27–45.

Panchal, R., Singh, A., & Diwan, H. (2021). Economic potential of recycling e-waste in India and its impact on import of materials. *Resources Policy*, *74*, 102264. DOI: 10.1016/j.resourpol.2021.102264

Pan, M., Linner, T., Pan, W., Cheng, H. M., & Bock, T. (2020). Influencing factors of the future utilization of construction robots for buildings: A Hong Kong perspective. *Journal of Building Engineering*, *30*, 10–20. DOI: 10.1016/j.jobe.2020.101220

Pannu, A. (2015). Artificial Intelligence and its Application in Different Areas.

Papagiannidis, E., Mikalef, P., & Conboy, K. (2025). Responsible artificial intelligence governance: A review and research framework. *The Journal of Strategic Information Systems*, *34*(2), 101885. DOI: 10.1016/j.jsis.2024.101885

Parthsarathy, P., & Bulbule, K. A. (2017). E-Waste Recycling: Best option for Mineral Substitution and Environmental Protection. *Journal of the Geological Society of India*, *89*(2), 221. DOI: 10.1007/s12594-017-0589-8

Patel, S., & Kumar, R. (2024). Analysing the draft Liquid Waste Management Rules 2024: Salient features and way forward. *National Law School of India University, Centre for Environmental Law and Policy.*

Paul, R. A. I., Palanisamy, M. A., Peramaiyan, P., Kumar, V., Bagavathiannan, M., Gurjar, B., Vijayakumar, S., Djanaguiraman, M., Pazhanivelan, S., & Ramasamy, K. (2024). Spray volume optimization with UAV-based herbicide application for effective droplet deposition and weed control in direct-seeded rice. *Frontiers in Agronomy*, *6*, 1491842. DOI: 10.3389/fagro.2024.1491842

Pearson, S., Camacho-Villa, T. C., Valluru, R., Gaju, O., Rai, M. C., Gould, I., Brewer, S., & Sklar, E. (2022). Robotics and autonomous systems for net zero agriculture. *Current Robotics Reports*, *3*(2), 57–64. DOI: 10.1007/s43154-022-00077-6

Pedersen, S. M., & Lind, K. M. (2017). Precision agriculture–from mapping to site-specific application. *Precision agriculture: Technology and economic perspectives*, 1-20.

Peretz-Andersson, E., Tabares, S., Mikalef, P., & Parida, V. (2024). Artificial intelligence implementation in manufacturing SMEs: A resource orchestration approach. *International Journal of Information Management*, *77*, 102781. DOI: 10.1016/j.ijinfomgt.2024.102781

Perey, R., Benn, S., Agarwal, R., & Edwards, M. (2018). The place of waste: Changing business value for the circular economy. *Business Strategy and the Environment*, *27*(5), 631–642. DOI: 10.1002/bse.2068

Perry, J., Klopfer, E., Norton, M., Sutch, D., Sandford, R., & Facer, K. (2008). AR gone wild: Two approaches to using augmented reality learning games in zoos. Proceedings of the 8th International Conference on International C01iference for the Learning Sciences, The Netherlands, (pp. 322-329).

Perucica, N., & Andjelkovic, K. (2022). *Is the future of AI sustainable? A case study of the European Union. Transforming Government: People*. Process and Policy., DOI: 10.1108/TG-06-2021-0106

Picatoste, A., Justel, D., & Mendoza, J. M. F. (2022). Exploring the applicability of circular design criteria for electric vehicle batteries. *Procedia CIRP*, *109*, 107–112. DOI: 10.1016/j.procir.2022.05.222

Pierson, M. E. (2001). Technology Integration Practice as a Function of Pedagogical Expertise. *Journal of Research on Computing in Education*, *33*(4), 413–430. DOI: 10.1080/08886504.2001.10782325

Piron, M., Wu, J., Fedele, A., & Manzardo, A. (2024). Industry 4.0 and life cycle assessment: Evaluation of the technology applications as an asset for the life cycle inventory. *The Science of the Total Environment*, *916*, 170263. DOI: 10.1016/j.scitotenv.2024.170263

Pollak, A., Gupta, A., & Göhlich, D. (2024). Optimized Operation Management With Predicted Filling Levels of the Litter Bins for a Fleet of Autonomous Urban Service Robots. *IEEE Access : Practical Innovations, Open Solutions*, *12*, 7689–7703. DOI: 10.1109/ACCESS.2024.3352436

Poornesh, K., Nivya, K. P., & Sireesha, K. (2020, September). A comparative study on electric vehicle and internal combustion engine vehicles. In *2020 International Conference on Smart Electronics and Communication (ICOSEC)* (pp. 1179-1183). IEEE. DOI: 10.1109/ICOSEC49089.2020.9215386

Popescu, S. M., Mansoor, S., Wani, O. A., Kumar, S. S., Sharma, V., Sharma, A., Arya, V. M., Kirkham, M. B., Hou, D., Bolan, N., & Chung, Y. S. (2024). Artificial intelligence and IoT driven technologies for environmental pollution monitoring and management. *Frontiers in Environmental Science*, *12*, 1336088. DOI: 10.3389/fenvs.2024.1336088

Pradhananga, P., ElZomor, M., & Santi Kasabdji, G. (2021) Identifying the challenges to adopting robotics in the US construction industry. *Journal of Construction Engineering and Management*, ; 147(5), pp. 05-021.

Prajapati, K. C., Patel, R., & Sagar, R. (2014). Hybrid vehicle: A study on technology. [IJERT]. *International Journal of Engineering Research & Technology (Ahmedabad)*, *3*(12), 8.

Prajapati, M. L., Gundaliya, P. J., & Sheth, A. (2020). Case study: Solar electric vehicles in India. *IOP Conference Series. Materials Science and Engineering*, *937*(1), 012033. DOI: 10.1088/1757-899X/937/1/012033

Prakash, J. (2023). Study of environmental sustainability and green manufacturing practices in the Indian automobile industry. *International Journal for Research Publication and Seminar*, *14*(2).

Preece, J., Sharp, H., & Rogers, Y. (2021). *Interaction design: Beyond human-computer interaction* (5th ed.). Wiley.

Preeti, M., & Sayali, A. (2021). Scientometric analysis of research on end-of-life electronic waste and electric vehicle battery waste. *Journal of Scientometric Research*, *10*(1), 37–46. DOI: 10.5530/jscires.10.1.5

Radanliev, P. (2025). AI Ethics: Integrating Transparency, Fairness, and Privacy in AI Development. *Applied Artificial Intelligence*, *39*(1), 2463722. DOI: 10.1080/08839514.2025.2463722

Radu, I. (2014). Augmented reality in education: A meta-review and cross-media analysis. *Personal and Ubiquitous Computing*, *18*(6), 1533–1543. DOI: 10.1007/s00779-013-0747-y

Rafael Araujo & João Mourão. (2023). *The Amazon Domino Effect: How Deforestation Can Trigger Widespread Degradation*. CPI. https://www.climatepolicyinitiative.org/publication/the-amazon-domino-effect-how-deforestation-can-trigger-widespread-degradation/

Ragel, R. G., Caldera, C. S., & Wijesundara, L. A. (2020). A review of the role of machine learning in IoT-based disaster management systems. *Computers & Electrical Engineering*, *89*, 106906.

Raghav, A., Singh, B., Jermsittiparsert, K., Raghav, R., & Yadav, U. (2024). Artificial Intelligence in Environmental and Climate Changes: A Sustainable Future. In *Maintaining a Sustainable World in the Nexus of Environmental Science and AI* (pp. 485-506). IGI Global. DOI: 10.4018/979-8-3693-6336-2.ch019

Rahman, M., & Parvin, R. (2024). Advancing Bangladesh's Digital Transition: From Digital Foundations to Technological Leadership.

Rahman, A.. (2022). AR-based tools for inclusive climate change education in urban classrooms. *Journal of Environmental Studies (Northborough, Mass.)*, *54*(4), 209–222.

Rahman, M. A., Hossain, M. A., & Kaiser, M. S. (2021). Deep Learning for Disaster Response: A Systematic Review. *IEEE Access : Practical Innovations, Open Solutions*, *9*, 144295–144312.

Raihan, A., Paul, A., Rahman, M.S., Islam, S., Paul, P., & Karmakar, S. (2024). Artificial Intelligence (AI) for Environmental Sustainability: A Concise Review of Technology Innovations in Energy, Transportation, Biodiversity, and Water Management. *Journal of Technology Innovations and Energy*.

Raihan, A. (2023). Artificial intelligence and machine learning applications in forest management and biodiversity conservation. *Natural Resources Conservation and Research*, *6*(2), 3825. DOI: 10.24294/nrcr.v6i2.3825

Raj, A. C., Anil, K. S., & Sastry, V. N. (2020). IoT based disaster monitoring and management system for real-time applications. *Procedia Computer Science*, *171*, 1738–1747.

Rajamanikam, A., & Solihin, M. I. (2019). Solid waste bin classification using Gabor wavelet transform. *Innovative Technology and Exploring Engineering*, *8*, 114–117.

Raj, M., Harshini, N. B., Gupta, S., Atiquzzaman, M., Rawlley, O., & Goel, L. (2024). Leveraging precision agriculture techniques using UAVs and emerging disruptive technologies. *Energy Nexus*, *14*, 100300. DOI: 10.1016/j.nexus.2024.100300

Rajumesh, S. (2024). Promoting sustainable and human-centric industry 5.0: A thematic analysis of emerging research topics and opportunities. *Journal of Business and Socio-economic Development*, *4*(2), 111–126. DOI: 10.1108/JBSED-10-2022-0116

Rakhio, A. (2024). Research advancements in recycling: YOLOV4 and darknet-powered object detection of hazardous items. *ParadigmPlus*, *5*(1), 1–11. DOI: 10.55969/paradigmplus.v5n1a1

Ramadan, M. N., Ali, M. A., Khoo, S. Y., Alkhedher, M., & Alherbawi, M. (2024). Real-time IoT- powered AI system for monitoring and forecasting of air pollution in industrial environment. *Ecotoxicology and Environmental Safety*, *283*, 116856. DOI: 10.1016/j.ecoenv.2024.116856 PMID: 39151373

Rame, R., Purwanto, P., & Sudarno, S. (2024). Industry 5.0 and sustainability: An overview of emerging trends and challenges for a green future. *Innovation and Green Development*, *3*(4), 100173. DOI: 10.1016/j.igd.2024.100173

Ramírez-Márquez, C., Posadas-Paredes, T., Raya-Tapia, A. Y., & Ponce-Ortega, J. M. (2024). Natural Resource Optimization and Sustainability in Society 5.0: A Comprehensive Review. *Resources*, *13*(2), 19. DOI: 10.3390/resources13020019

Rana, S., Goyal, P., & Bansal, P. (2021). IoT and Machine Learning-Based Disaster Management System. *Journal of King Saud University. Computer and Information Sciences.*

Randhawa, P. (2012). Delhi metro rail: Beyond mass transit. *Economic and Political Weekly*, •••, 25–29.

Rane, N (2023); Integrating Building Information Modelling (BIM) and Artificial Intelligence (AI) for Smart Construction Schedule, Cost, Quality, and Safety Management: Challenges and Opportunities. Cost, Quality, and Safety Management: Challenges and Opportunities (September 16, 2023).

Rane, N. L., Kaya, Ö., & Rane, J. (2024). *Artificial Intelligence, Machine Learning, and Deep Learning for Sustainable Industry 5.0.* Deep Science Publishing.

Rane, N., Kaya, Ö., & Rane, J. (2024). Digital Twins for Industry 4.0, 5.0, and Society 5.0. *Available atSSRN* 4918575. DOI: 10.2139/ssrn.4918575

Rani, B., Singh, U., Maheshwari, R.K., & Garh, L. (2011). Menace of Air Pollution Worldwide.

Rautela, K. S., & Goyal, M. K. (2024). Transforming air pollution management in India with AI and machine learning technologies. *Scientific Reports*, *14*(1), 20412. DOI: 10.1038/s41598-024-71269-7 PMID: 39223178

Rawat, A., Garg, C. P., & Sinha, P. (2024). Analysis of the key hydrogen fuel vehicles adoption barriers to reduce carbon emissions under net zero target in emerging markets. *Energy Policy*, *184*, 113847. DOI: 10.1016/j.enpol.2023.113847

Raworth, K. (2017). *Doughnut Economics: Seven Ways to Think like a 21st-Century Economist*. Chelsea Green Publishing.

Regona, M., Yigitcanlar, T., Xia, B., & Li, R. Y. M. (2022). Opportunities and adoption challenges of AI in the construction industry: A PRISMA review. *Journal of Open Innovation*, *8*(1), 45. DOI: 10.3390/joitmc8010045

Richter, J. L. (2022). A circular economy approach is needed for electric vehicles. *Nature Electronics*, *5*(1), 5–7. DOI: 10.1038/s41928-021-00711-9

Rishab Bailey & Jeanne Huang. (2024, June 18). *Briefing Note: AI Regulation in the European Union and China*. Digital Trade Alliance. https://dtalliance.org/2024/06/18/briefing-note-ai-regulation-in-the-european-union-and-china/

Rokadiya, S., & Bandivadekar, A. (2016). *Hybrid and electric vehicles in India: current scenario and market incentives. The International Council on Clean Transportation*. ICCT.

Romano, D., & Stefanini, C. (2025). Animal–robot interaction induces local enhancement in the Mediterranean fruit fly Ceratitis capitata Wiedemann. *Bioinspiration & Biomimetics*, *20*(3), 036009. DOI: 10.1088/1748-3190/adbb42 PMID: 40014926

Rosenbaum, E., Klopfer, E., & Perry, J. (2007). On location learning: Authentic applied science with networked augmented realities. *Journal of Science Education and Technology*, *16*(1), 31–45. DOI: 10.1007/s10956-006-9036-0

Roy, A., Tazyeen, S. M. F., Sarker, S. R., & Hossain, M. S. (2021). An IoT based real-time disaster monitoring system using deep learning. *Computers, Materials & Continua*, *69*(2), 1745–1762.

Roy, H., Alam, S. R., Bin-Masud, R., Prantika, T. R., Pervez, M. N., Islam, M. S., & Naddeo, V. (2022). A review on characteristics, techniques, and waste-to-energy aspects of municipal solid waste management: Bangladesh perspective. *Sustainability (Basel)*, *14*(16), 10265. DOI: 10.3390/su141610265

Roy, M. (2009). Planning for sustainable urbanisation in fast growing cities: Mitigation and adaptation issues addressed in Dhaka, Bangladesh. *Habitat International*, *33*(3), 276–286. DOI: 10.1016/j.habitatint.2008.10.022

Roy, S., Rautela, R., & Kumar, S. (2023). Towards a sustainable future: Nexus between the sustainable development goals and waste management in the built environment. *Journal of Cleaner Production*, *415*, 137865. DOI: 10.1016/j.jclepro.2023.137865

Rujanavech, C., Lessard, J., Chandler, S., Shannon, S., Dahmus, J., & Guzzo, R. "*Liam - An Innovation Story*," p. 8, 2016.

Rumayor, M., Margallo, M., Pinedo, J., & Albo, J. (2024). Learning by Doing using the Life Cycle Assessment tool: LCA Projects in collaboration with industries. *Education for Chemical Engineers*, *48*, 44–52. DOI: 10.1016/j.ece.2024.05.002

Rutqvist, D., Kleyko, D., & Blomstedt, F. (2020). An automated machine learning approach for smart waste management systems. *IEEE Transactions on Industrial Informatics*, *16*(1), 384–392. DOI: 10.1109/TII.2019.2915572

Sadraei, R., Biancone, P., Lanzalonga, F., Jafari-Sadeghi, V., & Chmet, F. (2023). How to increase sustainable production in the food sector? Mapping industrial and business strategies and providing future research agenda. *Business Strategy and the Environment*, *32*(4), 2209–2228. DOI: 10.1002/bse.3244

Saha, A., Ahmed, M., & Rahman, A. (2023). Flood forecasting by using machine learning: A study leveraging historic climatic records of Bangladesh. *Water (Basel)*, *15*(5), 887. DOI: 10.3390/w15050887

Saha, S., Bhattacharjee, S., Shit, P. K., Sengupta, N., & Bera, B. (2022). Deforestation probability assessment using integrated machine learning algorithms of Eastern Himalayan foothills (India). *Resources, Conservation & Recycling Advances*, *14*, 200077. DOI: 10.1016/j.rcradv.2022.200077

Said, Z., Sharma, P., Thi Bich Nhuong, Q., Bora, B. J., Lichtfouse, E., Khalid, H. M., Luque, R., Nguyen, X. P., & Hoang, A. T. (2023). Intelligent approaches for sustainable management and valorisation of food waste. *Bioresource Technology*, *377*, 128952. DOI: 10.1016/j.biortech.2023.128952 PMID: 36965587

Salem, K. S., Clayson, K., Salas, M., Haque, N., Rao, R., Agate, S., Singh, A., Levis, J. W., Mittal, A., Yarbrough, J. M., Venditti, R., Jameel, H., Lucia, L., & Pal, L. (2023). A critical review of existing and emerging technologies and systems to optimize solid waste management for feedstocks and energy conversion. *Matter*, *6*(10), 3113–p3684. DOI: 10.1016/j.matt.2023.08.003

Sales, S. (2024). Emerging Technologies in the Nigerian Construction Industry: The Role of Robotics. *Construction Innovation*, *19*(4), 351–369.

Sanchez-Calleja, I., Martínez-Gutiérrez, A., Ferrero-Guillén, R., Díez-González, J., & Perez, H. (2024). Contact system method for the precise interaction between cobots and mobile robots in smart manufacturing. *International Journal of Precision Engineering and Manufacturing*, *25*(2), 303–318. DOI: 10.1007/s12541-023-00907-3

Sanguesa, J. A., Torres-Sanz, V., Garrido, P., Martinez, F. J., & Marquez-Barja, J. M. (2021). A review on electric vehicles: Technologies and challenges. *Smart Cities*, *4*(1), 372–404. DOI: 10.3390/smartcities4010022

Sani, O. G., Azizi, A., & Malek, M. A. (2020). Assessing and mapping multi-hazard risk susceptibility using a machine learning technique. *Scientific Reports, 10*(1), 19620. DOI: 10.1038/s41598-020-76257-0 PMID: 33184353

Santos, B., Costa, R. L. C., & Santos, L. (2024). Cybersecurity in Industry 5.0: Open Challenges and Future Directions. *arXiv preprint arXiv:2410.09538*. DOI: 10.1109/PST62714.2024.10788065

Santos, R. C., M., & Cagica Carvalho, L. (. (2025). AI-driven participatory environmental management: Innovations, applications, and future prospects. *Journal of Environmental Management, 373*, 123864. DOI: 10.1016/j.jenvman.2024.123864 PMID: 39752951

Sarc, R., Curtis, A., Kandlbauer, L., Khodier, K., Lorber, K. E., & Pomberger, R. (2019). Digitalisation and intelligent robotics in value chain of circular economy oriented waste management – A review. *Waste Management (New York, N.Y.), 95*, 476–492. DOI: 10.1016/j.wasman.2019.06.035 PMID: 31351634

Sarkar, B. D., Shardeo, V., Dwivedi, A., & Pamucar, D. (2024). Digital transition from industry 4.0 to industry 5.0 in smart manufacturing: A framework for sustainable future. *Technology in Society, 78*, 102649. DOI: 10.1016/j.techsoc.2024.102649

Sarode, R., & Eadke, A. (2020). Electric Vehicle (Construction and Working principle). *Int. Res. J. Eng. Technol, 7*(10), 140–142.

Satheesh, S., & Shukla, K. (2024). Industry 5.0, Quality of Life, and Sustainable Adaptation to Climate Change: Challenges and the Transformational Way Forward. In *Quality of Life and Climate Change: Impacts, Sustainable Adaptation, and Social-Ecological Resilience* (pp. 200-219). IGI Global.

Saxena, P., Saxena, V., Pandey, A., Flato, U. a. P., & Shukla, K. (2023). Multiple aspects of artificial intelligence. https://doi.org/DOI: 10.60148/muasartificialintelligence

Schoch, M., Fournier De Lauriere, C., & Bernauer, T. (2025). Monitoring Urban Air Pollution in the Global South: Large Gaps Associated with Economic Conditions and Political Institutions. bioRxiv, 2025-03.

Secinaro, S., Brescia, V., Calandra, D., & Biancone, P. (2021). Towards a hybrid model for the management of smart city initiatives. *Cities (London, England), 116*, 103278. DOI: 10.1016/j.cities.2021.103278

Sermet, Y., & Demir, I. (2020). Virtual and augmented reality applications for environmental science education and training. In *New perspectives on virtual and augmented reality* (pp. 261–275). Routledge. DOI: 10.4324/9781003001874-17

Seyman-Guray, T., & Kismet, B. (2023). Drivers and barriers to implementing XR technologies in the construction industry in Turkey. *International Journal of Construction Management*, *120*, 1–16.

Shakirova, N., Berechikidze, I., & Gafiyatullina, E. (2023). The effects of immersive AR technology on the environmental literacy, intrinsic motivation, and cognitive load of high school students. *Education and Information Technologies*, *29*, 6121–9138.

Shakur, M. S., Lubaba, M., Debnath, B., Bari, A. B. M. M., & Rahman, M. A. (2024, March 05). Exploring the Challenges of Industry 4.0 Adoption in the FMCG Sector: Implications for Resilient Supply Chain in Emerging Economy. *Logistics (Basel)*, *8*(1), 27. DOI: 10.3390/logistics8010027

Shankar, A. U., Tewari, V., Rahman, M., Mishra, A., & Bajaj, K. K. (2023). Impact of Virtual Reality (VR) and Augmented Reality (AR) in education. Tuijin Jishu/ Journal of Propulsion Technology, 44(4), 1310-1318. DOI: DOI: 10.52783/tjjpt.v44.i4.1014

Sharma, A., Singh, M., & Chauhan, V. (2020). Big data and machine learning in disaster management: A survey. *Procedia Computer Science*, *167*, 561–570.

Sharma, H., Haque, A., & Blaabjerg, F. (2021). Machine learning in wireless sensor networks for smart cities: A survey. *Electronics (Basel)*, *10*(9), 1012. DOI: 10.3390/electronics10091012

Sharma, K., & Shivandu, S. K. (2024). Integrating artificial intelligence and internet of things (IoT) for enhanced crop monitoring and management in precision agriculture. *Sensors International*, *5*, 100292. DOI: 10.1016/j.sintl.2024.100292

Sharma, N., Singh, A., Dhyani, R., & Gaur, S. (2014). Emission reduction from MRTS projects–a case study of Delhi metro. *Atmospheric Pollution Research*, *5*(4), 721–728. DOI: 10.5094/APR.2014.081

Sharma, P., Vimal, A., Vishvakarma, R., Kumar, P., De Souza Vandenberghe, L. P., Kumar Gaur, V., & Varjani, S. (2022). Deciphering the blackbox of omics approaches and artificial intelligence in food waste transformation and mitigation. *International Journal of Food Microbiology*, *372*, 109691. DOI: 10.1016/j.ijfoodmicro.2022.109691 PMID: 35509146

Sharma, R., & Gupta, H. (2024). Harmonizing sustainability in industry 5.0 era: Transformative strategies for cleaner production and sustainable competitive advantage. *Journal of Cleaner Production*, *445*, 141118. DOI: 10.1016/j.jclepro.2024.141118

Sharma, R., & Gupta, H. (2024). Leveraging cognitive digital twins in industry 5.0 for achieving sustainable development goal 9: An exploration of inclusive and sustainable industrialization strategies. *Journal of Cleaner Production*, *448*, 141364. DOI: 10.1016/j.jclepro.2024.141364

Shet, H. N., & Moholkar, V. S. (2025). Comparative life cycle analysis of hydrogen fuel cell electric vehicles and battery electric vehicles: An Indian perspective. *International Journal of Hydrogen Energy*, *103*, 729–739. DOI: 10.1016/j.ijhydene.2025.01.254

Shi, Y., Sun, C., Yu, Y., & Chen, G. (2020). A hybrid machine learning model for flood prediction. *Applied Sciences (Basel, Switzerland)*, *10*(3), 1016.

Shojaei, R. S., & Burgess, G. (2022). Non-technical inhibitors: Exploring the adoption of digital innovation in the UK construction industry. *Technological Forecasting and Social Change*, *185*, 12–20. DOI: 10.1016/j.techfore.2022.122036

Shrestha, N. (2021). Factor analysis as a tool for survey analysis. *American Journal of Applied Mathematics and Statistics*, *9*(1), 4–11. DOI: 10.12691/ajams-9-1-2

Shreyas Madhav, A., Rajaraman, R., Harini, S., & Kiliroor, C. C. (2022, July). Application of artificial intelligence to enhance collection of E-waste: A potential solution for household WEEE collection and segregation in India. *Waste Management & Research*, *40*(7), 1047–1053. DOI: 10.1177/0734242X211052846 PMID: 34726090

Shrimali, G. (2021). Getting to India's electric vehicle targets cost-effectively: To subsidize or not, and how? *Energy Policy*, *156*, 112384. DOI: 10.1016/j.enpol.2021.112384

Shuford, J. (2024). Interdisciplinary Perspectives: Fusing Artificial Intelligence with Environmental Science for Sustainable Solutions. *Journal of Artificial Intelligence General science (JAIGS) ISSN:3006-4023.*

Singh, S. P., Sharma, N., Chandrakant, S. A., & Singh, S. P. (2021). Electric Vehicles in India: A Literature Review.

Singhal, A., Neveditsin, N., Tanveer, H., & Mago, V. (2024). Toward Fairness, Accountability, Transparency, and Ethics in AI for Social Media and Health Care: Scoping Review. *JMIR Medical Informatics*, *12*, e50048. DOI: 10.2196/50048 PMID: 38568737

Singh, D., & Satija, A. (2018). Prediction of municipal solid waste generation for optimum planning and management with artificial neural network—case study: Faridabad in Haryana State (India). *International Journal of System Assurance Engineering and Management*, *9*(1), 91–97. DOI: 10.1007/s13198-016-0484-5

Singh, R., Mishra, S., & Tripathi, K. (2021). Analysing acceptability of E-rickshaw as a public transport innovation in Delhi: A responsible innovation perspective. *Technological Forecasting and Social Change*, *170*, 120908. DOI: 10.1016/j.techfore.2021.120908

Singh, V., Singh, V., & Vaibhav, S. (2021). Analysis of electric vehicle trends, development and policies in India. *Case Studies on Transport Policy*, *9*(3), 1180–1197. DOI: 10.1016/j.cstp.2021.06.006

Singla, A., & Bansal, R. SUSTAINABILITY OF ELECTRIC VEHICLES.

Sipola, J., Saunila, M., & Ukko, J. (2023). Adopting artificial intelligence in sustainable business. *Journal of Cleaner Production*, *426*, 139197. DOI: 10.1016/j.jclepro.2023.139197

Sivapalan, M., Savenije, H. H., & Blöschl, G. (2012). Socio-hydrology: A new science of people and water. *Hydrological Processes*, *26*(8), 1270–1276. DOI: 10.1002/hyp.8426

Sjödin, D., Parida, V., & Kohtamäki, M. (2023). Artificial intelligence enabling circular business model innovation in digital servitization: Conceptualizing dynamic capabilities, AI capacities, business models and effects. *Technological Forecasting and Social Change*, *197*, 122903. DOI: 10.1016/j.techfore.2023.122903

Sláčala, J., Špačková, H., Menšík, M., Svobodová, P., & Barčák, T. "Use of robotic manipulation systems in automation of FORGING processes," presented at the METAL 2020, 2020, pp. 1200–1204. DOI: 10.37904/metal.2020.3662

Smith, J.. (2020). Augmented reality applications for enhancing recycling behavior: A pilot study. *Journal of Environmental Psychology*, *68*, 101382.

Song, C., & Cui, W. (2020). Review of underwater ship hull cleaning technologies. *Journal of marine science and application, 19*(3), 415-429.

Song, X. P., Liang, Y. J., Zhang, X. Q., Qin, Z. Q., Wei, J. J., Li, Y. R., & Wu, J. M. (2020). Intrusion of fall armyworm (*Spodoptera frugiperda*) in sugarcane and its control by drone in China. *Sugar Tech*, *22*(4), 734–737. DOI: 10.1007/s12355-020-00799-x

Soni, S., Kumar, C. K., Rawat, S. S., & Prajapat, Y. K. (2020). A REVIEW ON AIR POLUTION. *International Journal of Advance Research and Innovative Ideas in Education*, *6*, 460–464.

Sonone, S., & Dharme, A. (2019). A review paper on simulated intellect. *International Journal of Physics and Mathematics*.

Soori, M., Jough, F. K. G., Dastres, R., & Arezoo, B. (2024). AI-Based Decision Support Systems in Industry 4.0, A Review. *Journal of Economy and Technology, S2949948824000374*. Advance online publication. DOI: 10.1016/j.ject.2024.08.005

Sosunova, I., & Porras, J. (2022). IoT-enabled smart waste management systems for smart cities: A systematic review. *IEEE Access : Practical Innovations, Open Solutions, 10*, 73326–73363. DOI: 10.1109/ACCESS.2022.3188308

Spessert, B. M., Fischer, M., & Kühn, B. (2019, September). Combustion Engine Noise-a Historical Review. In *INTER-NOISE and NOISE-CON Congress and Conference Proceedings* (Vol. 259, No. 9, pp. 347-358). Institute of Noise Control Engineering.

Spinosa, L. M., & Costa, E. M. (2020). Urban innovation ecosystem & humane and sustainable smart city: A balanced approach in Curitiba. *Handbook of smart cities*, 1-23.

Sreenivas, A. G., Ranganath, , Wazid, , Vijayalakshmi, , & Shivayogiyappa, . (2024). Unmanned aerial vehicle (UAVs): A novel spraying technique for management of sucking insect pests in cotton ecosystem. *Journal of Biosystems Engineering, 49*(2), 103–111. DOI: 10.1007/s42853-024-00219-y

Srikanth, R. (2018). Role of electric mobility in a sustainable, and energy-secure future for India. *Current Science, 114*(4), 732–739. DOI: 10.18520/cs/v114/i04/732-739

Stahel, W. R. (2016). The circular economy. *Nature, 531*(7595), 435–438. DOI: 10.1038/531435a

Subbiah, S. S., & Paramasivan, S. K. (2024). Prediction of air quality pollutants using artificial intelligence techniques: A review. *WOMEN IN PHYSICS: 7th IUPAP International Conference on Women in Physics*. DOI: 10.1063/5.0183240

Subramanian, A. K., Thayalan, D., Edwards, A. I., Almalki, A., & Venugopal, A. (2021). Biomedical waste management in dental practice and its significant environmental impact: A perspective. *Environmental Technology & Innovation, 24*, 101807. DOI: 10.1016/j.eti.2021.101807

Sultana, M. T. N., Ahmed, F. U., & Islam, M. R. (2021). A review on disaster management using big data analytics and machine learning. *Journal of King Saud University. Computer and Information Sciences*.

Sundaralingam, S., & Ramanathan, N. (2023). A deep learning-based approach to segregate solid waste generated in residential areas. *Eng. Technol. Appl. Sci. Res., 13*(2), 10439–10446. DOI: 10.48084/etasr.5716

Sun, J., Guan, X., Zeng, Y., Zhang, J., Chen, X., & Zhan, X. (2024). Addressing pollution challenges for enterprises under diverse extreme climate conditions: Artificial intelligence- driven experience and policy support of top Chinese enterprises. *Frontiers in Public Health*, *12*, 1436304. Advance online publication. DOI: 10.3389/fpubh.2024.1436304 PMID: 39301513

Sun, P., Bisschop, R., Niu, H., & Huang, X. (2020). A review of battery fires in electric vehicles. *Fire Technology*, *56*(4), 1361–1410. DOI: 10.1007/s10694-019-00944-3

Sun, W., Bocchini, P., & Davison, B. D. (2020). Applications of artificial intelligence for disaster management. *Natural Hazards*, *103*(3), 2631–2689. DOI: 10.1007/s11069-020-04124-3

Sutherland, I. (1968). A head-mounted three-dimensional display. *Proceedings of the AFIPS Fall Joint Computer Conference, 33*, 757–764.

Sutikno, T., Fathurrahman, H. I. K., Wahono, T., & Handayani, L. (2024). Robotics Technologies in Urban Smart Waste Management. *High-tech and Innovative Series*, *1*, 43–54.

Suto, K., & Oie, Y. (2019). Anomaly detection for disaster management using mobile social networking services. *Sensors (Basel)*, *19*(6), 1362. PMID: 30893769

Swapan, M. S. H., Zaman, A. U., Ahsan, T., & Ahmed, F. (2017). Transforming urban dichotomies and challenges of South Asian megacities: Rethinking sustainable growth of Dhaka, Bangladesh. *Urban Science (Basel, Switzerland)*, *1*(4), 31. DOI: 10.3390/urbansci1040031

Szpilko, D., de la Torre Gallegos, A., Jimenez Naharro, F., Rzepka, A., & Remiszewska, A. (2023). Waste management in the smart city: Current practices and future directions. *Resources*, *12*(10), 115. DOI: 10.3390/resources12100115

Szpilko, D., Naharro, F. J., Lăzăroiu, G., Nica, E., & Gallegos, A. D. L. T. (2023). Artificial Intelligence in the Smart City—A Literature Review. *Engineering Management in Production and Services*, *15*(4), 53–75. DOI: 10.2478/emj-2023-0028

Taherdoost, H. (2019). What Is the Best Response Scale for Survey and Questionnaire Design; Review of Different Lengths of Rating Scale / Attitude Scale / Likert Scale'. *International Journal of Academic Research in Management*, *8*(1), 1–10.

Tang, M., & Liao, H. (2021). multi-attribute large-scale group decision making with data mining and subgroup leaders: An application to the development of the circular economy. *Technological Forecasting and Social Change*, *167*, 120719. DOI: 10.1016/j.techfore.2021.120719

Taseer, A., & Han, X. (2024). Advancements in variable rate spraying for precise spray requirements in precision agriculture using Unmanned aerial spraying Systems: A review. *Computers and Electronics in Agriculture*, *219*, 108841. DOI: 10.1016/j.compag.2024.108841

Teófilo, M., Lucena, V. F., Nascimento, J., Miyagawa, T., & Maciel, F. "Evaluating accessibility features designed for virtual reality context," *2018 IEEE International Conference on Consumer Electronics (ICCE)*, 2018, pp. 1-6. DOI: 10.1109/ICCE.2018.8326167

Thakur, J., de Almeida, C. M. L., & Baskar, A. G. (2022). Electric vehicles batteries for a circular economy: Second life batteries as stationary storage. *Journal of Cleaner Production*, *375*, 134066. DOI: 10.1016/j.jclepro.2022.134066

Thew H, Graves C, Reay D, Smith S, Petersen K, et al. Mainstreaming climate change education in UK higher education institutions 2021.

Thirumal, S., Udawatta, N., Karunasena, G., & Al-Ameri, R. (2024). Barriers to Adopting Digital Technologies to Implement Circular Economy Practices in the Construction Industry: A Systematic Literature Review. *Sustainability (Basel)*, *16*(8), 3185. DOI: 10.3390/su16083185

Tijani, B., & Feng, Y. (2021). Social impacts of adopting robotics in the construction industry: A systematic literature review. *InProceedings of the 23rd International Symposium on Advancement of Construction Management and Real Estate*; pp. 668-680. Springer Singapore. DOI: 10.1007/978-981-15-3977-0_51

Tonle, F. B., Niassy, S., Ndadji, M. M., Tchendji, M. T., Nzeukou, A., Mudereri, B. T., Senagi, K., & Tonnang, H. E. (2024). A road map for developing novel decision support system (DSS) for disseminating integrated pest management (IPM) technologies. *Computers and Electronics in Agriculture*, *217*, 108526. DOI: 10.1016/j.compag.2023.108526

Trevelyan, J. P., Kang, S.-C., & Hamel, W. R. (2008). Robotics in Hazardous Applications. In Siciliano, B., & Khatib, O. (Eds.), *Springer Handbook of Robotics* (pp. 1101–1126). Springer Berlin Heidelberg., DOI: 10.1007/978-3-540-30301-5_49

Tripathy, A. K., & Kumar, S. (2021). IoT and cloud-based disaster management system: A review. *Materials Today: Proceedings*, *47*, 1908–1913.

Tripathy, A. K., Sahoo, A., Mohanty, R., & Satapathy, S. (2021). An integrated approach to disaster management using IoT and machine learning. *International Journal of Information Management*, *57*, 102196.

Tyagi, S. (2024). A CRITICAL ASSESSMENT OF THE DIGITAL PERSONAL DATA PROTECTION ACT IN LIGHT OF THE GDPR FRAMEWORK - Sunidhi Tyagi. *Ijalr.* https://ijalr.in/volume-4-issue-3/a-critical-assessment-of-the-digital-personal-data-protection-act-in-light-of-the-gdpr-framework-sunidhi-tyagi/

UNDP. (2024). *The Role of International Cooperation in the Responsible Use of AI Usage in Serbia and Globally | United Nations Development Programme*. https://www.undp.org/serbia/news/role-international-cooperation-responsible-use-ai-usage-serbia-and-globally

UNESCO. (2014). *Roadmap for implementing the global 760 action programme on education for sustainable development*. UNESCO.

United Nations. (2015). *SDGs- Sustainable Development Goals*. United Nations.

United Nations. (2023). *The Sustainable Development Goals Report 2023*. United Nations.

Upadhyay, A., Mukhuty, S., Kumar, V., & Kazancoglu, Y. (2021). Blockchain technology and the circular economy: Implications for sustainability and social responsibility. *Journal of Cleaner Production*, *293*, 126130. DOI: 10.1016/j.jclepro.2021.126130

Uwaoma, P., Eleogu, T., Okonkwo, F., Farayola, O., Kaggwa, S., & Akinoso, A. (2024). AI's Role in Sustainable Business Practices and Environmental Management. *International Journal of Research and Scientific Innovation*. https://doi.org/.DOI: 10.51244/IJRSI.2023.1012029

Vacchi, M., Siligardi, C., & Settembre-Blundo, D. (2024). Driving manufacturing companies toward industry 5.0: A strategic framework for process technological sustainability assessment (P-TSA). *Sustainability (Basel)*, *16*(2), 695. DOI: 10.3390/su16020695

Valeriya, G., Kirola, M., Gupta, M., Bharathi, P., & Acharya, P. (2024). Evaluating the Impact of AI-Based Sustainability Measures in Industry 5.0: A Longitudinal Study. In *BIO Web of Conferences* (Vol. 86, p. 01058). EDP Sciences.

Van Tam, N., Toan, N. Q., & Van Phong, V. (2024). Investigating potential barriers to construction digitalization in emerging economies: A study in Vietnam. *International Journal of Information Management Data Insights*, *4*(1), 10–26. DOI: 10.1016/j.jjimei.2024.100226

Veas, E., Grasset, R., Ferencik, I., Grünewald, T., & Schmalstieg, D. (2013). Mobile augmented reality for environmental monitoring. *Personal and Ubiquitous Computing*, *17*(7), 1515–1531. DOI: 10.1007/s00779-012-0597-z

Venigandla, K., Vemuri, N., & Aneke, E. N. (2024). Empowering Smart Cities with AI and RPA: Strategies for Intelligent Urban Management and Sustainable Development. *Valley Int. J. Digit. Libr., 2024*(4), 1117–1125. DOI: 10.18535/ijsrm/v12i04.ec02

Verma, V. K., Gupta, A. K., Singh, A. K., Raghav, A., Singh, B. P., & Singh, A. K. (2024, October). Real-time Plant Disease Detection and Pesticide Application System Using IoT and CNN. In *2024 International Conference on Computing, Sciences and Communications (ICCSC)* (pp. 1-6). IEEE. DOI: 10.1109/ICCSC62048.2024.10830417

Vishnu, S., Ramson, S. J., Rukmini, M. S. S., & Abu-Mahfouz, A. M. (2022). Sensor-based solid waste handling systems: A survey. *Sensors (Basel), 22*(6), 2340. DOI: 10.3390/s22062340 PMID: 35336511

Vishnu, S., Ramson, S. J., Senith, S., Anagnostopoulos, T., Abu-Mahfouz, A. M., Fan, X., Srinivasan, S., & Kirubaraj, A. A. (2021). IoT-enabled solid waste management in smart cities. *Smart Cities, 4*(3), 1004–1017. DOI: 10.3390/smartcities4030053

Vora, C., Aryal, A., Willoughby, S., & Wang, C.; (2024) Investigating Stakeholder Perception and Developing a Decision Framework for Robot Adoption in Construction. *Journal of Construction Engineering and Management,*;150(4), pp. 04-24.

Wakchaure, M., Patle, B. K., & Mahindrakar, A. K. (2023). Application of AI techniques and robotics in agriculture: A review. *Artificial Intelligence in the Life Sciences, 3*, 100057. DOI: 10.1016/j.ailsci.2023.100057

Wang, C., Qin, J., Qu, C., Ran, X., Liu, C., & Chen, B. (2021). A smart municipal waste management system based on deep-learning and Internet of Things. *Waste Management (New York, N.Y.), 135*, 20–29. DOI: 10.1016/j.wasman.2021.08.028 PMID: 34461487

Wang, H. (2024). Application of new features based on artificial intelligent robot technology in medium-scale urban design pedigree and intelligent management and control. *Intelligent Systems with Applications, 22*, 200379. DOI: 10.1016/j.iswa.2024.200379

Wan, J., Li, X., Dai, H. N., Kusiak, A., Martinez-Garcia, M., & Li, D. (2020). Artificial-intelligence-driven customized manufacturing factory: Key technologies, applications, and challenges. *Proceedings of the IEEE, 109*(4), 377–398. DOI: 10.1109/JPROC.2020.3034808

Waqar, A., Othman, I., Falqi, I. I., Almujibah, H. R., Alshehri, A. M., Alsulamy, S. H., & Benjeddou, O. (2023). Assessment of barriers to robotics process automation (RPA) implementation in safety management of tall buildings. *Buildings (Basel, Switzerland)*, *13*(7), 1663. DOI: 10.3390/buildings13071663

Wei, Z., Li, H., & Zhou, Z. (2023). Human-Robot Collaboration in the Construction Industry: A New Era. *Automation in Construction*, *142*, 121–132.

WHO. (2021). *Ethics and Governance of Artificial Intelligence for Health: WHO Guidance* (1st ed.). World Health Organization.

Whyman, B. (2023). *AI Regulation is Coming- What is the Likely Outcome?*https://www.csis.org/blogs/strategic-technologies-blog/ai-regulation-coming-what-likely-outcome

Wilk-Jakubowski, G., Harabin, R., & Ivanov, S. (2022). Robotics in crisis management: A review. *Technology in Society*, *68*, 101935. DOI: 10.1016/j.techsoc.2022.101935

Wolniak, R., & Stecuła, K. (2024). Artificial Intelligence in Smart Cities—Applications, Barriers, and Future Directions: A Review. *Smart Cities*, *7*(3), 1346–1389. DOI: 10.3390/smartcities7030057

Wong, C., Wood, J., & Paturi, S. (2021). Managing waste in the smart city of Singapore. *Tropical Constrained Environments and Sustainable Adaptations: Businesses and Communities*, 225-241.

Wu, H. K., Lee, S. W. Y., Chang, H. Y., & Liang, J. C. (2013). Current status, opportunities and challenges of augmented reality in education. *Computers & Education*, *62*, 41–49. DOI: 10.1016/j.compedu.2012.10.024

Wuni, I. Y., Shen, G. Q., Osei-Kyei, R. (2020). Quantitative evaluation and ranking of the critical success factors for modular integrated construction projects, International Journal Construction Management; 22 (11), 2108–2120, .DOI: 10.1080/15623599.2020.1766190

Wu, P. H.. (2020). Gamification of environmental education with augmented reality applications: Enhancing critical thinking. *Interactive Learning Environments*, *28*(7), 895–911.

Wu, T.-W., Zhang, H., Peng, W., Lü, F., & He, P.-J. (2023). Applications of convolutional neural networks for intelligent waste identification and recycling: A review. *Resources, Conservation and Recycling*, *190*, 106813. DOI: 10.1016/j.resconrec.2022.106813

Wu, W., Lee, S. W., Chang, H., & Liang, J. (2020). Current status, opportunities, and challenges of augmented reality in education. *Computers & Education*, *62*(1), 41–49.

Wu, X., Freese, D., Cabrera, A., & Kitch, W. A. (2015). Electric vehicles' energy consumption measurement and estimation. *Transportation Research Part D, Transport and Environment*, *34*, 52–67. DOI: 10.1016/j.trd.2014.10.007

Xu, C., & Xue, Z. (2024). Applications and challenges of artificial intelligence in the field of disaster prevention, reduction, and relief. *Natural Hazards Review*, *4*(1), 169–172. DOI: 10.1016/j.nhres.2023.11.011

Xu, J., & Moreu, F. (2021). A Review of Augmented Reality Applications in Civil Infrastructure During the 4th Industrial Revolution. *Frontiers in Built Environment*, *7*, 640732. DOI: 10.3389/fbuil.2021.640732

Yadav, V., & Karmakar, S. (2020). Sustainable collection and transportation of municipal solid waste in urban centers. *Sustainable Cities and Society*, *53*, 101937. DOI: 10.1016/j.scs.2019.101937

Yamane, T. (1967). *Statistics: An Introductory Analysis* (2nd ed.). Harper and Row.

Yamaoka, S. (1986). HEALTH EFFECTS OF AIR POLLUTION. *Chemistry & Industry*, 827–830.

Yamauchi, B. M. (2004). In Gerhart, G. R., Shoemaker, C. M., & Gage, D. W. (Eds.), *PackBot: A versatile platform for military robotics* (pp. 228–237)., DOI: 10.1117/12.538328

Yang, J., Blount, Y., & Amrollahi, A. (2024). Artificial intelligence adoption in a professional service industry: A multiple case study. *Technological Forecasting and Social Change*, *201*, 123251. DOI: 10.1016/j.techfore.2024.123251

Yan, H., Yin, Q., Peng, J., & Bai, B. (2019). Multi-functional tugboat for monitoring and cleaning bottom fouling. *IOP Conference Series. Earth and Environmental Science*, *237*(2), 1–6. DOI: 10.1088/1755-1315/237/2/022045

Yetilmezsoy, K., Ozkaya, B., & Cakmakci, M. (2011). Artificial intelligence-based prediction models for environmental engineering. *Neural Network World*, *21*(3), 193–218. DOI: 10.14311/NNW.2011.21.012

Yigitcanlar, T., & Cugurullo, F. (2020). The sustainability of artificial intelligence: An urbanistic viewpoint from the lens of smart and sustainable cities. *Sustainability (Basel)*, *12*(20), 8548. DOI: 10.3390/su12208548

Yi, L., & Hiroatsu, F. (2022). Incentives for innovation in robotics and automated construction: Based on a tripartite evolutionary game analysis. *Sustainability (Basel)*, *14*(4), 2475. DOI: 10.3390/su14042475

Yilmaz, S., & Goktas, Y. (2021). The role of gamified augmented reality in climate change education. *Journal of Interactive Learning Research*, *32*(1), 91–108.

Yoon, S. A., Anderson, E., Lin, J., & Elinich, K. (2017). How augmented reality enables conceptual understanding of challenging science content. *Educational Technology Research and Development*, *65*(4), 611–633.

Yuan, Q., & Chen, T. (2025). Holding AI-Based Systems Accountable in the Public Sector: A Systematic Review. *Public Performance & Management Review*, •••, 1–34. DOI: 10.1080/15309576.2025.2469784

Zafar, M. H., Langås, E. F., & Sanfilippo, F. (2024). Exploring the synergies between collaborative robotics, digital twins, augmentation, and industry 5.0 for smart manufacturing: A state-of-the-art review. *Robotics and Computer-integrated Manufacturing*, *89*, 10–27. DOI: 10.1016/j.rcim.2024.102769

Zaidan, E., & Ibrahim, I. A. (2024). AI Governance in a Complex and Rapidly Changing Regulatory Landscape: A Global Perspective. *Humanities & Social Sciences Communications*, *11*(1), 1121. DOI: 10.1057/s41599-024-03560-x

Zaidan, M. A., Motlagh, N. H., Boor, B. E., Lu, D., Nurmi, P., Petäjä, T., Ding, A., Kulmala, M., Tarkoma, S., & Hussein, T. (2023). Virtual Sensors: Toward High-resolution Air Pollution Monitoring Using AI and IoT. *IEEE Internet of Things Magazine*, *6*(1), 76–81. DOI: 10.1109/IOTM.001.2200103

Zhang, A., Venkatesh, V. G., Liu, Y., Wan, M., Qu, T., & Huisingh, D. (2019). Barriers to smart waste management for a circular economy in China. *Journal of Cleaner Production*, *240*, 118198. DOI: 10.1016/j.jclepro.2019.118198

Zhang, H. Y., Khan, M. K., & Wang, L. (2021). Blockchain-based disaster management system with situation awareness and decision support. *Computers & Electrical Engineering*, *93*, 107217.

Zhang, K., & Herrmann, G. (2024). Control practice for robotic applications in challenging environments. *Annual Reviews in Control*, *57*, 100956. DOI: 10.1016/j.arcontrol.2024.100956

Zhang, M., Xu, R., Wu, H., Pan, J., & Luo, X. (2023). Human-robot collaboration for on-site construction. *Automation in Construction*, *150*, 10–28. DOI: 10.1016/j.autcon.2023.104812

Zhang, Q., Yang, Q., Zhang, X., Bao, Q., Su, J., & Liu, X. (2021). Waste image classification based on transfer learning and convolutional neural network. *Waste Management (New York, N.Y.)*, *135*, 150–157. DOI: 10.1016/j.wasman.2021.08.038 PMID: 34509053

Zhang, Y. W., Liu, M. N., Chen, D., Xu, X. M., Lu, J., Lai, H. R., Wen, C., & Yin, Y. X. (2025). Development and Testing of Row-Controlled Weeding Intelligent Robot for Corn. *Journal of Field Robotics*, *42*(3), 850–866. DOI: 10.1002/rob.22457

Zhao, X., Zhang, L., & Li, H. (2024). Performance, emissions, and economic analyses of hydrogen fuel cell vehicles. *Renewable & Sustainable Energy Reviews*, *168*, 112849.

Zhao, Z., Hicks, Y., Sun, X., McGuinness, B. J., & Lim, H. S. (2025). A novel self-supervised method for in-field occluded apple ripeness determination. *Computers and Electronics in Agriculture*, *234*, 110246. DOI: 10.1016/j.compag.2025.110246

Zhen, Z., & Yao, Y. (2024). The confluence of digital twin and blockchan technologies in Industry 5.0: Transforming supply chain management for innovation and sustainability. *Journal of the Knowledge Economy*, *16*(1), 1–27. DOI: 10.1007/s13132-024-02151-0

Zhu, L., Tian, Z., & Du, J. (2023). Spatial–temporal redundancy evaluation of the municipal solid waste incineration treatment capacity: The case study of China. *Environmental Science and Pollution Research International*, *31*(7), 9948–9963. Advance online publication. DOI: 10.1007/s11356-023-26989-0 PMID: 37072590

Zoumpoulis, P., Konstantinidis, F. K., Tsimiklis, G., & Amditis, A. (2024). Smart bins for enhanced resource recovery and sustainable urban waste practices in smart cities: A systematic literature review. *Cities (London, England)*, *152*, 105150. DOI: 10.1016/j.cities.2024.105150

Zyoud, S., & Zyoud, A. (2024). Internet of things supporting sustainable solid waste management: Global insights, hotspots, and research trends. *International Journal of Environmental Science and Technology*, 1–30.

About the Contributors

Mourade Azrour received his PhD from Faculty of sciences and Techniques, Moulay Ismail University of Meknes, Morocco. He has received his MS in computer and distributed systems from Faculty of Sciences, Ibn Zouhr University, Agadir, Morocco in 2014. Mourade currently works as computer sciences professor at the Department of Computer Science, Faculty of Sciences and Techniques, Moulay Ismail University of Meknès. His research interests include Authentication protocol, Computer Security, Internet of things, Smart systems, Machine learning and so ones. Mourade is member of the member of the scientific committee of numerous international conferences. He is also a reviewer of various scientific journals. He has published more than 60 scientific papers and book chapters. Mourade Has edited 2 scientific books "IoT and Smart Devices for Sustainable Environment" and "Advanced Technology for Smart Environment and Energy". Finally, he has served as guest editor in journals "EAI Endorsed Transactions on Internet of Things", "Tsinghua Science and Technology", "Applied Sciences MDPI" and "Sustainability MDPI."

Imran Hossain is a dedicated academician and researcher with a profound commitment to sustainable development and environmental governance. Currently, Imran serves as a lecturer in Political Science at Varendra University, Bangladesh. Beyond his academic role, Imran has delved into extensive research on the SDGs, collaborating with both national and renowned international organizations. Imran's research portfolio is diverse and impactful, encompassing projects on various aspects of sustainable development, including environmental governance, waste management, climate change, water management, and sustainable cities. Notably, he is at the forefront of exploring the intersection between sustainable development and Industry 4.0 tools, aiming to utilize technology to achieve sustainable solutions. His research interests span sustainable governance, smart technology, sustainable waste management, and IoT-based solutions for environmental sustainability. Imran actively engages in scholarly discourse by participating in national and international conferences. As a testament to his expertise, he serves as a reviewer for various scientific journals indexed in Scopus. Imran has published a number of research articles and book chapters indexed in Scopus and other renowned databases. Imran's dedication to advancing knowledge extends to editorial roles, where he served as the editor of several scientific books published by Springer Nature, IGI Global International, CRC Press, Taylor and Francis Group. Beyond academia, Imran is a prolific writer and columnist for national dailies, where he shares insights and perspectives on sustainable development, environmental governance, and smart technology.

A.K.M. Mahmudul Haque is a Professor at the Department of Political Science, University of Rajshahi. He did his PhD at the Institute of Bangladesh Studies, University of Rajshahi where he examined the policies, practices, and challenges of environmental governance of Bangladesh. Haque is a Visiting Scholar at the University of Victoria, Canada. He also attended a PhD Internship Program at the Centre for Asia Pacific Initiatives (CAPI) of the University of Victoria, Canada. His areas of specialization are environmental governance, climate change, sustainable development, and local governance. His key contributions focus on evaluating local government's environmental governance. Haque's dedication to advancing knowledge extends to editorial roles, where he currently serves as the editor of several scientific books published by IGI Global International and CRC Press, Taylor, and Francis Group. Dr. Haque has already carried out a number of research projects funded by different national and international organizations. Dr. Haque published more than 60 scientific papers in various national and international journals. He frequently contributes columns to national dailies in both Bengali and English. He supervised/is supervising the research of thirteen students, including three PhD fellows (already awarded their PhD). He has delivered numerous lectures mainly on environmental governance and sustainable development in about 20 conferences and seminars held at various universities domestically and overseas.

***

Clinton Aigbavboa research primarily addresses digital transformation and sustainability in the construction industry, particularly within the global south. Since joining University of Johannesburg, he has authored and coauthored over 1300 accredited journal articles, conference papers and book chapter.

Abdullah Al Bassam is an emerging and enthusiastic researcher in the fields of machine learning, artificial intelligence, full-stack engineering, IoT, and robotics.

Yet in his final year of his Bachelor's degree in Science, Bassam is a researcher who crosses disciplines while holding computational intelligence in both theoretical and typical use. Solid in data science and digital signal processing, Bassam is keen on optimizing advanced methods for how AI automates and creates smart systems. His interests range to optoelectronics and engine combustion analysis, in which he seeks alternatives to an efficient, precise approach. He also studies AI-powered customer service automation, using natural language processing (NLP) for effective multilingual services. He is also a Full Stack Engineer who codes up scalable software while creating links between the intelligent front end of the system with the backend architectures. Apart from that, he is working in robotics and IoT projects which would take the systems to autonomous and industrial automation. Bassam is a professionally working Office Engineer cum Design Engineer for funded project doing technical design and execution of project. Besides that, to promote programming literacy among native speakers, he is authoring a complete JavaScript book in Bangla. He is driven by deep commitment to innovation and new technologies, and keeps pursuing high-impact research areas that will play a role in tomorrow's AI, data science, and intelligent automation.

John Aliu, PhD, is currently a Post-Doctoral Research Associate with the Institute for Resilient Infrastructure Systems at the College of Engineering, University of Georgia. Dr. Aliu's research interests lie in the areas of curriculum development, sustainability, sustainable construction, skills development, employability studies, diversity, equity and inclusion. He has authored and co-authored several publications in top journals in the field of construction digitalization, sustainable construction, and engineering education. John is a regular reviewer for different peer-reviewed journals.

Anisha Biswas is an experienced academic and healthcare professional with 2 years of teaching experience as an Assistant Professor in the Hospital Management Department at Brainware University. She is currently pursuing her PhD in Management, focusing on Healthcare Marketing. Before transitioning to academia, she worked for 5 years as a Clinical Administrator at TATA Medical Centre, gaining valuable insights into hospital operations and healthcare management. Anisha completed her Master's in Hospital Administration from NSHM Knowledge Campus. Throughout her teaching career, she has actively engaged in academic seminars and presented research papers, further contributing to the field of healthcare management.

Pallab Chowdhury is an educator with experience teaching cutting-edge topics like Machine Learning (ML), Deep Learning (DL), Natural Language Processing

(NLP), Data Analytics, Embedded Systems, the Internet of Things (IoT), and Semiconductor Research. Proven ability to serve in both technical and leadership capacities; creates an inspiring atmosphere for student growth through teamwork and innovation. Enthusiastic about teaching, motivating students, and fostering creativity.

Indrajit Ghosal presently serves as an Associate Professor & Research Coordinator at Brainware University, Barasat, Kolkata (India), and Dept. of Management. Dr. Indrajit was awarded a Ph.D. in Online Marketing (Management) in 2020. He was awarded FRPM in 2017. He has done his masters in MCA and an undergraduate with BSc.IT. He has more than 16+ years of experience in teaching. He has published 75 research papers in many reputed journals with indexing such as ABDC, ABS, Scopus, WOS, UGC, and many more, and published 6 patents. He has presented papers in 36 International, 7 national, and 4 case studies in conferences. He is the author of 6 Books and 5 edited books published by Taylor & Francis, Nova Science, IGI Global, and published Book chapters. He assists 3 Ph.D. scholars, and four MPhil scholars already awarded under his supervision. He reviews many reputed journal indexes in Scopus, WOS, ABDC, ABS, and SSCI. As a Keynote Speaker/ Resource person/Chairperson, he has been invited to 30 Conferences/workshops/seminars. He is associated as an Examiner for PhD. DRC seminar in Seacom Skills University, West Bengal, Mewar University, Rajasthan. He is a Member of ACM (Meerut Chapter), SSRN, MTTF (Bhatinda, Punjab), Life Member and executive member of the Education Research and Development Association (ERDA), Member of IARA (Guwahati), IARA (Tiruchirappalli) & World Leadership Academy. His area of research is Digital Marketing, Consumer behavior, Online Marketing, Entrepreneurship, and E-commerce.

Oluwadamilare Olamide Ilesanmi is a researcher and a member of the Sustainable Infrastructure Management Plus (RG-SIM+) research group, also a Quantity Surveyor, Data Analyst, and Construction Project Manager. He graduated from the Department of Quantity Surveying, Federal University of Technology, Akure Ondo-state, Nigeria, with a first-class honor. He won the awards for the Best Graduating Student in Information Communication Technology (ICT) and Construction Management in the department. He also won an award for the most influential male in entrepreneurship skills in the School of Environmental Technology of the same university. His research interests include construction automation and robotics, digitalization and sustainability in construction, construction economics, etc

Babasaheb Jadhav is an Associate Professor in the area of Financial Management and International Business Management at Dr. D. Y. Patil Vidyapeeth's,

Global Business School and Research Centre, Pune, Maharashtra, India. He has more than 10 years of experience in Industry, Research and Academics. He has completed his PhD from Savitribai Phule Pune University (SPPU) Formerly Pune University in Financial Management. He has a Dual Master's Degree in Business Administration and Bachelor's Degree in Business Administration. He has in his credit ongoing Funded Research Projects, 7 Books, more than 50 Research Papers, Case Studies in various National and International Indexed Journal's with high Impact Factors, and 4 Articles in reputed Magazines as well as National Newspapers. He is also working as Editorial Board Member and Advisory Board Member for various Management Journals or Conferences. He has invited as resource person for Guest Lectures, Workshops and Conferences. His areas of interest and research are Financial Management, Taxation, Economics, General Management and International Business Management.

K. Murugasridevi, Ph.D., is an accomplished Agricultural Entomologist with expertise in biological control, chemical ecology, and molecular entomology. She holds a doctorate from Tamil Nadu Agricultural University, Coimbatore, and has received prestigious awards including the DST-INSPIRE Fellowship and multiple gold medals for academic excellence. With over five years of teaching and research experience, she has authored numerous peer-reviewed publications, contributed to international conferences, and served as a reviewer and editor in reputed journals. Her research primarily focuses on sustainable pest management, parasitoid-host interactions, and entomopathogenic fungi. Currently, she serves as Assistant Professor at Amrita School of Agricultural Sciences, Coimbatore.

Dinesh Komarasamy presently working as a Assistant Professor (SRG) in the department of computer science and engineering, Kongu Engineering College, perundurai. He received the Bachelor of Engineering and Master of Engineering in Computer Science and Engineering under Anna University, Chennai, India in 2010 and 2012 respectively. He has received his Doctorate of Philosophy (Ph.D) under the title of "A Framework For Scheduling Of Heterogeneous Workloads In Cloud Management System" in the year 2019. He has around 3.6yrs. of research experience in College of Engineering, Anna University, Guindy. He has completed his research in optimal scheduling of jobs in cloud environment. He has published many Papers which are indexed by SCI and Scopus. His area of research includes cloud computing and machine learning. Presently, he is working as a Assistant Professor in the department of computer science and engineering in Kongu Engineering College, perundurai for the past 8 years. He has currently doing research projects funded by Indian Council of Social Science Research (ICSSR). He also presented papers in several national and international conferences.

Taskia Haq Lyric is an Assistant Professor at the Department of History, University of Dhaka and a PhD researcher at the School of History, University of Leeds, UK. She completed her Bachelors and Masters from the Department of History, University of Dhaka. Previously, she was Faculty Member at Stamford University Bangladesh. She is an author of numerous research articles and book chapters. She has participated and presented papers in various national and international seminars, symposiums and workshops in Bangladesh and the United Kingdom. Her research interests are—Climate and Disaster History, Social History, Economic History, Ancient History, Historical Geography and South Asian History. Her ORCID ID: 0009-0003-4505-044X

M. Kavyadarshini is currently pursuing her second year of B.E. Computer Science and Engineering at Kongu Engineering College (Autonomous), Perundurai, Erode. Her areas of interest include full stack development, and she is actively learning and building projects using technologies like JavaScript and Node.js. She is passionate about exploring new technologies and aims to become a skilled and confident developer.

M. R. Prasanndh Raaju is currently pursuing a bachelor's degree in computer science and engineering (3rd year) at Kongu Engineering College (Autonomous), Perundurai, Erode. His areas of interest include Data Science, IoT, Machine Learning, and he has actively participated in various academic projects, technical workshops and doing a government sanctioned project Tomatix (for Tomato Grand Challenge). He have also contributed in four book chapters under various domains in which one have got published. As an enthusiastic learner, he aspires to contribute to innovative solutions in the field of Computer Science.

Mageshen holds a Ph.D. in Soil Science, specializing in soil fertility and soil health, with a focus on the interactions between soil nutrients, microorganisms, and plant growth. Over the course of his academic journey, he has gained expertise in various aspects of soil science, including the chemistry and biology of soils, nutrient cycles, and the environmental impacts of soil degradation. Currently he is working as an Assistant Professor (Soil Science) in Amrita School of Agricultural Sciences, Coimbatore, Tamil Nadu, India

Navonita Mallick is a doctoral candidate at KIIT Law School, Bhubaneswar, specializing in Child Rights with a particular emphasis on legal frameworks addressing child sexual abuse in India. She is actively engaged in research, publications, and presentations at national and international conferences, addressing critical issues of child protection, digital transformation, and legal reform. She has authored multiple books and scholarly articles published by renowned international publishers, addressing diverse legal themes ranging from criminal law to sustainable governance. Navonita's analytical and communication skills drive her commitment to fostering impactful legal and social change.

Dorcas Titilayo Moyanga is a Lecturer in the Department of Quantity Surveying, Federal University of Technology, Akure, Ondo State, Nigeria. Her research interest includes quantity surveying practice, construction management and economics and construction economics.

Shiblee Noman is an Assistant Professor at the Department of Japanese Studies, University of Dhaka and a PhD researcher at the School of Earth and Environment, University of Leeds, UK. He completed his Bachelors and Masters from the Department of Political Science, University of Dhaka. Previously, he was a Faculty Member at the Department of Political Science, Jagannath University and earlier at Brac University. He is co-author of a research book Brexit Referendum: Democratic Dividend or Deficit (2016) and author of numerous research articles and book chapters. He has participated and presented papers in various national and international seminars, symposiums and workshops in Bangladesh, the United Kingdom, Indonesia, Japan, India and Thailand. His research interests are—Climate Change Governance and Diplomacy, Public Policy, Bangladesh Politics, Japan and East Asian Politics, Geo-politics and Political Economy.

Ayodeji Emmanuel Oke is an associate professor at the Department of Quantity Surveying, Federal University of Technology Akure, Nigeria, and a Research Fellow at the University of Johannesburg, South Africa. He founded the Research Group on Sustainable Infrastructure Management Plus (RG-SIM+) at the Federal University of Technology, Akure, and he is currently the team leader. Prof. Oke's research interest is in sustainable infrastructure management (SIM), emphasizing sustainable construction, value management, quantity surveying, and construction in the digital era. He is among the top 2% most-cited Scientists in the world in 2023 by Stanford University. With more than 400 publications, including several research books, he is well suited to author The Circular Economy Business Model: A Framework for Transforming Construction Organizations.

Poornima Ramesh is an Assistant Professor of Environmental Science at Vanavarayar Institute of Agriculture, Tamil Nadu. She holds a Ph.D. from Tamil Nadu Agricultural University, where her research focused on mitigating tropospheric ozone stress in rice cultivars. Her academic interests span environmental pollution, composting technology, climate-resilient agriculture, and soil and water remediation. She has published over 20 peer-reviewed articles and contributed to several book chapters on topics including microplastics in agriculture, wastewater management, and bioremediation. Dr. Poornima has also served in research and teaching roles at RVS Agricultural College and as a Senior Research Fellow in an ISRO-funded project.

Mohemmed Yousuf R received the M.E. and B.E. degrees in CSE from Anna University. He is currently working as an assistant professor at the Bannari Amman Institute of Technology, Sathyamangalam, Erode, Tamil Nadu, India. His research interests include cybersecurity, IoT, and machine learning.

Mohana Saranya Sellappan earned her master's degree from the College of Engineering Guindy (CEG), Anna University, Chennai. She is currently an Assistant Professor at Kongu Engineering College in Erode, with a focus on research in deep learning and computer vision. She has secured a project grant from the Ministry of Education (MoE) and has successfully completed a student project funded by AICTE. Additionally, she has organized two national-level seminars sponsored by CSIR and NBHM. Her academic contributions include publishing nine research papers in international journals and presenting 36 articles at international conferences. She is also a lifetime member of the Computer Society of India.

Shashwata Sahu, a distinguished research scholar currently pursuing her Ph.D. in Law at KIIT University, Bhubaneswar, embodies an unwavering passion for the field of law. With a profound interest in Constitutional Law, she stands as a self-driven scholar, firmly rooted in the principles of altruism. With an impressive literary portfolio, Shashwata has authored nine impactful books, each delving into crucial aspects of law and society. Her works include "The Growing Significance of Artificial Intelligence," "The role of Indian media in covering crimes: a critical analysis," "The use of Amicus Curiae Briefs at the International Criminal Tribunals," "Assessing the legal landscape of bio-medical waste management in India," "Journey of Law: From Deities to Decrees," "Evolution of Local-Self Government in India", "The Invisible Manual Scavengers of India", "Law and Justice in a Globalizing World: An LL.M. Companion" and "Comparative Public Law: An LLM. Companion." Beyond her authored books, Shashwata's contributions extend to numerous publications in both international and national journals indexed in Scopus and Web of Science.

Sunita Sharma is working as Professor of Physics adn Associate Dean in School of Basic Sciences, Galgotias university, Greater Noida. Dr Sharma has 21 years of research and 11 years of teaching experience. She has completed her PhD from Guru Jambheshwar University of Science and Technology, Hisar. She has done postdocs from Raman Research Institute, Bangalore, Inter University Accelerator Centre, Delhi and Indian Institute of Technology, Delhi. Her research interests are nanomaterials, thin films, solar cells, lasers and hydrogen generation.She is an international speaker in online global classroom teaching in collaboration with Universiti Teknologi Malaysia. She has guided several PhD students in the area of thin films and solar cells. She has published 72 papers in journals of national and international repute. Dr Sharma is a member of Materials Research Society of India and Photonics Society of India.

Bhupinder Singh working as Professor at Sharda University, India. Also, Honorary Professor in University of South Wales UK and Santo Tomas University Tunja, Colombia. His areas of publications as Smart Healthcare, Medicines, fuzzy logics, artificial intelligence, robotics, machine learning, deep learning, federated learning, IoT, PV Glasses, metaverse and many more. He has 3 books, 139 paper publications, 163 paper presentations in international/national conferences and seminars, participated in more than 40 workshops/FDP's/QIP's, 25 courses from international universities of repute, organized more than 59 events with international and national academicians and industry people's, editor-in-chief and co-editor in journals, developed new courses. He has given talks at international universities, resource person in international conferences such as in Nanyang Technological University Singapore, Tashkent State University of Law Uzbekistan; KIMEP University Kazakhstan, All'ah meh Tabatabi University Iran, the Iranian Association of International Criminal law, Iran and Hague Center for International Law and Investment, The Netherlands, Northumbria University Newcastle UK,

Abdus Sobhan is currently pursuing a Bachelor's degree in Computer Science and Engineering (CSE) at Varendra University. His research interests include Artificial Intelligence (AI), Machine Learning (ML), and Deep Learning. In addition to his academic pursuits, Abdus Sobhan writes tech blogs on various topics, sharing insights on cutting-edge technologies and their applications.

Tanvir Islam Sourav is a third-year student in the Department of Computer Science and Engineering at a private institution in Bangladesh researching the use of microcontrollers and sensors to collect weather data in real time in Rajshahi.

Index

A

B

C

D

E

F

H

I

M

N

P

R

S

www.ingramcontent.com/pod-product-compliance
Lightning Source LLC
LaVergne TN
LVHW080344011026
849161LV00009B/2186